# ENCYCLOPEDIA OF ORGANIC CHEMISTRY

# ENCYCLOPEDIA OF ORGANIC CHEMISTRY

## *Vol. 8*

*By*
**Sananda Chatterjee**
*DAE, DCA, CMCNet*
*Scientific Consultant*
*Institute for Natural Sciences*
*Kolkata*
*(West Bengal)*

D P H

**DISCOVERY PUBLISHING HOUSE PVT. LTD.**
**NEW DELHI-110 002**

*Published by:*
**Tilak Wasan**
**DISCOVERY PUBLISHING HOUSE PVT. LTD.**
4383/4B, Ansari Road, Darya Ganj
New Delhi-110 002 (India)
*Phone* : +91-11-23279245, 43596064-65
*Fax* : +91-11-23253475
*E-mail* : parul.wasan@gmail.com
discoverypublishinghouse@gmail.com
*web* : www.discoverypublishinggroup.com

***First Edition:* 2012**

**ISBN: 978-81-8356-875-3 (Set)**

**Encyclopedia of Organic Chemistry**



***Printed at:***
***Shree Balaji Art Press***
***Delhi***

# Preface

Organic Chemistry is that branch of chemistry, which deals with the reactions related to the life form, or one of the source of the life form, carbon (we can say the end of it).

This carbon mainly conjugates with different elements such as nitrogen, hydrogen and oxygen sometimes metals and other non-metals also). Organic chemistry studies mainly the extraction, structure, properties, of these compounds.

In this book chapters deals with the different methods of purification, extraction of compounds, some processes were used and devised by Alchemists.

The extraction preservation of natural dyes are examples of alchemical work, these been discussed in the chapter "dyes indicators and pigments".

Dyes is the mal-pronunciation of the damathi word dyum meaning colour, it was, the ancient Egyptians who use different natural pigments to extract different colours.

In this book the organic compounds has been mainly classified into two broad groups, the aromatic and the aliphatic compounds. The extraction of alcohol is definitely an alchemical process'.

The destructive distillation of wood is one of the finest processes devised by the alchemists for the preparation of methanol is one of the important topics of this book.

Looking ahead in to the modem world, the world of future, the subject comes in vision is the biochemistry, the base of which is organic chemistry, the chapter containing carbohydrates, and the chapter, chemical compounds of biological and biochemical interest is the best element of focus for the students stepping foot in the advanced staircase of biochemistry.

The last chapter, chapter 41, discusses the different organic reactions and their suitably mechanisms, this will help those students who are very much afraid of organic formulas and reactions. This chapter also deals with the chirality one of the interest of Dr. Palash Gangopadhyay (author's friend).

Organic Chemistry is not a mere subject, it is the rhythm of life tuned in the essence of carbon, which looks black but carries the dazzleness of diamond in its heart.

**SANANDA CHATTERJEE**

# Contents

CHAPTER

# 38

# Plastics, Resin and Condensates

## Chemistry of Resin Formation

Amino resins and plastics, like other synthetic resins, are formed by the process of polymerization. This consists of building one large molecule from many small ones. Since water is formed in the process, it is known as condensation polymerization. The materials which polymerize the methylol (hydroxy methyl) compounds formed by addition of formaldehyde to the amine:

$$RNH_2 + CH_2O \rightarrow RNHCH_2OH$$

These methylol compounds are not resins, but many molecules condense to form a resin. A molecule of water is evolved as each molecule of methylolamine condenses to increase the size of the molecule:

$$2\,RNHCH_2OH \rightarrow NHRCH_2NRCH_2OH.$$

The molecule formed from two molecules of methylol-amine can condense with a third molecule to form a larger molecule: $NHRCH_2NRCH_2NRCH_2OH$. This compound still contains a methylol group and con-denses further. There is considerable evidence that the hydrogen and hydroxyl in this molecule are eliminated as water, thus resulting in the formation of a cyclic compound (1). If only one methylol group were present, the molecule formed would

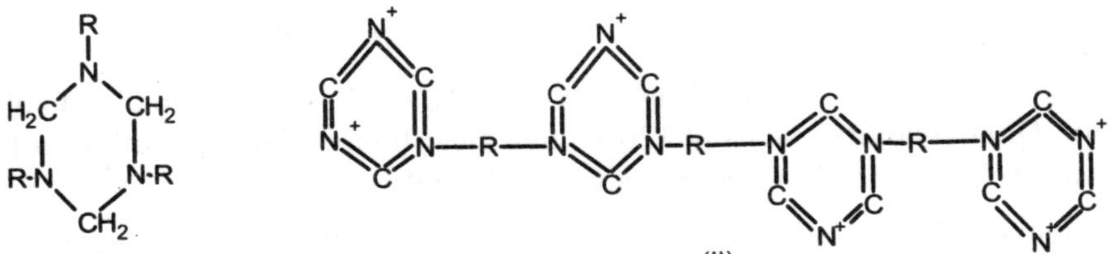

(I)

(II)

not be very large, nor would it have particularly interesting properties as a resin. However, in the methylolureas and methylolmelamines there is more than one methylol group present in each molecule, and a structure of connected rings may be formed as the condensation proceeds. This results in the complicated network represented by (II). Since all these rings may be attached to other rings, very large molecules are formed. The insolubility of many amino resins is consistent with such a structure.

Condensation polymerization may be stopped and started again almost at will. Thus many of the amino resins are partially polymerized when made, and polymerization is later completed, as in the molding of amino plastics.

Since the methylolamino derivatives are not readily soluble in organic solvents, they cannot be used with many film-forming materials. The introduction of an alkyl group, if sufficiently large, forms ethers that are soluble in a variety of solvents. Butyl alcohol is often used for this purpose. If a methylol derivative is heated with butyl alcohol in acid solution, an ether, which is soluble, is formed: RNH

$$CH_2OH + C_4H_9OH \rightarrow RNHCH_2OC_4H_9.$$

Other alcohols may be used; the resulting resin solution is useful in imparting hardness to other film-forming resins.

## Urea-formaldehyde Condensates

The simplest reaction products of urea and formaldehyde are the methylolureas. Einhom and Hamburger have described a method of preparation which consists of stirring 1 mole of urea with 2 moles of 37% formalin at 25-30°C. In alkaline solution until aldehydes no longer is present. Monomethylolurea can be made in the same manner using but 1 mole of formalin to 1 of urea and cooling the reaction vessel with ice. Formalin (37%) is added to a 50% aqueous solution of urea (eqs. 1 and 2).

$$O{=}C(NH_2)_2 + H{-}CHO \longrightarrow O{=}C(NH{-}CH_2{-}OH)(NH_2) \quad (1)$$

1-(hydroxymethyl)urea

$$O{=}C(NH_2)_2 + H{-}CHO \longrightarrow O{=}C(NH{-}CH_2{-}OH)_2 \quad (2)$$

1,3-bis(hydroxymethyl)urea

Monomethylolurea is a white crystalline solid melting at 111°C. It is soluble in cold water and in warm methanol. Dimethylolurea melts at 126°C. to a clear liquid which solidifies on further heating. Dimethylolurea is soluble in cold water and in warm alcohol.

The behaviour of the methylolureas on heating has been studied with a view to clarifying the mechanism of the reaction. When dimethylolurea is heated, 0.5 mole of formaldehyde and 1 mole of water are evolved. When monomethylolurea is heated, 1 mole of water is liberated hut no formaldehyde. If alkalies are present, less water is evolved from dimethylolurea, and less formaldehyde is liberated.

The residue on heating dimethylolurea is clear and resinous if heating is done under pressure, while only amorphous powders are obtained by heating monomethylolurea.

Products which have been assu [illegible] when methylolureas are condensed are the methyleneureas, as (III). A substituted methylene group occurs in Schiff bases (RN = CRR'). The existence of the methylene urea structure in amino plastics has never been definitely established. The methylene compounds are unstable and Polymerize as they are formed. Acid solutions promote the formation of methylene compounds, while the methylol groups are more stable in alkaline solution.

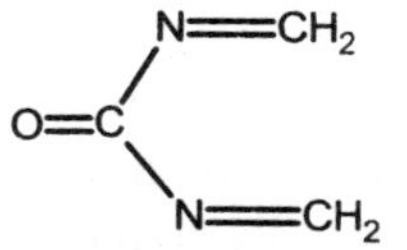

1,3-bis(methylene)urea

(III)

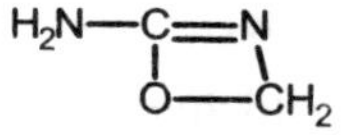

2*H*-1,3-oxazet-4-amine

(V)

The structure of methyleneureas polymers has usually been assumed to be linear. If this is the case, the chains cannot be very long, since the condensation products of monomethylolurea are not useful plastics. Walter has presented evidence for a monomeric cyclic methyleneurea (formula IV), which is water-insoluble and has resinous properties. This is formed only in strongly acid solution.

Much of the early literature on urea-formaldehyde condensates assumed the formation of linear C-X-C-N-C-N chains by polymerization of methylene compounds. However, most Schiff bases polymerize to the cyclic trimers:

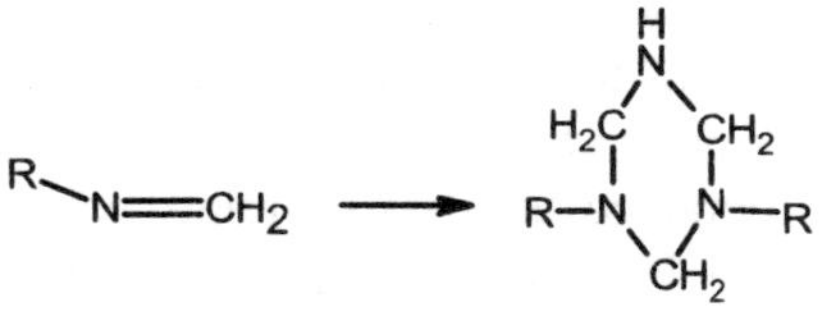

It has been suggested that the methyleneureas polymerize to form such cyclic trimers. Thus, monomethyleneurea might be expected to form a ring with three $NH_2$, groups attached which are capable of reacting with formaldehyde. The methylene groups can again form rings leading to a highly cross-linked structure, which is necessary for the insolubility associated with thermosetting resins:

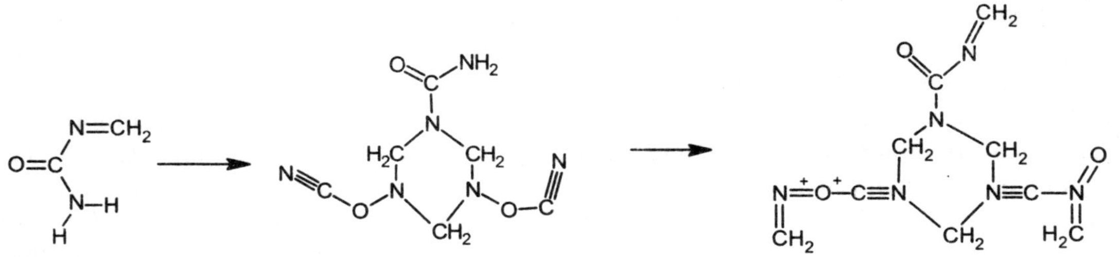

1-methyleneurea

3,5-bis(cyanooxy)-1,3,5-triazinane-1-carboxamide

Marvel investigated the reaction of other amides with formaldehyde and found that cured condensates, much like those from urea, can be made from glycin-amide, $NH_2CH_2CONH_2$. The resin cures to an insoluble resin which in the cured state contains 1.5 moles of formaldehyde to 1 mole of urea. When t-aminocapro-amide, $NH_2(CH_2)_5CONH_2$, was condensed with formaldehyde and cured, the same molar ratio of formaldehyde to amide was found. The cured resin was apparently rather more flexible than the cured urea-formaldehyde resin. However, with the N-methylamide of glycine, $NH_2CH_2CONHCH_3$, the trimer was formed, but polymerization could not be carried further. It was thus shown that substitution of any of the hydrogen atoms in the urea molecule will prevent the formation of a cross-linked molecule when condensed with formaldehyde.

In all the study of urea-formaldehyde condensates, the separation of large soluble polymers has never been accomplished. The resins, just before curing occurs, have average molecular weights of less than 1000. These same low-molecular materials are rapidly converted to cured, insoluble resins of large molecular size. This behaviour is consistent with the cross-linked nature of the structure formed, and may be further support of the cyclic structure, for once the cyclic trimers begin to tie together, large insoluble molecules may be formed readily.

## Thiourea-formaldehyde Condensates

The chemistry of the reaction of formaldehyde with thiourea is, in general, believed to be much the same as with urea. However, the condensates at low stages are rather less soluble in water than comparable condensates from urea. Pollak has found that condensates of thiourea with formaldehyde made in alkaline solution do not readily give precipitates with silver nitrate solution. He attributed this behaviour to the addition of the methylol group to the sulfur atom. When thiourea and formaldehyde were allowed to react at 50°C without the addition of alkalies, crystals were deposited on cooling which melted at 97-98°C and gave a precipitate with silver nitrate test solution. When magnesium oxide was present, the precipitate was obtained only on standing, while condensation in the presence of sodium hydroxide gave no precipitate. In many patents, urea and thiourea are considered as equivalents, and substantially the same conditions are recommended for the formation of thiourea as for urea condensates with formaldehyde.

## Melamine-formaldehyde Condensates

The conditions of the reaction of melamine with aqueous formalin are somewhat different from the reactions of urea, because of the low solubility of melamine in water. Reactions are usually conducted at temperatures of 80-100°C to bring the melamine into solution more readily. The amino groups in melamine can each add two methylolgroups, while in urea apparently only one mole of formaldehyde adds to each $NH_2$ group. Hexamethylolmelamine is formed by heating melamine at 90°C with an excess of neutral formaldehyde, or at room temperature for 15 hours. The hexa-methylolmelamine crystallizes with one molecule of water of crystallization. The tri-methylol compound is formed at ordinary temperatures by using 3 moles of formalin to 1 of melamine. It is much less stable than the hexamethylolmelamine and forms a resin if heated. This behavior seems to favour the normal structure for melamine if the analogy to urea holds, for only in the $NHCH_2OH$ structure does the dehydration to the $N = CH_2$ structure occurs readily. The relative stability of the hexamethylolmel-amine is understandable since it cannot polymerize until formaldehyde has been eliminated.

The structure of melamine-formaldehyde polymers has almost always been interpreted in terms of formation of linear -$CH_2N$- chains cross-linked with the same type of structure. The formation of a six-membered ring by addition of three $RN{=}CH_2$ molecules seem as plausible an explanation of the polymerization reaction as it is in the case of urea-formaldehyde polymers. The structure of the polymer would be highly branched, and this is in agreement with the insolubility of the polymers formed:

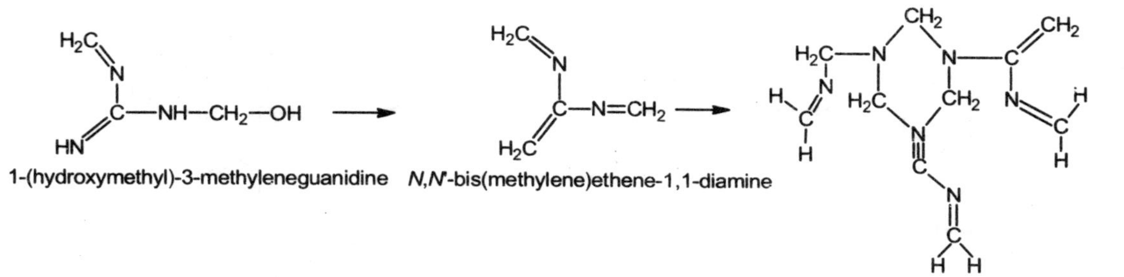

1-(hydroxymethyl)-3-methyleneguanidine *N,N'*-bis(methylene)ethene-1,1-diamine

Until more information is available, it is not possible to state definitely what the structure of cured melamine-formaldehyde polymers is. It is entirely possible that both cyclic and linear polymers are formed.

## Sulphonamide-formaldehyde Condensates

The molecular weight of sulphonamide-formaldehyde polymers has been studied by a variety of methods. The crude polymer was fractionated to separate the more soluble from the less soluble fractions. None of the fractions was appreciably higher than the trimer, while the most soluble fractions indicated the dimmer. This evidence suggests quite strongly the formation of a cyclic trimer. If linear polymers were formed, it would be probable that higher polymers than the trimer would be formed, and would be separated in the fractionation procedure. The sulphonamide-formaldehyde products, since they form low polymers, are usually soft and thermoplastic, and are readily soluble in solvents:

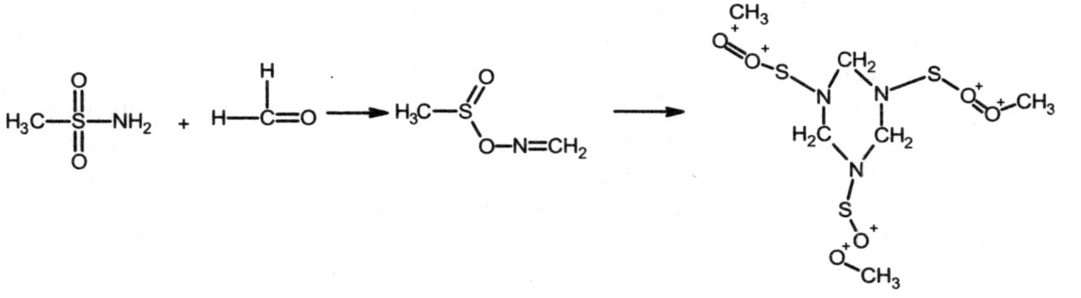

(*Z*)-1,1',1"-[1,3,5-triazinane-1,3,5-triyltris(thio)]tris(methyldioxidenediium)

If disulphonamides are used, the situation is quite different, since cross linking can occur, and hard insoluble resins are formed. m-Benzenedisulphonamide is typical of such bifunctional derivatives.

## Aniline-formaldehyde Condensates

Aniline can react with formaldehyde in two ways, either to add to the $NH_2$ group to form a methylene group, or to replace a hydrogen in the benzene ring to form a methylene bridge in the polymer structure. It is possible that the formaldehyde first adds to the $NH_2$ group and the methylene ($CH_2$) group subsequently rearranges to the ring:

aniline + H—C(H)=O ⟶ N-methyleneaniline ⟶ 1,3,5-triphenyl-1,3,5-triazinane

aniline + H—C(H)=O ⟶ 4,4'-methylenedianiline

As the condensation is carried out in acid solution, compounds containing methylol groups are not isolated, because, if they are formed, they soon lose water to form the methylene derivatives. The diphenylmethane derivative is free to react with formaldehyde to form links between molecules, resulting in a rather highly branched structure. Aniline-formaldehyde polymers usually do not become as hard or as insoluble as urea- or melamine-formaldehyde resins; this suggests a much less branched structure.

## Mixed Polymers

Since formaldehyde reacts with many types of organic molecules, it is not surprising that mixtures of various materials have been condensed to produce new resins and plastics. Most of the amino derivatives considered here have been used together. It is common practice to use melamine and urea with formaldehyde. Because urea is much lower in cost than melamine, cost can be reduced by its judicious use, often without too great sacrifice in useful properties. Urea and thiourea have often been condensed together. Urea and phenol have been condensed with formaldehyde in an attempt to improve the water resistance of the urea resin. Since the rate of condensation of the two materials is not the same, one component may be less thoroughly utilized than the other. Aniline has been condensed with formaldehyde along with phenol to produce condensates that are somewhat less brittle than phenol-aldehydes plastics. Since the monosulphonamides form rather soft resins, the incorporation of some sulphonamide with urea and formaldehyde in the formation of urea plastics has been effected to reduce the reactivity of urea resins and to plasticize them.

The above examples are conservative and conventional when compared to the mixtures that have been tried to form new plastics. It is only natural that two materials should be used together in an attempt to obtain all the desirable properties of both starting materials. It often happens that the resin has all the deficiencies of both materials. It is, however, sometimes more effective to blend two materials and condense them with formaldehyde than to blend the condensates.

Dicyanodiamide, the starting material for melamine, can be condensed with formaldehyde; it can be mixed with melamine and the mixture condensed with formaldehyde. Often much more dicyanodiamide than melamine is used. Phenol and dicyanodiamide have also been used to form condensates with formaldehyde and used for making molding powders.

Biuret has also been used as the starting material for amino plastics. Dimethylolbiuret (m.p., 139-140°C.) can be made by refluxing biuret with 37% formalin which has been neutralized to pH 7.6 with triethanol-amine. Molding powders or lacquers can be made from the dimethylol compound.

Melamine and phenol have been condensed with formaldehyde. In one case, 3 moles of phenol and 2 moles of melamine were condensed with 11 moles of formalin. The mixture was made alkaline (to pH 9.1) and refluxed 1 hour, a small amount of acid was added (to pH 7.3) and reflux continued. A hard resin was obtained on dehydrating in vacuum.

β-Hydroxyethylureas can be produced by heating ethanolamine with urea at 130° until the theoretical amount of ammonia has been evolved. The ethanol urea can be condensed with formaldehyde in acid solution without formation of insoluble gels. β-Hydroxyethylurea melts at 90° when pure. Urea and di(β-hydroxyethyl)urea can be condensed with formaldehyde to form thermoplastic resins.

To produce water-soluble urea-phenol condensates sulphanilic acid, urea, phenol, and formaldehyde are condensed in alkaline solution. The sulphonic acid group of sulphanilic acid imparts water solubility, and equimolar quantities of urea and sulphanilic acid are employed.

A mixed urea-silicon resin has been suggested using ethyl orthosilicate with a methylated urea in butyl alcohol solution. Water is added to hydrolyze the ethyl silicate and initiate polymer formation. Acid catalysts may be added to affect the cure of the urea resin.

The addition of a salt of an aliphatic diamine and a dibasic acid to urea and formalin in the condensation reaction leads to formation of a resin with improved resistance to water and better resistance to impact.

Amides of fatty acids may also be condensed with thiourea or with urea and formaldehyde. Since these amides are bifunctional, they do not tend to form insoluble condensates. The amides of hydrogenated fish oil acids and stearamide have been suggested as materials to be used with thiourea.

Because proteins contain amide groups which condense with formaldehyde, it is possible to use the natural proteins, such as casein with urea, thiourea, or mel-amine resins. It is also possible to mix the urea resin with protein; the formaldehyde liberated in curing the urea-formaldehyde resin is available for condensation with the protein material.

Aliphatic as well as aromatic diamides may be used with urea or thiourea in forming condensates with formaldehyde. Diamides have two groups which can condense with formaldehyde and therefore do not reduce the possibility of cross linking as do amides of monobasic acids. However, diamides are less active than urea, and the second hydrogen atom is much less active.

Most of the literature is concerned with the reactions of formaldehyde with amino compounds. The higher aldehydes do not react nearly as readily, nor do the reaction products form insoluble resins readily. As we have seen earlier, the possibility of the formation of a methylene compound, $-N=CH_2$, seems a necessary requirement to formation of active resins. With higher aldehydes, this structure cannot be formed.

There has been some interest in condensates with unsaturated aldehydes, particularly with acrolein. Here the unsaturated groups in the aldehyde molecule may possibly polymerize, and a resin may be formed by addition polymerization. Thio-urea has been used with acrolein to form resinous condensates. An aryl carboxamide or sulphonamide may also be used with acrolein.

Whife melamine is the only cyclic amine to have achieved commercial importance, many other amino diazines and amino triazines have been condensed with formaldehyde to produce reactive, heat-hardening resins.

Triaminopyrimidine (formula V) can be condensed with formaldehyde to form a condensate which can be hardened with heat. 2,4-Diaminoquinazoline (formula VI) forms a dimethylol compound which forms an insoluble resin on further heating. Tetraminopyrimidine (formula VII) condenses readily with formaldehyde

(V)

quinazoline-2,4-diamine (VI)

pyrimidine-2,4,5,6-tetramine (VII)

to form a very waterproof resin. Hydroxydiaminopyrimidine reacts readily with formaldehyde and can be condensed with formaldehyde in the presence of urea, phenol, and sulphonamide, as can the other cyclic amines listed above.

The list of possible mixed condensates is by no means complete. The literature of the subject is large, and the examples have been chosen to illustrate the possibili-ties that exist in forming mixed condensates. Few of these materials have as yet achieved any commercial acceptance, except the melamine-urea condensates with formaldehyde, which offer certain advantages in some cases and economies in others where the less expensive urea can be employed to replace melamine.

## Formation of Ethers

In discussing the reactions of amines with formaldehyde, the reactions leading to the formation of polymers have been considered. There is another reaction which, although it decreases the extent of polymer formation, is exceedingly useful in modifying the properties of urea resins. This reaction is acetal formation, and proceeds readily when an alcohol and an aldehyde are heated in acid solution:

$$CH_2O + 2ROH \rightarrow CH_2(OR)_2$$

The same type of reaction can be carried out with a methylol compound, except that only one mole of alcohol reacts, forming an ether:

$$RNHCH_2OH + R'OH \text{'}\rightarrow RNHCH_2OR'.$$

As long as the alcohol group remains, a methylenamino ($N{=}CH_2$) group will not form, nor will a polymer be formed at this part of the molecule. If part of the metliylol groups reacts, the others are free to form the polymer, and hardening of the resin will occur. When higher alcohols, butyl, capryl, and other monohydric alcohols are used, the ethers are soluble in organic solvents, particularly the alcohols from which they are made, and these solutions can be diluted with hydrocarbon solvents; usually aromatic solvents are required. These solutions, if dried and baked, tormvery hard films which are somewhat brittle. They can be modified, however, by incorporation of flexible resins. These resins often con-tain alcohol groups which may condense with the urea resin to form acetals, the resulting combination having the desired flexibility. It is often necessary to choose plasticizing resins with free hydroxyl groups, to insure reaction between the amino resin and the plasticizing resin, otherwise the urea resin becomes insoluble on curing, and the desired plasticizing action is not effected. In many instances the condensation of the urea resin with the plasticizing resin is carried out in making the urea resin, hut this step is not always necessary.

It seems possible that reactions of this type occur with many materials with which amino plastics are used. Cellulose has long been used successfully as a filler for amino plastics, and there is a possibility that a bond is established between the amino resin and the alcohol groups in cellulose.

Methods of forming the methyl and higher ethers have been described and consist of condensing urea with an excess of aqueous formalin at 90° for. 20 minutes and concentrating to a viscous solution, which is taken up in methanol. Ethyl alcohol or butyl alcohol is used for the higher ethers. Concentrated hydro-chloric acid is added, and the mixture allowed to stand 1 hour. The solution is neutralized, concentrated, and distilled and gives the dimethyl ether of dimethyloluron (VIII). The ethyl and butyl ethers have also been prepared and can be distilled *in vacuo*.

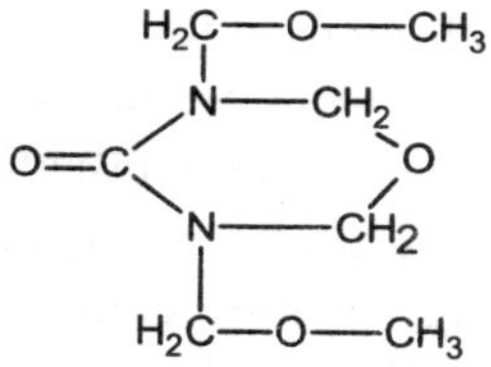

3,5-bis(methoxymethyl)-1,3,5-oxadiazinan-4-one

(VIII)

Methanol reacts very readily to form ethers. The higher alcohols react less rapidly, and conditions are more difficult since the higher alcohols are not soluble in water or methylolamine solutions. In many such cases the methylolamine is first modified with butyl alcohol, and the solution in butyl alcohol is heated with the water-insoluble alcohol. Such treatment has been suggested to utilize the glycols formed by reduction of dimer fatty acids formed on heat-bodying. On heating, butyl alcohol may be volatilized and replaced by the less volatile glycol. Resins may also be incorporated by using the butyl-alcohol-soluble amino resin. Polyesters from glycerol, phthalic anhydride, and castor oil can be used, the hydroxy groups in the castor oil acids being available for reaction with the amino resin. Resins, such as maleic-rosin glyceride, may also be present with the castor oil alkyd.

## Accelerators for Amino Resins

Amino resins are usually supplied in a reactive state as molding powders, adhesives, and other forms, which require further condensation to achieve the desired properties. It is, however, highly desirable that the uncured materials possess good stability and do not advance appreciably on being stored for months. Such stability on storage is not compatible with rapid cure, so catalysts are incorporated, with the resins or added just prior to use. Dry catalysts may, however, be incorporated with dry resins, which are stable until treated with water. The literature on accelerators for amino plastics and resins is extensive indeed, and reflects a great deal of ingenuity in devising compounds that will promote activity under conditions of use. Ammonium salts, particularly ammonium chloride, have been used to the greatest extent. The action has been explained by the hydrolysis to ammonia and to hydrochloric acid. The ammonia is taken up by formaldehyde, thus leaving the free acid to accelerate the further condensation of the amine and the aldehyde.

The acidity of the medium is the usual means of controlling the extent and rate of condensation between an amine and formaldehyde. In alkaline solutions, a complete cure to an insoluble resin is probably never achieved. As acidity is increased, the rate of cure is increased proportionally, and the hydrogen-ion concentration (pH) usually affords a good index of the rate at which condensation between the amine and formaldehyde will occur. In the case of liquid applications, such as adhesives, it is convenient to add a solution of an accelerator, but in the case of moulding powders, addition of an accelerator just prior to use is not easily accomplished. In such cases, latent accelerators are usually employed, that is, materials which become active or liberate activating acids on the application of heat. Organic halogen derivatives are favourite materials in this class since there are many such compounds which liberate the hydrogen halide on heating. Bromohydrocinnamic acid is a typical example of this type of material. On heating to molding temperatures, hydrogen bromide is liberated; this causes rapid curing of the resin. Cinnamic acid has also been used for this purpose. It is not particularly soluble at ordinary temperatures, but becomes soluble at the temperature of molding. Since it is a much weaker acid than the mineral acids, its accelerating effect is not pronounced.

Among the halogen derivatives that are useful as latent accelerators is (3-chloroethylurea), $ClCH_2CH_2NHCONH_2$. The corresponding bromine compound is also effective, as well as the β-dibromo- and tribromoethylureas. These all depend on liberation of hydrochloric or hydrobromic acid on heating to cure the amino resin rapidly. Esters of phosphoric acid are also used, particularly where solubility in organic solvents is desired. From 0.1 to 2% of the weight of the dry resin is recommended, and the methyl esters and butyl esters of phosphoric acid are effective. Hydrolysis makes the esters active, and the acid esters, or partially hydrolyzed esters of phosphoric acid, have also been used.

Amides or imides which, on hydrolysis, liberate organic acids such as phthalic or benzoic acid have been recommended as accelerators for urea-formaldehyde molding powders. Among the imides that have been suggested are N-benzoyl-succinimide, N-phthaloyldiphthalimide, and

N-propionylphthalimide. Typical of the amides recommended are N-benzoyltoluenesulphonamide and N-benzoylsaccharin.

If the resins contain an excess of formaldehyde or reactive methylol groups, the addition of small amounts of urea may hasten the cure. Materials which com-bine readily with aldehydes such as active phenols or resorcinol, may also be used as accelerators.

Materials to stabilize amino plastics until they are used are also added, particularly in solution form. The most effective method is to adjust the pH to the neutral point 7. Buffer salts have been used, but often added electrolytes are undesirable. The lower alcohols particularly methanol, have a pronounced stabilizing effect. The nonvolatile glycols and glycerols also have a stabilizing effect, but they also retard the final cure of the resin.

## Manufacture of Amino Resins and Plastics

Amino resins are usually produced from aqueous formaldehyde. In the case of the alcohol-modified resins, preparation from the anhydrous paraformaldehyde is somewhat simpler; however, due to its greater cost, it is not greatly used for this type of resin. The conditions of manufacture differ - considerably with the type of resin to be produced, and this depends on the application in which it is to be used. Amino resins are supplied in water or alcohol solutions, and as dry powders and as filled powders for moulding.

Equipment for the manufacture of amine-aldehyde resins consists chiefly of kettles for carrying out the condensation of the amine with formaldehyde, evaporators for concentrating the resin, and some type of dryer. For the reaction kettle, nickel has been recommended as the best material of construction. Twenty per cent nickel-clad steel has been found to be satisfactory for the inner shell of the reactor. The horseshoe-type agitator is used since high-speed agitators are not useful with highly viscous resins. Five hundred feet per minute is the maximum velocity of the outside edge of the agitator, and the clearance between the agitator and the kettle should be about 1/4 inch.

## Urea- and Melamine-formaldehyde Resins

The simplest condensates are the methylolureas or methyloln1elamines, and these are usually prepared by mixing the amine and formaldehyde for a short time. A typical low-stage resin is formed when urea is mixed with 2 moles of formalde-hyde (as 37% solution) or a slight excess. The mixture is adjusted to a slightly alkaline reaction, pH 7.6-8. The solution cools on addition of urea, which dissolves with the absorption of heat. The mixture is kept below 25°C. and preferably below 15 a for 2 days. The precipitate which forms is filtered off and washed with alcohol.

Since melamine is not readily soluble in water or formalin at room temperature, heating to 80-90°C. for a short time is usually employed to form methylol compounds. An example of such a method comprises heating 1 mole of melamine with 8 moles of neutral formalin until solution of the melamine occurs, and for 10 minutes more. The mixture is cooled and allowed to stand for 2 days, then filtered and washed with alcohol. Analysis indicates that hexamethylolmelamine is formed.

A crude trimethylolmelamine is prepared by heating 3 moles of melamine with 10 moles of 30% formaldehyde at pH 9 for about 30 minutes.

Adhesive resins are usually carried somewhat further than the methylol stage, and are offered as syrups or dry powders. Such a resin can be prepared from urea by heating 2 moles of formalin with 1 mole of urea, adjusted to pH 4.5-5.0 by addition of sodium hydroxide solution. The mixture is refluxed 1 hour, the amount of excess formaldehyde is titrated with sodium sulfite, sufficient ammonia is added to combine with this excess to form hexamethylenetetramine, and refluxing is con-tinued for an hour. The solution may then be evaporated in vacuum. Many of the urea resins for adhesives are spray-dried to 2-3% moisture content.

Methods used in Germany for forming urea resins for adhesives have recently been disclosed. One method consists in condensing 1 mole of urea with 2 moles of 30% formalin. The resin is çoncentrated at 65°C at pH 7 to 55% resin solids. A minor amount of wood flour, rye, or potato flour is added before spray drying.

Adhesive resins from melamine may be made by heating melamine with neutral formalin at 80°C. Heating is continued until a sample becomes turbid on dilution with 3 parts of water. The solution is then cooled, and evaporated *in vacuo* to a thick syrup.

Resins for molding require little more processing than do adhesive resins. German methods are possibly not the most advanced, but are illustrative of methods for this type of plastic material. Two moles of urea are condensed with 3 moles of 30% formalin, using ammonia to adjust the reaction at pH 7 throughout the condensation, grinding, and mixing. Alpha-cellulose was used as filler and zinc stearate as mold lubricant. Hexamethylenetetramine and zinc oxide were also incorporated in the mix. α-Dichlorohydrin (0.8%) was added to the molding powder as the accelerator. Wood flour was used in place of the commercial alpha-cellulose, using equal weights of resin and filler. However, the stability of this wood-flour-milled material was not good. In other formulations, 20-30% of the urea was replaced with thiourea. Zinc sulfate was used as the hardener where molded parts came in contact with hot liquids.

Another method for producing a urea molding powder consists in con-densing 1 mole of urea and 2 moles of 37% formalin in alkaline solution using 1/2 mole of ammonia with a small amount of fixed alkali. This solution can be evaporated, fillers added, and the mixture dried and ground. Accelerators are added to neutralize the alkalinity and create an acid medium; 0.1 to 0.2% of *b*-bromohydrocinnamic acid may be used. Lowęr ratios of formaldehyde to urea may be employed. The reaction may be carried out at 25-28°, and the water is largely removed by evaporating at 70°C. or below. The syrup so obtained is added to sulfite pulp and dried in a tunnel dryer, between 40° and 700 to below 3% moisture.

The dried material is ground in a ball mill. Benzoyl peroxide may also be used as an accelerator, since, on heating, benzoic acid is formed; the latter acts as an acid catalyst for the curing of the resin. Thiourea molding materials are prepared by similar methods. Thiourea is condensed with 37% formalin in alkaline solution using sodium hydroxide solution. The condensation is carried out for several hours, the reaction product is kneaded with fibrous fillers and dried. A mixed urea-melamine compound may be prepared from 1 mole each of mel-amine and urea, with 3 moles of 30% formalin. The pH is adjusted to 6.8 with ammonia and then heated at 800 for 20 minutes. The solution is filtered, and the pH adjusted to 7.0 with sodium hydroxide solution. Cellulose and zinc stearate (as a mold lubricant) are added. The material is then dried and ground. It is claimed that this resin is nearly equal in performance to one made from melamine alone.

In the recently developed melamine-formaldehyde resins for treatment of paper, it has been found that resins formed in strongly acid solution, or aged in strongly acid solutions, acquire a positive charge and are more readily taken up by the fibers. To prepare positively charged resin solutions, tri- or hexamethylol-melamine is dissolved in water to about 15% solids, and 1 mole of hydrochloric acid per mole of methylolmelamine is added, and the solution is heated to 50 DC. Free formaldehyde is liberated, and in both cases about 2.5 moles of formalin remain combined with each mole of melamine.

## Alcohol-modified Resins

The formation of alcohol-soluble resins depends on forming ethers of the methyl-olureas in the presence of acid catalysts. Since acid catalysts are used, which also will accelerate the condensation of the amine aldehyde resins, conditions must be controlled to prevent precipitation of the resin. It is possible to produce rather low-stage modified resins, or to polymerize the resins to somewhat more viscous materials.

The methyl ethers may be prepared by condensing 2 moles of formaldehyde with 1 mole of urea at pH 4.5-5.0 under reflux for 1 hour. The solution is then neutralized and spray-dried. Sixty parts of the dried powder is added gradually to 40 parts of methanol containing 1 part of maleic acid at 60°. Methyl ethers of urea-formaldehyde condensates have been suggested as plasticizers for amino-formalde-hyde molding resins.

The butyl-alcohol-modified urea resins have been prepared by a variety of methods. The usual method consists of forming the condensation product of about 2 moles of aqueous formalin with 1 mole of urea in a slightly alkaline solution, and drying the product at moderate temperatures. Dimethylolurea is the principal component of these resins and may be used to form the alcohol-soluble resins. The dry condensate is boiled with butyl alcohol in the presence of acid catalysts and the water of condensation is removed with the boiling alcohol. More butyl alcohol may be added to complete the condensation. The reaction is usually carried on until the solution obtained can be diluted with toluene without resulting in precipitation.

It is not essential to dry the urea-formaldehyde condensate, since a butyl-alcohol-modified resin can be prepared by refluxing 60 parts of urea with 243 parts of 37% formalin in the presence of 5 parts ammonia water (sp.gr. 0.88) for 3 hours. Two moles of formaldehyde were found to have been combined in this treatment; 148 parts of butanol and 2.5 parts of 75% phosphoric acid were added, and the mixture was dehydrated *in vacuo*. Butyl alcohol was added to replace the alcohol lost in the dehydration.

One gram mole of urea may be condensed at pH 8-10 with 2 to 2.5 moles of formalin for 1 hour at 85°C. The resin is dried at 55-60 ° at 26-29 in. vacuum and added to 225 grams of butyl alcohol containing 1 gram of 75% phosphoric acid, and heated at 90-95°C until a clear solution is obtained. The solution is then distilled to remove water formed in the condensation and to obtain the desired solids content.

Melamine-formaldehyde resin solutions in butyl alcohol may be obtained in a somewhat similar manner. Hexamethylolmelamine is stirred for an hour with twice its weight of butyl alcohol containing a small amount of 85% phosphoric acid. The mixture is then heated to its boiling point and the butyl alcohol-water azeotrope distills off and is condensed. The water is separated, and the alcohol layer returned to the reaction. The solution is evaporated to 50% resin content when dehydration is complete.

Higher alcohols can be used in place of butyl alcohol; lower alcohols are often added to hold the resin in solution until the ether of the higher alcohol is formed. Urea and three times its weight of 37% formalin are heated at reflux temperature with β-ethylhexyl alcohol and methanol in the presence of phosphoric acid.

More methanol is added if necessary to form a clear solution. The wet methanol which distills off is condensed and an equal volume of dry methanol added to the reaction vessel until dehydration is complete. More β-ethylhexyl alcohol is added and the solution is evaporated *in vacuo* until it contains less than 0.5% methanol. Butyl Carbitol can be used in place of β-ethylhexyl alcohol.

## Sulphonamide Resins

Sulphonamide resins can be made by refluxing equimolar amounts of the sulphonamide and 37% aqueous formaldehyde for 3 hours, and subsequently dehydrating to form brittle, thermoplastic resins. These resins may be used as plasticizers, or the sulphonamides may be condensed with formaldehyde in the pres-ence of urea to produce internally plasticized resins.

## Aniline-formaldehyde Resins

Since aniline is readily soluble in formalin in acid solution, condensation is carried out in the presence of a strong acid, usually hydrochloric acid. The reaction between aniline and formaldehyde is exothermic. If 1 mole of formaldehyde is used per mole of aniline, the condensates are soluble and readily fusible. As more formaldehyde is condensed with aniline, the condensate becomes progressively less soluble and nearly insoluble condensates can be obtained. Fillers, either cellulose or mineral fillers, are added to the solution of the resin, and the resin is precipitated by addition of alkali. The precipitate is washed thoroughly with water and dried. The resins are thermoplastic but do not have sufficiently easy flow for use in injec-tion molding.

## Applications of Amino Resins and Plastics

The number of applications for amino resins has increased greatly in recent years. Uses for urea-formaldehyde resins have shown an especially great expansion in this period. During the war period, they replaced phenolic moldings for civilian uses to some extent, but the use of urea resins in war materials expanded greatly, and their use for civilian purposes was largely shut off. The fields of application may be roughly classified as those in paper manufacturing, in textiles, in adhesives, in molding, and in enamels and lacquers. Urea resins have been used in all of these fields and melamine resins are rapidly moving into these same fields, often in a somewhat different part of the field. The plastics from aniline have not been used so extensively and are chiefly used in molded articles, while the sulphonamide resins are used chiefly in lacquers.

## Commercial Resins and Plastics

Table 38.1 includes many of the common commercial resins and plastics classified according to their use. No attempt has been made to include air the commercial varieties, but, as far as possible, the best-known materials in each class have been included. The producers and the trade names under which the particular resins are sold are included.

**Table 38.1: Common Commercial Resins and Plastics**

| Trade name | Type | Producer[a] |
|---|---|---|
| | **Papers Resins** | |
| Parez | Melamine-formaldehyde | Am. Cy. |
| Uformite | Urea-formaldehyde | Res. Prod. |
| | **Textile Resins** | |
| Aerotex | Melamine- and urea-formaldehyde | Am. Cy. |
| Lanaset | Melamine-formaldehyde | Am. Cy. |
| Resloom | Melamine-formaldehyde | Monsanto |
| Rhonite | Urea-formaldehyde | Rohm & Haas |
| Lacet | Melamine-formaldehyde | Am. Cy. |
| | **Adhesive and Laminating Resins** | |
| Melmac | Melamine-formaldehyde | Am. Cy. |
| Melurac | Urea- and melamine-formaldehyde | Am. Cy. |
| Bakelite | Urea-formaldehyde | Bakelite |
| Plaskon | Urea-formaldehyde | Plaskon |
| Uformite | Urea-formaldehyde | Res. Prod. |
| Urac | Urea-formaldehyde | Am. Cy. |
| Cascamite | Urea-formaldehyde | Casein |

| Trade name | Type | Producer[a] |
|---|---|---|
| | **Molding Powders** | |
| Beetle | Urea-formaldehyde | Am. Cy. |
| Cibanite | Aniline-formaldehyde | Ciba |
| Melmac | Melamine-formaldehyde | Am. Cy. |
| Plaskon | Urea-and melamine-formaldehyde | Plaskon |
| Resimene | Melamine-formaldehyde | Monsanto |
| | **Resin Solutions for Lacquers** | |
| Beckamine | Urea- and melamine-formaldehyde | Reichhold |
| Beetle | Urea-formaldehyde | Am. Cy. |
| Melamac | Melamine-formaldehyde | Am. Cy. |
| Santolite | Sulfonamide-formaldehyde | Monsanto |
| Plskon | Urea-formaldehyde | Plaskon |
| Uformite | Urea-and melamine-formaldehyde | Res.Prod |

[a]Am. Cy., American Cynamid Co., Plastics Division, N.Y.; Casein, Casein Company of America, N.Y.; Ciba, Ciba Products Corp., Hoboken, N.J.; Monsanto, Monsanto Chemical Co., Plastics Division, Springfield, Mss; Plaskon Plaskon Division, Libbey-Owens-Ford Glass Co., Toledo; Reichhold, Reichoold Chemicals, Inc., Detroit; Res. Prod., TheResinous Products & Chemical Co., Philadelphia; Rohm & Haas, Rohm & Hass Co., Inc., Philadelphia.

## Resins for Textiles

The urea- and melamine-formaldehyde resins are being actively developed in the treatment of textiles. The urea resins were the first to be used, since they have been available for a much longer time. The resins are either soluble or readily dispersible in water, and therefore they can be readiy and economically applied. The amino-formaldehyde resins are usually laid down inside the fibres,and thus the treated fabrics often appear to be unchanged by the treat-ment. Since the cured resins are hard and brittle, their presence as a film on the surface of the fabric would render it hard and stiff. The amount of resin that can be taken up by the fibers, before saturation and resulting surface hardness, will vary with the type of resin, method of application, and the type of fiber that is treated. Table 35.2 shows the effect of use of various amounts of a water-soluble urea formaldehyde resin on the properties of cotton and spun-rayon fabrics. It will be seen that the spun rayon takes up a considerably larger amount of resin without becoming much stiffer. In many applications, methylolurea or methylmelamine is employed, although the ethers from the lower alcohols or from glycol are also employed. A wetting agent is often added to ensure wetting out of the fabric, and a catalyst to accelerate the setup of the resin. Ammonium chloride may be used for this purpose. A more highly polymerized resin often is less effective in improving abrasion resistance. The low-molecular methylol derivatives are more effective in improving the shrink resistance of cotton, wool, and rayon. In a wide variety of fabrics including cotton, rayon, and woolen fabrics, the shrinkage was reduced 80% by treatment with methylolmelamine.

**Table 38.2: Urea-Formaldehyde Resin in Fabrics**

| Resin content, per cent of fibre | Stiffness | | Tensile strength, p.s.i. | |
|---|---|---|---|---|
| | Cotton | Spun rayon | Cotton | Spun rayon |
| 0 | 2.0 | 2.0 | 118 | 120 |
| 1 | 2.1 | 2.2 | 114 | 124 |
| 3 | 2.1 | 2.4 | 112 | 127 |
| 8 | 2.4 | 2.5 | 98 | 130 |
| 20 | 9.4 | 3.6 | 86 | 114 |

Methylolurea and methylolmelamine, when applied to cotton fabrics and cured, cannot be removed without destroying the fabric. It is possible that the methylol derivatives combine with the cellulose. Since cellulose is an alcohol, it may combine with the amino resins, although it has not been possible to measure the extent of combination.

The methylolureas and methylolmelamines are soluble in water on heating and are stable at high dilution, and their ethers with the lower alcohols are also readily soluble in water. The fabrics are impregnated with a 12% solution in water. Diammonium phosphate or ammonium chloride may be used as the catalyst. To cure the resins, 30 minutes is required at 225°, but the cure is not entirely complete at this temperature. At 250°, 10 minutes is required while, at 275°,4 to 5 minutes is sufficient. The treated fabrics take up certain dyes more slowly than untreated materials, and this behavior can be used to indicate whether the resin is fully cured. The treatment does not appreciably affect the appearance or feel of the fabrics, but shrinkage on washing is greatly reduced.

Fabrics are also treated with amino resins, in combination with other resins, to improve their water repellency. In this type of treatment, a coating of resins is applied to the fabrics. Urea and melamine resins modified with alcohols have been used with other resins to harden and reduce their stickiness at high temperatures. Polyvinyl butyral has been largely used for waterproofing fabrics, and the use of alcohol-modified urea and melamine resins has greatly improved the heat resistance of the coating.

## Paper-treating Resins

One use of urea and melamine resins, somewhat related to their use in textiles, has been in the treatment of paper pulp. Since the resin must usually be completely adsorbed on the pulp for economical operation, special resins have had to be developed for the papermaking operation. The use of the resins has made possible the manufacture of papers of much greater wet strength. The urea resin for such applications may be supplied as a dry powder or as a sirup with high solids content. The resin is best applied in the beater. Hydrochloric acid, alumi-num chloride, or alum may be used as the catalyst; the alum is satisfactory and may be used in conjunction with hydrochloric acid. Rosin size may be used with the urea resin to improve water resistance, and asphalt emulsions have been used with urea resins in making fiberboard. About 2-3% resin (on the weight of paper) may be used to give good results; larger amounts do not greatly increase the wet strength. The urea resin size solution (5%) with catalyst (ammonium sulfate or aluminum chloride) is stable for several days at 49°C but cures in 1 to 2 minutes on drying the paper.

Special resins have been developed which can be added directly to the beater and are largely taken up by the paper pulp. The wet strength of the resin-treated paper is greatly increased and the dry strength is also higher. Map paper for the army is usually resin-treated.

Bag paper, blueprint and chart paper, and food wrappings have also been resin-treated to improve their wet strength.

The preparation of melamine resins for use with paper has been carefully studied. Resins from melamine and formaldehyde have been developed which can be applied in the beater or the headbox, and which are adsorbed by the paper pulp. The early resins required overnight aging in acid solution, but the aging time has been reduced to 3 hours. The resin is supplied as a fine powder containing not over 3% of moisture. This powder is readily soluble in water up to 20% solids content. Hydrochloric acid, 0.8 mole per mole of resin, is added and the solution is aged 3 hours at 80-100°F. The resin is more effective if added just before forming the sheet. The pH should be adjusted to 4. The aging process results in positively charged resin particles which are readily adsorbed on the paper. The treated paper improves in properties for several days, particularly when rolled hot and allowed to stand in the roll. The resin is largely adsorbed by the finer fibers of the pulp. The treated paper is greatly improved in dry and wet tensile strength; in one case over four time the untreated wet strength was obtained. The folding and breaking strengths are also greatly improved by the resin treatment. During the war, thou-sands of tons of paper were resin-treated for the armed forces.

## Adhesives

Amino resins have been widely used as adhesives and, as in many other applications, urea-formaldehyde resins have achieved the largest volume, while melamine resins have more recently been adopted, wherever their unusual properties make them desirable. One definite advantage of the urea-formaldehyde adhesives has been the speed with which they can be cured, and it is possible to effect a cure at ordinary temperatures. Since they are applied in water solution, and can be extended with inexpensive fillers, they are fairly low-cost adhesives.

A variety of urea-formaldehyde adhesives are commercially available. Most are in the form of dry powders, but resin syrups are also available. Catalysts are used to cure the resins; often a choice of accelerators is offered, depending on the temperature at which the cure is to be accomplished. Some powdered resins are available which contain the catalyst, and which are activated by putting the resin in, solution in water. The catalyst is also supplied separately, either as a dry powderor as a solution, and a small amount is added to the resin before use. This is often advantageous, since the life of a catalyzed resin solution is often only a few hours. Catalysts consist of ammonium salts-ammonium chloride is often used or acids or salts which liberate acids by hydrolysis or on heating.

Urea resins have fairly good resistance to water even when extended with starch; melamine resins have rather better resistance to water and are used where this resistance is required. Phenolic resins are often more water-resistant, but their dark color and the higher temperatures usually required to cure them have limited their use. Fiberboard containers have been successfully laminated with urea resins where better strength when wet is required. Dextrin is often used to extend the urea resin, and this is dissolved in water at 88ºC.; the urea-formaldehyde resin is added after cooling to 72ºC, and the catalyst is added at 50°C or below.

For laminated wood, the veneers are coated with a solution containing about 60% water-soluble urea-formaldehyde resin containing about 2% catalyst (on the weight of resin). Fifty pounds of wheat flour may be added to each one hundred pounds of resin without greatly reducing the water resistance of the adhesive. The glue solution usually has a useful life of about 24 hours. The adhesive is spread on the veneers with grooved rubber rolls, using 30-40 lb. of glue per 1000 sq.ft. The veneers should be assembled within 30 minutes of the time they are coated. The adhesive is cured by heating in a hot press at 220-240 of. Temperatures as high as 300° may be used for thick sections, and relatively high pressures are usually required (200-300 p.s.i.).

Resins are also employed which can be used for plywood at ordinary tempera-tures. The life of the resin solution after the catalyst has been incorporated is rather short, usually about 4 hours. Low pressures are sufficient (150 p.s.i.), and at 3.5°C the cure is nearly complete in 2 hours. Higher temperatures will give shorter curing cycles. A somewhat similar urea resin is used for gluing lumber which will set up in 30 minutes at 3.5°C, and at 49° in 30 minutes when extended with wheat flour.

Resins from urea and melamine with formaldehyde are also available and yield the relatively rapid cure of the urea resins with the improved water resistance of the melamine resins. Mixtures of melamine-formaldehyde with urea-formaldehyde resins are also employed for somewhat similar purposes. Adhesives from melamine and formaldehyde are available as dry powders and are used in much the same way as the urea-formaldehyde resins. Adhesives have been developed using special catalysts. which cure at temperatures as low as 60°C and which are resistant to boiling water when cured. Water-soluble melamine-formaldehyde condensates have been found useful in rendering cellulose ethers water-insoluble; 5 to 15% of resin is required. The melamine-formaldehyde resins are more effective than the urea-formaldehyde resins.

## Laminating Resins

Phenolic resins have been widely employed in producing laminates with paper and fabrics. The amino resins are employed in this field principally in applications where their improved color is required. Melamine-formaldehyde resins have superior arc resistance, which indicates their use in some electrical applications.Water-soluble urea resins are usually used, but the alcohol-modified resins also find application. The fabric or paper is dipped in the resin solution, and, after squeezing or scraping off the excess, is dried under controlled conditions to partially set up the resin. The sheets are cut to size, laid up to the desired thickness, and consolidated with heat and pressure. Table and counter tops are familiar ex-amples of such laminates. Translucent laminates from urea-formaldehyde resins are used for light fixtures. Certain grades can be shaped by heating at 250-300°C to give curved shapes.

Melamine resins have recently been used for laminated articles with better heat resistance and surface hardness. It has been possible to obtain a heavy surface layer of cured resin, resulting in improved resistance to abrasion and better resist-ance to moisture. The sheets are cured at 140-165°C. at 1000 p.s.i. for a few minutes to several hours, depending on the thickness of the piece. The use of glass fabrics with melamine resins has made possible fabrication of laminates with excellent are resistance and heat resistance. A water-soluble melamine-formaldehyde resin is used for impregnating the glass fabric, but 5% of butyl alcohol is added to the water to improve wetting of the glass fibres. About 40% of resin is added to the fibres; more resin results in crazing, while the strength is much less as the resin content is decreased.

## Moulding Resins

Amino plastics derive their name from their use in molding operations. As we have seen from the applications considered above, they are now used in a variety of ways, all closely related to, but not strictly as, plastics. Application as moulding materials continues to be one of the important fields in which they are used. Urea plastics were first developed as clear, transparent castings. Stability of solid cast articles was never satisfactory, and amino plastics are now supplied with fillers. Cellulose is the usual filler; although wood flour has been used, good stability was not achieved through its use. Paper pulps are widely used as fillers for urea and melamine resins. Such pulps should have at least 88% alpha cellulose content and less than 4% material soluble in 1% sodium hydroxide solution, for best results in moldings. A minimum degradation of pulp furnishes the most desirable filler. Of the amino plastics, the urea is the most important in terms of volume. Melamine plastics are used where their better heat resistance or greater resistance to water is desired. Aniline-formaldehyde resins, and aniline-phenol-formaldehyde resins, are used in special applications, particularly in molding electrical equipment.

Urea plastics are supplied in granules or as a fine powder. These granules are about 50% urea resin with cellulose filler. Pigments and catalysts are also included. They are availab le in a variety of degrees of plasticity where slow or easy flow is desired. The less plastic granules can be moulded at somewhat higher temperatures than the softer grades. For large shapes with a long draw, a long-flowing type of molding material is offered. The molding compound is supplied dry and should be kept tightly sealed, especially in humid atmospheres. It should be stored as far as possible below 25ºC, since prolonged, storage at high, temperature than other soft gradesnm in further condensation of the resin, and reduces its plasticity. The usual methods of molding thermosetting plastics are applicable to urea, plastics. The granules are preformed to give pills or tablets of the desired weight, having a density approaching that of the finished piece. Moulding of urea resins is carried out at 138-160ºC, which can be reached at steam pressures from 40-120 p.s.i. Pressures of 1 to 4 tons per sq.in. are used. Since urea resins overcure or "burn" readily, a temperature of 320°C should not be maintained for more than 2 minutes. By preheating the molding material, considerable time can be saved in the molding operation, and lower pressures can be employed. Heating of preforms to 80ºC before placing in the mold is effective in this respect. If higher temperatures are used, the time of preheating must not be too long, or the resin may harden and fail to mold satisfactorily. The time required for curing the resin depends largely en the thickness and size of the piece and may vary from less than a minute to ten minutes. Urea moldings have many different uses; familiar ones include buttons, clock cases, display boxes, closures, lighting reflectors, and tableware.

**Table 38.3: Properties of Amino Plastics[a]**

| Properties | Urea-formaldehyde Cellulose | Melamine-formaldehyde | | | Aniline-formaldehyde Unifilled |
|---|---|---|---|---|---|
| | | Cellulose | Fabric | Mineral | |
| Sp. gr. | 1.48–1.50 | 1.49 | 1.40-1.45 | 1.7-2.0 | 1.22-1.25 |
| Heat distortion pt., °F | 260-280 | 385 | — | 266 | 210-245 |
| Compresion ratio | 2-3 | 2-2.7 | 10-15 | 2.1-2.5 | 2.5-3 |
| Moulding temp., °F | 280-325 | 280-340 | 300-370 | 280-360 | 300-340 |
| Moulding pressure, tons/sq. in | 1-3 | 1-3 | 2-3 | 1-2 | 0.75-1 |
| Transfer moulding pressure, tons/sq. in | — | 2-10 | 5-15 | 2-10 | — |
| Heat resistance continuance, °F | 170 | 210 | 230 | 300-400 | 180-190 |
| Mould shrinkage, mils/in. | 6-11 | 8-15 | 3-5 | 4-7 | 2-3 |
| Tensile strength, 1000 p.s.i. | 8-13 | 8-13 | — | 5-7 | 8-12 |
| Impact strength (notched Izod), ft.-lb. | 0.28-0.32 | 0.24-0.35 | 05.-0.8 | 0.3-0.4 | 0.32 |
| Water absorption (24 hr.), % | 1-3 | 0.3-0.6 | 0.1-0.2 | 0.08-0.14 | 0.7-4.5 |
| Burning rate | Very slow | Nil | Nil | Nil | Slow |
| Dielectric strength, v./mil | 300-400 | 340 | 250-350 | 390 | 600-650 |
| Power factor (60 cycles), % | 4-4.9 | 3.7 | — | 4.5 | 0.2 |
| Power factor ($10^6$ cycles), % | 2.65 | 2.9-4.5 | 4.1 | 2.8 | 0.6-0.8 |
| Dielectric const. (60 cycles) | 7.6-8.2 | 7.9 | — | 7.7 | 3.8 |
| Dielectric const. ($10^6$ cycles) | 6.6 | 7.0 | 7.2 | 5.7 | 3.5 |
| Tracking resistance | — | Exc. | Fair | Exc. | — |
| Arc resistance, sec. | 120 | 125 | 111 | 130-190 | — |

Melamine resin moulding compounds are available in several varieties, depending on the filler employed-cellulose-filled, mineral-filled, and chopped-fabric-filled. The cellulose-filled variety is moulded under much the same conditions as urea resins, except that the temperature of molding can be carried to 340°F. Momentary release of pressure toward the end of the cycle to relieve gas pressure may be necessary. When good electrical properties are required, a mineral-filled melamine moulding compound is employed. Mica, asbestos, and to a less extent talc, are typical mineral fillers. These mineral-filled compounds are supplied in three degrees of flow-soft, medium, and hard. The soft variety flows at about one-half the pressure required by the usual soft phenolic or urea molding materials. The mineral-filled materials have a rather long curing cycle - 2 to 15 minutes may be required. Transfer molding may be used to advantage where close tolerances must be met. The fabric-filled melamine resins are used when high resistance to shock is required; their molding is comparable to that of other bulky molding materials. High pressures, 5-15 tons per sq.in., are needed to mold these materials.

Aniline-formaldehyde moulding materials are thermoplastic, but are not plastic enough to be used for injection molding. The moldings are translucent and are reddish-brown. Aniline-formaldehyde resins are notable for good electrical properties, and the low-loss properties are not greatly decreased on prolonged exposure to moisture. Aniline-formaldehyde plastics are available in sheets 1/16 to 1 in thick and in rods. Tubes are formed by heating the sheets and forming. Varnishes of aniline-formaldehyde resins are also available.

Some of the properties of molded amino plastics are shown in Table 38.3. For details on the tests reported, Specifications have been established for three types of urea-formaldehyde and four types of melamine-formaldehyde plastics. The impact strength of urea moldings has been found to show con-siderable loss on aging 1 year (18). Soaking in water lowers the impact strength about 25%. After holding urea moldings 4 weeks at 65°, the moisture resistance was slightly lowered.

## Coating Resins

Amino resins are seldom used alone in coatings *(q.v.)* since they are too brittle for most applications. They find their chief use as hardening agents for softer, more flexible coating materials, Alkyd resins are chiefly used for this purpose; the plasticizing and drying type of alkyds can both be employed. The color is lighter with nondrying alkyds, and the faster-drying types of alkyds usually are found to impart more color. Not all alkyd resins are applicable with amino resins, but several alkyds have been developed for use with amino resins.

Of the amino resins the urea- and melamine-formaldehyde resins modified with butyl alcohol or with capryl alcohol are usually used. Resins modified with alcohols lower than propyl alcohol cannot be readily blended with the alkyds, nor can they be diluted with hydrocarbon solvents. The capryl-alcohol-modified resins are miscible with a greater variety of resins, and can be used with a few oleo-resinous vehicles. Melamine resins bear their usual relation to the urea resins, being more expensive at the present time, but having certain superior characteristics, particularly greater hardness and better heat resistance. In most vehicles in which these resins are used, they are present in a minor amount, and often represent from 20 to 33% of the resin content.

Since alkyd resins often contain hydroxyl groups and can be used as alcohols, it is possible to condense the methylolureas with the alkyd resin to obtain alkyd -modified urea resins. The conditions for producing homogeneous vehicles are rather limited. These alkyd-modified urea resins can be used alone or may be used in conjunction with other alkyd resins.

The urea-formaldehyde resins modified with butyl alcohol are supplied in solu-tion. in butyl alcohol, or in butyl alcohol-xylene mixtures, and have a solids content of 50-60%. The capryl-alcohol-modified resins usually are supplied in solution in capryl alcohol. The alkyd resin solution is usually used as the grinding vehicle to disperse the pigments, although some of the urea solutions may be used. The ratio of alkyd resin to urea resin solution varies principally with the type of alkyd resin used. With the nondrying alkyds, such as castor oil-phthalic alkyds, and the sebacic acid alkyds, equal amounts of urea and alkyd resin may be employed. This is usually the upper limit, and more alkyd than urea resin is often used. With the drying type of alkyds, 2 to 4 parts of alkyd resin may be used to 1 part of urea resin. The resins are usually baked to dry the alkyd resin and complete the condensation of the urea resin. Phosphoric acid esters have been suggested to effect hardening of urea resins at ordinary temperatures. The usual baking schedules begin at 42°C., at which temperature 60 minutes is required. At 66°C the baking cycle is 20 minutes. Higher temperatures will lead to some discolouration in white enamels, but in colors, 175°C can be used and 15 minutes is sufficient. Greater hardness can be achieved with the use of higher temperatures, and enamels baked at 300°C will develop considerably greater hardness than those baked at lower temperatures.

Melamine-formaldehyde resins, when modified with butyl alcohol or capryl alcohol, are used in very much the same way as the urea resins. They impart greater hardness with alkyds at the same ratio of amino resin to alkyd, and can be used to give the same hardness with lower amounts of melamine resin. Higher temperatures can be employed in baking the melamine resins without causing dis-coloration of white enamels. Alcohol-modified melamine-aldehydes resins are miscible with a wide variety of natural and synthetic resins, including dammar, rosin resins, indene polymers, and cellulose ethers. Melamine alcohol-modified resins are generally somewhat more miscible than the urea resins. Acid catalysts have been used with melamine resins, but the acidity of the alkyd resin is often sufficient to affect a satisfactory cure.

Finishes employing amino resins are characterized by unusual hardness, mar resistance, and stability on outdoor exposure. They are often used for automobiles, refrigerators, washing machines, machinery, kitchen cabinets, and many household and industrial appliances.

The sulphonamides have been used as plasticizers for nitrocellulose, for proteins, for cellulose acetate, and for shellac. Mixtures of o- and p-toluene sulphonamides, and a mixture of o- and p-ethyltoluenesulphonamides, are used for this purpose. When the toluenesulphonamides are condensed with formaldehyde, resins are obtained which vary from soft, sticky balsams to low-melting, hard resins. They may be used with cellulose esters and ethers, and are often advantageously used with plasti-cizers, where they may improve the retention of the plasticizers. They are useful in heat-sealing lacquers, particularly in conjunction with nitrocellulose.

## Miscellaneous Applications

It has not been possible to consider all the methods of utilizing amino resins under the headings considered above, and many new applications for these resins are being developed.

The use of melamine-formaldehyde and other amine-formaldehyde resins has been reported for treating water to remove anions. Resins are formed and rendered water-insoluble; they contain free $NH_2$ groups, which combine with the acid radicals in the water that is treated. It is thus possible, by using anion-exchanging resins together with cation-exchanging resins, to produce ion-free water by filtering through beds of the resins. The resins are revivified by treatment with an alkali, often sodium hydroxide, which removes the acidic constituents from the resin. If sulfates are present in the water, they are taken up by the resin and sodium sulfate is formed on regenerating the resin. Melamine-formaldehyde resins have been de-veloped for these types of uses; resins from aliphatic diamines are also used.

Water-soluble urea-formaldehyde resins have been used with plaster of Paris to improve the set of the plaster; parts can be molded under pressure from mixtures of the urea resin and plaster. Cores for the molding of metal parts have recently been found to be improved by use of water-soluble urea-formaldehyde resins; the resin is mixed with sand and baked; when the hot metal is poured into the mold, much of the resin is destroyed, leaving a surface that collapses readily and allows ready withdrawal of the molded part.

Foamed urea resins have been suggested as insulating materials. Water-soluble resins are beaten to foam of desired density and cured by addition of suitable catalysts. Wood flour or other fillers have been added to improve the strength. The foam sets as the resin cures and is carefully dried. If the density of the dry foam is about 0.05, the material develops cracks on drying, and materials of lower density than 0.05 are rather fragile and easily broken. However, it has been found that low-density, foamed urea resins can be collapsed under pressure to improve their strength. Foamed urea resins have been used in aircraft to fill dead air spaces in which explosive mixtures might collect. Since cured urea resins are insoluble in organic solvents, they can be used to seal asphalt and prevent it from staining painted surfaces.

Dimethylolurea is a commercial material and is the first step in formation of urea resins. It is desirable in many cases to use more highly polymerized materials than dimethylolurea, but, in cases where penetration of fibrous materials is desired, the smaller the molecule, the better the penetration. Thus, dimethylolurea is well adapted for such uses. It has been found that aqueous solutions of dimethylolurea of about 25% concentration will readily penetrate wood. Plywood may be treated in this way, and thicker sections are also impregnated. Pressure may be used to effect deeper penetration of the solution, and alternate vacuum and pressure are even more effective. The wood is swollen by the saturation and may double its weight. The solution should be held close to neutral, and should not be allowed to become more acid than pH 6. Often dimethylolurea as a higher content of methylol groups than is required, and 1 to 2 parts of urea is used with 6 parts of dimethylolurea the treated wood is dried carefully; temperatures above 88°C are required to render the resin insoluble. The resulting wood is much harder than untreated wood and is somewhat heavier. It is remarkably water-resistant and swells but slightly on prolonged exposure to water. The treatment has been recommended for veneers, furniture, flooring, boats, and office equipment.

## Identification of Amino Resins and Plastics

Amino plastics are used in such a wide variety of ways and so often where other types of resins may be employed that it is often desirable to identify the material quickly. Tests by heating a sample of the plastic material are often used for rapid characterization. Urea, thiourea, and melamine plastics are thermosetting and are not softened by heat. Aniline-formaldehyde resins are readily softened by heat, and burn readily giving off black smoke and the odor of formaldehyde. Urea and melamine plastics burn with difficulty and the odor of formaldehyde is notice-able from both materials. A sodium fusion with a positive test for nitrogen will indicate the presence of amino resins and, if sulfur also is found present, thiourea resins are indicated if the resin is not thermoplastic.

The combination of the weak acid cation exchange with degasification reduces both the calcium and magnesium levels as well as the alkalinity level in the raw water.

(2) ***Dealkalization-softening uses.*** This pretreatment should be investigated when pH adjustment of the raw water by an acid addition is indicated for the desalination process. Weak acid resins use about 10-percent more acid than that required for pH adjustment alone and will reduce the calcium and magnesium concentration as an additional advantage.

In brackish waters containing essentially only calcium, magnesium, and alkalinity, the use of weak acid cation resins with degasification could be considered as a possible desalination process. Since some types of weak acid cation resins also permit the efficient removal of sodium bicarbonate, the process becomes applicable as a desalination process when the raw water contains mainly sodium and alkalinity.

## Desalination

Ion exchange can be used as a desalination process in the production of potable water.

*a.* ***Requirements***. There are several basic requirements for the ion-exchange process to be used economically for the desalination of brackish waters. The ion-exchange resins should operate at high capacities. The ion-exchange resins should be regenerated close to the stoichiometric equivalence capacity. The acid and base regerants should be low cost. The waste regenerants should be rinsed from the ion-exchange resins with a minimum of water, so that the capacity of the resin is not exhausted significantly. Regenerant waste volumes should be minimized, and unused regenerants should be recovered and reused to reduce the waste disposal volume.

*b.* ***Limitations***. The use of ion exchange in the desalination of brackish water has several limitations. The volume of water treated is inversely proportional to the ionic concentration in the water. Regenerant consumption per unit volume of treated water is high and becomes higher as the salinity of the brackish water increases. The size of the ion-exchange equipment follows the same rationale-the more saline the water, the larger the ion-exchange equipment. A low salinity water, usually product water, is required for regeneration of the ion-exchange resins.

*c.* ***Treatment processes***. The treatment processes employed have either been on a pilot plant scale or have been used in a limited number of full-size installations. The processes have generally utilized weak acid cation and weak base anion resins. These resins have higher capacities and require less acid and base regenerants than strong acid cation and strong base anion resins.

## Process and the RDI Process

(1) ***Desal process***. The Desal Process has several variations, but the main thrust of the process is the use of the weak base anion resins in the bicarbonate form.

(2) ***RDI Process***. The RDI Process is a three unit system using four different resins. The water first passes through a strong base anion resin where the strong acid anions, such as chloride, sulfate, and nitrate, are replaced with the bicarbonate ion from the resin. The water then moves through a layered ion exchange unit of weak acid cation and strong acid cation resins, where the calcium, magnesium, and sodium are removed, the bicarbonates are converted into carbonic acid, and the neutral salt leakage from the previous anion unit is converted into free mineral acidity, i.e.,sulphuric, hydrochloric, and nitric acids.

Then, the water travels through a weak base anion resin, where the free mineral acidity is adsorbed but the carbonic acid passes through unaffected. The water is then degasified, which removes the dissolved carbon dioxide. The weak acid cation and strong acid cation resins are regenerated with either sulfuric or hydrochloric acid, first through the strong acid cation resin and then through the weak acid cation resin. The strong base anion and weak base anion resins are regenerated in series with sodium bicarbonate, first through the strong base anion resin and then throughthe weak base anion resin. The RDI Process is shown in Fig. 38.2.

*d.* ***Three-unit variation.*** In the three-unit variation, the strong acid anions in the water, such as chloride, sulfate, and nitrate, are replaced with the bicarbonate ion from a weak base anion resin in the bicarbonate form. The process then employs a weak acid cation resin that replaces the calcium, magnesium, and sodium in water with the hydrogen ion from the resin. The carbonic acid that is formed is adsorbed by a second weak base anion resin in the free-base form. When the system has exhausted its treating capability, the lead weak base anion resin is regenerated with ammonia, caustic, or lime, the weak acid cation resin is regenerated with sulfuric, hydrochloric, nitric, or sulfurous acid, and the tail-end weak base anion is not regenerated. The lead weak base anion resin is now in the free-base form and the weak acid cation resin in the hydrogen form. After its adsorption of carbonic acid in the previous service cycle, the tail-end weak base anion is in the bicarbonate form. The service flow direction is reversed for the next service cycle, with the former tailend weak base anion in the lead position and the former lead weak base anion in the tail-end position. The direction of service flow is reversed on each succeedingservice cycle after regenerating only the weak acid cation and the former lead weak base anion. This three-unit variation of the Desal Process is shown in Fig. 38.3 with the following sequence of operation: Service-A followed by Regeneration-B, Regeneration-B followed

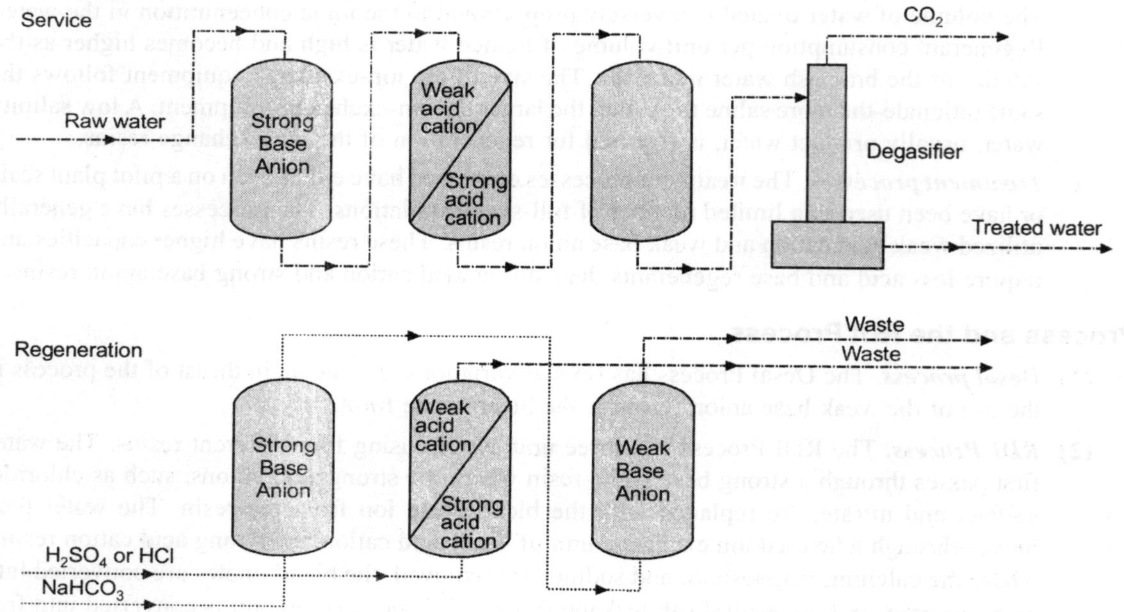


**Fig. 38.2. RDI Process**

by Service-C, Service-C followed by Regeneration-D, Regeneration-D followed by Service-A, Service-A followed by Regeneration-B, etc., in a repeating sequence. *e. Two-unit variation.* In the two-unit variation, carbon dioxide is fed to the raw water. The carbon dioxide in the water (carbonic acid) converts the weak base anion resin in the lead unit to the bicarbonate form and the strong acid anions in the water, such as chloride, sulfate, and nitrate, are replaced with the bicarbonate ion from the resin.

The process then employs a weak acid cation resin, in the same manner as the three-unit variation, which replaces the calcium, magnesium, and sodium in the water with the hydrogen ion from the resin. The carbonic acid or dissolved carbon dioxide that is formed is now removed by a degasifier.

Ammonia, caustic, or lime can be used to regenerate the weak base anion resin and sulfuric, hydrochloric, nitric, or sulfurous acid can be used to regenerate the weak acid cation resin. The two unit variation of the Desal Process is shown in Fig. 38.4.

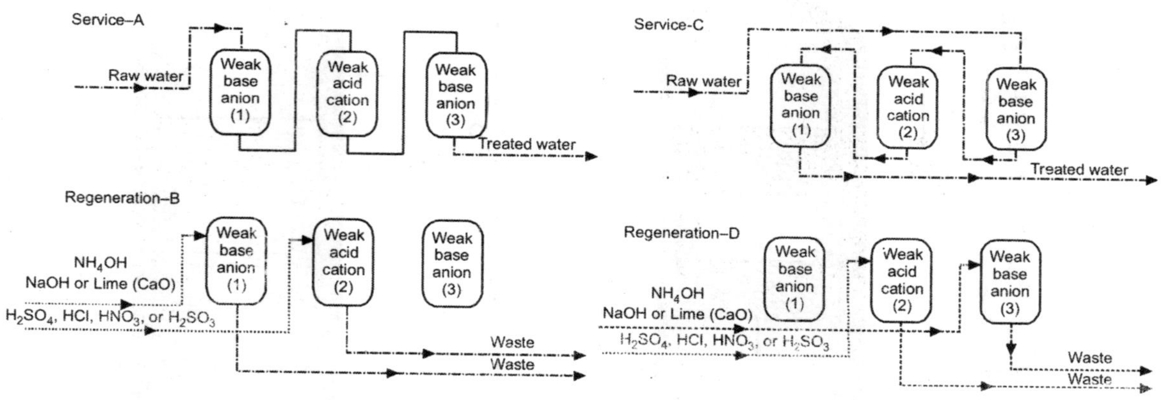


**Fig. 38.3. Three-unit variation Desal Process**

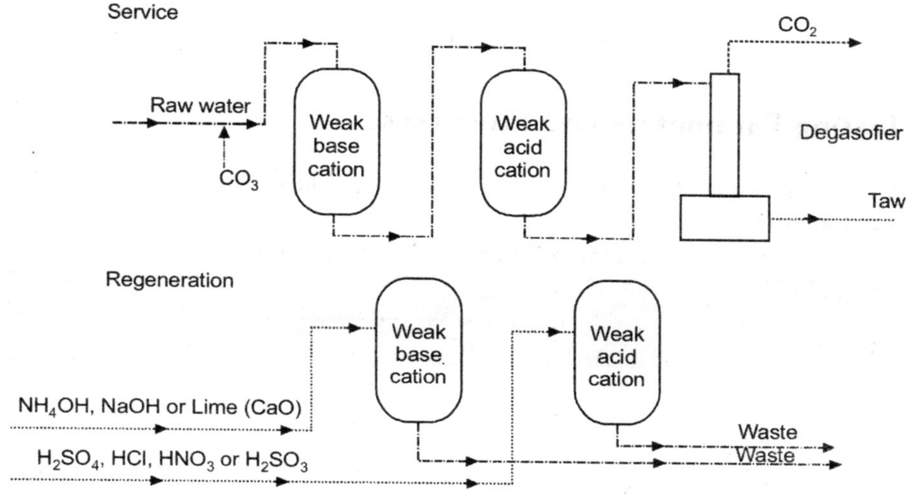


**Fig. 38.4. Two-unit variation Desal Process**

**Demineralization**. No other demineralization or desalination technique can, in a single pass, produce water as pure as does ion exchange. In the production of steam, it is sometimes necessary to use water with a lower level of total dissolved solids. Ion exchange should be considered if water with less than approximately 300 milligrams per liter of total dissolved solids must be purified further.

A typical cation-anion two-bed demineralization flow sheet is shown in Fig. 38.5. The cost of ion-exchange regeneration, including regeneration waste disposal, is directly related to the amount of dissolved solids to be removed. For many small users, such as laboratories, replaceable mixed-bed ion-exchange cartridges are the most economical method used to obtain ultrapure water.

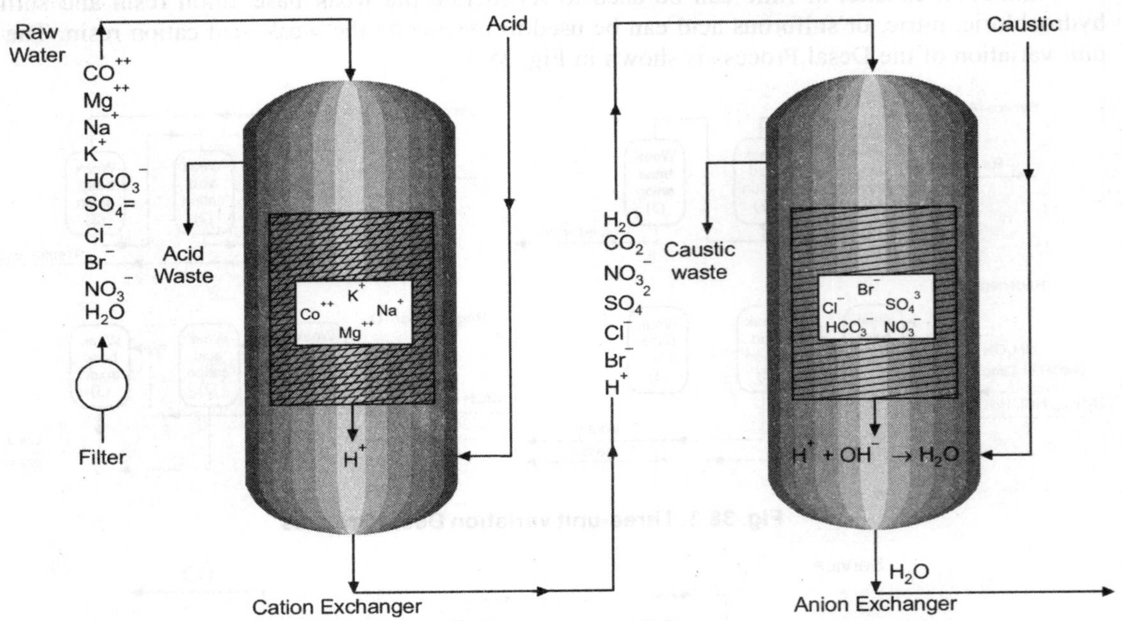


## Water Testing Parameters and Processess

- Water is an essential resource for living systems, industrial processes, agricultural production and domestic use.

**Saturated Dissolved Oxygen (DO) Levels**

| Temperature (°C) | Saturated level of DO (ppm) | |
|---|---|---|
| | Freshwater | Sea water |
| 10 | 10.9 | 9.0 |
| 20 | 8.8 | 7.4 |
| 30 | 7.5 | 6.1 |
| 40 | 6.6 | 5.0 |

- The principal factors that are taken into consideration when determining water quality are:
- dissolved oxygen content (DO)
- turbidity
- acidity and alkalinity
- trace elements and nutrients such as nitrogen, phosphorus, halogens (chloride and fluoride ions), alkali metals (sodium and potassium ions), calcium and magnesium ions.
- microorganisms

**Water Quality of a Typical Natural Aquatic System**

| Substance or Quality | River Water | Sea Water |
|---|---|---|
| pH | 6.8 | 8.0 |
| Dissolved Oxygen | 6-8 ppm | 6-8 ppm |
| $Na^+$ | 6.7 ppm | $1.1 \times 10^4$ ppm |
| $K^+$ | 1.5 ppm | 380 ppm |
| $Ca^{2+}$ | 17.5 ppm | 400 ppm |
| $Mg^{2+}$ | 4.8 ppm | $1.3 \times 10^3$ ppm |
| $Cl^-$ | 4.2 ppm | $1.9 \times 10^4$ ppm |
| $SO_4^{2-}/HSO_4^-$ | 17.5 ppm | $2.6 \times 10^3$ |
| $CO_3^{2-}/HC_3^-$ | 33.0 ppm | 142 ppm |
| $Hg^{2+}$ | < 1 ppb | 0.03 ppb |
| $Cd^{2+}$ | < 1 ppb | 0.1 ppb |
| $Pb^{2+}$ | < 1 ppb | 4-5 ppb |

| Test | Method | Reason for testing |
|---|---|---|
| **Temperature** | Use an alcohol thermometer in a hard plastic cover | Temperature influences the amount of dissolved oxygen in water which in turn influences the survival of aquatic organisms (raising the temperature of a freshwater stream from 20 to 30°C will decrease the dissolved oxygen saturation level from about 9.2 ppm to 7.6 ppm.). Increasing temperature also increases the rates of chemical reactions taking place in the water. Increases in temperature are often associated with hot water discharge from power stations and industries that use water as a coolant. |
| **pH** | Use a pH meter in a hard plastic cover, pH paper or Universal Indicator solution. | pH measures the acidity or alkalinity of water. pH of rain water is about 5.5-6.0. Typically, natural water has pH 6.5-8.5. A pH<5 (acidic water) is most damaging to eggs and larvae of aquatic organisms. Most aquatic life (except for some bacteria and algae) cannot survive pH<4.Natural alkalinity is due to $CO_2(g)$, $HCO_3^-$, $CO_3^{2-}$ and $OH^-$, carbonate rocks such as limestone and |

| Test | Method | Reason for testing |
|---|---|---|
| | | dolomite increase alkalinity. Alkalinity is increased by caustic substances from industry (KOH, NaOH), soil additives in agriculture such as lime $Ca(OH)_2$, superphosphate which is mixture of $Ca(H_2PO_4)_2$ and $CaSO_4$, and soaps and detergents. Natural acidity is due to $CO_2(g)$, $HPO_4^{2-}$, $H_2PO_4^-$, $H_2S$, $Fe^{3+}$, other acidic metal ions, proteins & organic acids. Increases in acidity can be due to acids used in industry, acid mine drainage, acid rain. |
| **Turbidity** | Use a Sechi disc or 500mL of water in a measuring cylinder standing on paper marked with a black cross. | Turbidity is a measure of water clarity. Suspended solids in water can stop light reaching submerged plants and can raise water temperature. Suspended solids often present in water are mud, clay, algae, bacteria and minerals such as silica, calcium carbonate and ochre (iron oxide). Suspended solids can be increased by the discharge of wastes (domestic sewage, industrial and agricultural effluents), leaching of wastes (from mines), and agitation (dredging or shipping). |
| **Total Dissolved Solids (TDS)** | Use an appropriate TDS meter. Freshwater meters: 0-1990 ppm (parts per million). Dual range brackish water meters: 0-19,900 ppm. Salt-water meters: to above 35,000 ppm. | This is a conductivity test of available ions in the water, including $Ca^{2+}$, $Na^+$, $K^+$, $Fe^{2+}$, $Fe^{3+}$, $HCO_3^-$ and ions containing P, S & N. High levels of $Na^+$ is associated with excessive salinity and is found in many minerals. Potassium is incorporated into plant material and is released into water systems when plant matter is decayed or burnt. |
| **Dissolved Oxygen (DO)** | • Winkler Titration Method-Halve the water sample. Place one sample in the dark (for BOD analysis).<br>- To the other sample add 2mL $MnCl_2(aq)$ (4g $MnCl_2$ dissloved in 10mL distilled water) + 2mL alkaline iodide solution (3.3g NaOH + 2.0g Ki dissolved in 10mL distilled water). Shake sample.<br>- Add 2 mL concentrated HCl. Shake. The iodine formed is directly proportional to the dissolved oxygen.<br>- Titrate 50.0mL of the above solution with 0.0125M sodium thiosulfate solution using starch as an indicator. The end-point is reached when the blue-black colour disappears. | The Dissolved Oxygen test measures the current oxygen levels in the water. The Do level varies with temperature. DO levels are highest in the afternoon due to photosynthesis and lowest just before dawn. DO is lowered by an increase in temperature (as from a discharge of hot water form a power station), increases in aerobic oxidation (due to increases in organic matter from sewage or due to inorganic fertilisers such as phosphates and nitrate with overstimulate algal growth). Water with DO<1ppm is dead. |

| Test | Method | Reason for testing |
|---|---|---|
| | • Colorimetric Method (A field kit is available using a 'Smart' colorimeter) Collect 2 water samples, 1 for DO test, 1 for BOD test. Sample must be collected under water to ensure there are no trapped air bubbles. | |
| **Biochemical Oxygen Demand (BOD)** | The first water sample from above is kept in the dark for 5 days at the temperature at which the sample was collected. Then the dissolved oxygen is determined using the Winkler titration method as above. Subtract the mass of oxygen obtained an day 5 from mass of oxygen on day 1 to determine the BOD (mg/L). Unpolluted natural waters have BOD<5mg/L. Treated sewage can have BOD 20-30 mg/L. | BOD measures the rate of consumption of oxygen by organisms in the water over a 5 day period. Increases in BOD can be due to animal and crop wastes and domestic sewage. Untreated domestic sewage BOD~350 ppmWaste water from breweries BOD~550 ppmWaste water from petroleum refineries BOD~850 ppmAbattoir wastes BOD~2,600 ppmPulpmill wastes BOD~25,000 ppm |
| **Salinity** | Titrate a known volume of the water sample with silver nitrate solution (2.73g $AgNO_3$ per 100mL distilled water) using $K_2CrO_4$ as indicator. The end-point of the titration is given by the reddening of the silver chloride precipitate ($AgCl_{(s)}$). Volume of $AgNO_3$ used = chloride content in g/L. | Many aquatic organisms can only survive in a narrow range of salt concentrations since salt controls their osmotic pressure. |
| **Total Phosphate test** | - acid digestion using concentrated $H_2SO_4$ and ammonium persulfate.<br>- Titrate using NaOH and phenolphthalein as indicator-Use a few drops of $H_2SO_4$ to turn the solution clear again.<br>- Add ammonium molybdate solution then solid ascorbic acid.<br>- An intense blue complex of molybdenum blue is formed which can be measured colorimetrically.<br>Absorbency is measured at 882nm. (A field kit is available using a 'Smart' colorimeter) | Total Phosphate is used as an indicator of pollution from run-off in agricultural areas or domestic sewage. Concentrations of 0.2mg/L are common. Concentrations of 0.05mg/L indicate the possibility of eutrophication (increased nutrient concentrations) and algal blooms are likely. Natural phosphate is due to decayed organic matter and phosphate minerals. |
| **Total Nitrogen test** | • Kjeldahl digestion-digestion with concentrated sulfuric acid, converting the nitrogen into | Total Nitrogen is an important indicator of eutrophic waters, especially for those contaminated by animal wastes, fertiliser run- |

| Test | Method | Reason for testing |
|---|---|---|
| | ammonia sulfate - Solution is then made alkaline-liberated ammonia is distilled, and the amount determined by titration with standard acid.<br>• distillation titration method for samples containing >1mg/L<br>• Add Nessler's reagent (100g mercuric iodide + 70g KI in 100mL distilled water, then add 160g NaOH in 700mL distilled water, then dilute to 1L) for samples containing <1mg/L and measure absorbancy colorimetrically at 425nm. | off and domestic sewage. Aquatic nitrogen is essential for the growth of organisms and is produced in natural processes including decay of proteins, the action of lightning, and the action of nitrogen-fixing bacteria on ammonia. |
| **Hardness** | • Calcium Ions, $Ca^{2+}$-complexometric titration using EDTA (ethylenediaminetetraacetic acid) at a pH of 12-13 (at this pH $Mg^{2+}$ is precipitated and not complexed with EDTA)-OR potentiometric techniques using selective electrodes-OR Atomic Absorption Spectroscopy (AAS)-OR Gravimetric Method - measure the amount of $CaCO_3(s)$ precipitated by a known volume of 0.02M $Na_2CO_3$ -OR by Flame Test - $Ca^{2+}$ flame test a brick-red colour in a non-luminous Bunsen flame<br>• $Mg^{2+}$-complexometric titration using EDTA at pH=10 (both $Ca^{2+}$ and $Mg^{2+}$ will complex with EDTA at this pH, $[Mg^{2+}]$ can be found by subtracting the results of this titration from the results of the first titration.)-OR potentiometric techniques using selective electrodes-OR Atomic Absorption Spectroscopy (AAS) | Calcium ions are a major contributor to water hardness and are due to water running through rocks containing minerals such as gypsum ($CaSO_4.2H_2O$), calcite ($CaCO_3$), dolomite ($CaMg(CO_3)_2$). Hard water has a noticeable taste, produces precipitates with soaps which inhibits lathering and forms precipitates (scale) in boilers, hot water systems and kettles. Temporary hardness (or 'bicarbonate hardness') is due to $Ca(HCO_3)_2$ which deposits $CaCO_3(s)$ as scale on boiling the water. Magnesium ion levels are often high in irrigation water and can cause scouring in stock. $Ca^{2+}$ and $Mg^{2+}$ can combine with $Cl^-$ and/or $SO_4^{2-}$ causing permanent hardness which can't be removed by boiling. Water can be softened by an ion exchange process using a solid material such as a resin or clay that is capable of exchanging $Na^+$ or $H^+$ for $Ca^{2+}$ and $Mg^{2+}$. |
| **Micro-organisms** | Microorganisms in a water sample are counted under a microscope Method for finding the number of coliform organisms in a water sample:<br>• Filter a known volume of water sample through a filter that retains microorgnisms<br>• Transfer the filter to a sterile petri dish containing appropriate agar and incubate at 35°C for 20-40 hours. | Many protozoa, bacteria, viruses, algae and fungi are found in natural water systems. Some are pathagenic (typhoid, cholera and amoebic dysentery can result from water-borne pathogens). The excessive growth of algae (called 'algal bloom') can degrade water quality because it lowers dissolved oxygen levels thereby killing other living things. The level of bacterial contamination of water due to animal waste is measured by determining the number of coliform organisms such as *E. coli*. |

| Test | Method | Reason for testing |
|---|---|---|
| | Also incubate a control plate with agar only. Colonies will develop on the filter wherever bacteria are retained.<br>• Either visually, or using a microscope, count the number of coliform colonies. Express these values as CFU/100mL. | |
| **Heavy Metals** | • $Ni^{2+}$ + dimethylglyoxime in ethanol turns pink-red<br>• $Fe^{3+}$ + ammonium thoicyanate turns blood-red<br>• $Cu^{2+}$ + dithizone in 1,1,1-trichloroethane turns yellow-brown<br>• $Cd^{2+}$ + dithizone in 1,1,1-trichloroethane turns blue-violet<br>• $Pb^{2+}$ + dithizone in 1,1,1-trichloroethane turns brick-red$Pb^{2+}$ + 2KI(aq)$\rightarrow$ yellow precipitate of $PbI_2$(s)<br>• $Zn^{2+}$ + dithizone in 1,1,1-trichloroethane turns pink | Heavy metals in concentrations above trace amounts are generally toxic to living things.Trace amounts (<0.05 mg/L) of Zn, Cu and Mn are present in most natural waters. Zn and Cu may be present in higher levels in irrigation areas due to the use of galvanised iron, copper and brass in in plumbing fixtures and for water storage. In irrigation areas, acceptable levels are 0.2 mg/L for $Cu^{2+}$, and 2.0 mg/L for $Zn^{2+}$ and $Mn^{2+}$. |
| **Other Ions** | • $Al^{3+}$ + aluminon $\rightarrow$ pink-red<br>• $Mg^{2+}$ + magneson I $\rightarrow$ light blue<br>• $Na^{+}$ flame test $\rightarrow$ yellow flame<br>• $K^{+}$ flame test $\rightarrow$ lilac flame<br>• $NH_4^{+}$ & $NH_3$ + Nessler's reagent (100g mercuric iodide + 70g KI in 100mL distilled water, then add 160g NaOH in 700mL distilled water, then dilute to 1L) $\rightarrow$ yellow-brown<br>• $NO_3^{-}$ + conc $H_2SO_4$ + $FeSO_4$ $\rightarrow$ brown ring forms at junction<br>• $S^{2-}$ + lead acetate solution $\rightarrow$ black deposit of PbS<br>• $SO_4^{2-}$ + $BaCl_2$(aq) $\rightarrow$ white precipitate of $BaSO_4$(s) | |

CHAPTER

39

# Terpenes and Camphors

## Introduction

The fragrant oil which occurs in fruits flower and leaves and stems of plants are grounded together under the name of essential oil, in order to distinguish them from the fixed oil and fats. The essential oils are volatile with steam and compromise a large number of products of different kinds used in pharmacy and perfumery. Among them occur, chiefly in the 'coniferæ' and 'citrus' family a number of unsaturated hydrocarbon called **terpens**, with the empirical formula $C_{10}H_{16}$ and are cyclic compounds derived by partial reduction from is to render the cyclic nucleus unsaturated in character. The monocyclic terpens possess, two olefinic bonds and two side chains, the bicyclic terpens have only one double bonds, but two ring structures, since the 6 – membered ring is bridged by the attachment of the isopropyl group to a second point, as in pinene. In addition to the terpens proper are **sesquiterrpenes**, $C_{15}H_{24}$, the **polyterpenes**, $(C_5H_8)_n$) and the open chain olefin isoprene, $C_5H_8$, which is obtained by dry distillation of rubber, and is apparently the parent substance of other terpens which may all be regarded as polymers of isoprene. The camphors are derivatives of terpens containing oxygen and are mostly, alcohols, aldehydes or ketones. Among the more important members of the group of terpens and camphors are pinine, limonene, terpinol menthol, bornerol camphor and the olefinic terpene aldehyde general or citral, which are described below.

## Nomenclature of Terpenes

Terpens, $C_8H_{16}$, are all unsaturated hydrocarbon containing either one or two ethylene bonds in the molecule according to which they are again divided as follows:

*1. Monocyclic Terpenes*: Containing two double bonds. These are di hydro cymes, and may therefore be regarded as partially reduced benzene derivatives. Included in this class are limonene (dipentane) slyvestrene, terpinene and terpinolene. Owing to the presence of two double bonds they are capable of adding on two or four monovalent atoms or groups.

*2. Dicyclic Terpenes:* Containing one double bond in the molecule. These are built up with two ring systems e.g. a combination of a cyclohexane with a cyclobutane ring. Compounds of this type are pinene, camphene and fenchene, each of which can unite with two monovalent atoms or groups.

Camphors and other terpene derivatives can be classified in a similar manner. According to the recent works, the sesquiterrpenes are related to naphthalene in the same manner as the terpenes to benzene. Comparatively little is known of the chemistry of this group.

We will now discuss in detail about these compounds.

# LIMONENE

## Introduction

This compound exists in a dexo – lævo and inactive forms, as would be expected from the presence of an asymmetric atom in the molecule.

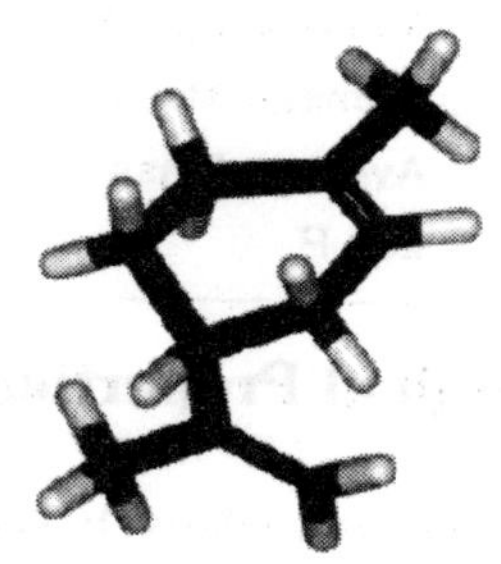

**Dilimonene** Also known as carvene citrine or hesperidene, is the chief constituent of the oil of orange rind, dill oil and oil of cumin. It occurs together with pinene in oil of lemons, L – Limonene is found with l – pinine in pine needle oil, it may be prepared from carvone or from perillaldehyde. Both forms of limonene are liquids of boiling point 175°C, with strong smell of lemon they yield crystalline tetrabromides $C_{10}H_{16}Br_4$ m.p 104 – 105, which are also distinguished by rotation of opposite sign. Both limonenes unite nitrosyl chloride, each forming two compounds differing in specific rotation. On treatment with alkali these part with hydrochloric acid and re transformed in to the oxime of carvone.

di Limonene, di pentane cinene, is produced by mixing equal amounts of d an dl limonenes, and is found associated with cineol, $C_{10}H_{18}O$, in oleum cinæ. It is formed by the elimination of water from terpineol and from pinene or camphene by heating at 250°C to 270°C.

Limonene is formed from geranyl pyrophosphate, via cyclisation of a neryl carbocation or its equivalent as shown. The final step involves loss of a proton from the cation to form the alkene.

OPP
geranyl pyrophosphate
$-H_+$
limonene

## Physcial Properties

|  |  |
|---|---|
| Molecular Formula | = $C_{10}H_{16}$ |
| Formula Weight | = 136.23404 |
| Composition | = C(88.16%) H(11.84%) |
| Melting Point | = –95.2 °C |
| Boiling Point | = 176 °C |
| Molar Refractivity | = 45.35 $cm^3$ |
| Molar Volume | = 163.2 $cm^3$ |
| Parachor | = 368.2 $cm^3$ |
| Index of Refraction | = 1.467 |
| Surface Tension | = 25.8 dyne/cm |
| Density | = 0.834 g/$cm^3$ |
| Dielectric Constant | = 2.44 |
| Polarizability | = 17.98 $cm^3$ |
| Monoisotopic Mass | = 136.125201 Da |
| Nominal Mass | = 136 Da |
| Average Mass | = 136.234 Da |
| Log P | = 4.45 |

## Chemical Properties

Limonene is a relatively stable terpene, which can be distilled without decomposition, though it forms isoprene when passed over a hot metal filament. It is easily oxidised in moist air to carveol and carvone.[2] Oxidation using sulphur leads to p-cymene and a sulphide. Limonene occurs naturally as the (*R*)-enantiomer, but it can be racemised to dipentene simply by heating at 300 °C. When warmed with mineral acid, limonene forms the conjugated diene terpinene, which can itself easily be oxidised to p-cymene, an aromatic hydrocarbon. Evidence for this includes the formation of Diels-Alder α-terpinene adducts when limonene is heated with maleic anhydride. It is possible to effect reaction at one of the double bonds selectively. Anhydrous hydrogen chloride reacts preferentially at the disubstituted alkene, whereas epoxidation with MCPBA occurs at the trisubstituted alkene. In both cases the second C=C double bond can be made to react if desired. In another synthetic method Markovnikov addition of trifluoroacetic acid followed by hydrolysis of the acetate gives terpineol.

## Uses

As the main odour constituent of citrus (plant family Rutaceæ), *d*-limonene is used in food manufacturing and some medicines, e.g., bitter alkaloids, as a flavouring, and added to cleaning products such as hand cleansers to give a lemon-orange fragrance. See: orange oil.

Limonene is increasingly being used as a solvent for cleaning purposes, such as the removal of oil from machine parts, as it is produced from a renewable source (citrus oil, as a byproduct of orange juice manufacture.) Limonene works as paint stripper when applied to painted wood. The (*R*)-enantiomer is also used as botanical insecticide.

The (*S*)-enantiomer, also known as *l*-limonene, is used as a fragrance in some cleaning products. In contrast to the citrus (orange-lemon) scent (see above) of *d*-limonene, the enantiomer *l*-limonene has a piney, turpentine-like odor.

Limonene is very common in cosmetic products.

## TERPINOLENE

### Introduction

This compounds is another terpene which is obviously important in its group.

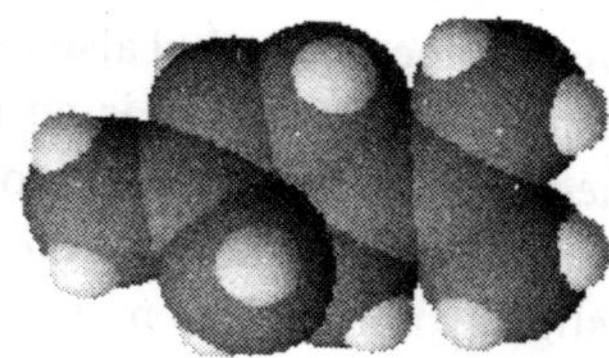

### Preparation

It is obtained by the reaction of terpinol and oxalic acid, it boils at 183°C and on further treatment with acids it produces terpinene.

$CH_3$ ... $H_3C$ $CH_3$ —oxalic acid→ $CH_3$ ... $H_3C$ $CH_3$

This compound is present in the cardamom oil and is optically inactive, it smells like cymene and boils at 179°C. Terpene is formed when dipentane or phellandrene is boiled with dilute sulphuric acid. Hence it is also obtained when terpin hydrate, cineol, terpinol or Dihydro – carvenol is heated with dilute sulphuric acid. Terpene is conveniently prepared by shaking pinene (oil of turpentine) with conc. sulphuric acid. It forms a nitrosite of mp 155°C does not give definite addition products with bromine or hydrogen halides. Ordinary terpinene consists mainly of $\Delta^{1:4}$ dihydrogen – cyumene and contains also some $D^{1:4}$ dihydro – cymene.

***Sylvestrene,*** $C_{10}H_{16}$ is the limonene of the m – cymene series, and has been prepared from Swedish and Russian terpentines Simonsen and Rao have shown that it does not occur naturally in these sources but is formed from the carene present by secondary charges due to the treatment of the oils with hydrogen chloride.

***Phellandrene, $C_{10}H_{16}$*** **This compound was isolated from water fennel (Phallandrium Aquaticum) from which takes its name. It also occurs in other ethereal oils and exists in *d* and *l* forms, neither of which has yet been obtained in the pure state. With acid it very readily under goes change.**

# MENTHOL

## History and Occurrence

**There is evidence that menthol has been known in Japan for more than 2000 years, but in the West it was not isolated until 1771, by Hieronymous David Gaubius. Early characterizations were done by Oppenheim, Beckett, Moriya and Atkinson (-)-Menthol (also called *l*-menthol or *(1R,2S,5R)*-menthol) occurs naturally in peppermint oil (along with a little menthone, the ester menthyl acetate and other compounds), obtained from *mentha x piperita*. Japanese menthol also contains a small percentage of the 1-epimer, (+)-neomenthol. Menthol is an organic compound made synthetically or obtained from peppermint or other mint oils. It is a waxy, crystalline substance, clear or white in color, which is solid at room temperature and melts slightly above. The main form of menthol occurring in nature is (-)-menthol, which is assigned the (1R,2S,5R) configuration. Menthol has local anesthetic and counterirritant qualities, and it is widely used to relieve minor throat irritation.**

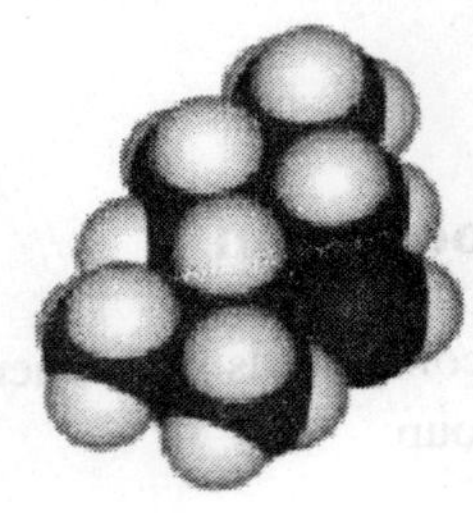

## Structure

**Natural menthol exists as one pure enantiomer, nearly always the *(1R,2S,5R)* form (bottom left of diagram below). The other seven stereoisomers are:**

(+) Menthol | (+)-Isomenthol | (+)-Neomenthol | (+)-Neoisomenthol

(–) Menthol | (–)-Isomenthol | (–)-Neomenthol | (–)-Neoisomenthol

In the natural compound, the isopropyl group is oriented *trans*- orientation to both the methyl and alcohol groups. Thus it can be drawn in any of the ways shown:

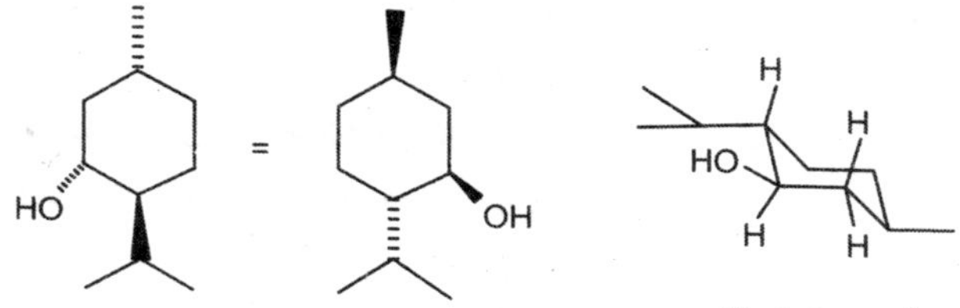

(–)-Menthol

Chair form of (–)-Menthol

In the ground state all three bulky groups in the chair are equatorial, making (-)-menthol and its enantiomer the most stable two isomers out of the eight.

There are two crystal forms for racemic menthol; these have melting points of 28°C and 38°C. Pure (-)-menthol has four crystal forms, of which the most stable is the á form, the familiar broad needles.

## Synthesis

As with many widely-used natural products, the demand for menthol greatly exceeds the supply from natural sources. Menthol is manufactured as a single enantiomer (94% ee) by Takasago International Co. on a scale of 400,000 tonnes per year. The process involves an asymmetric synthesis developed by a team led by Ryoji Noyori:

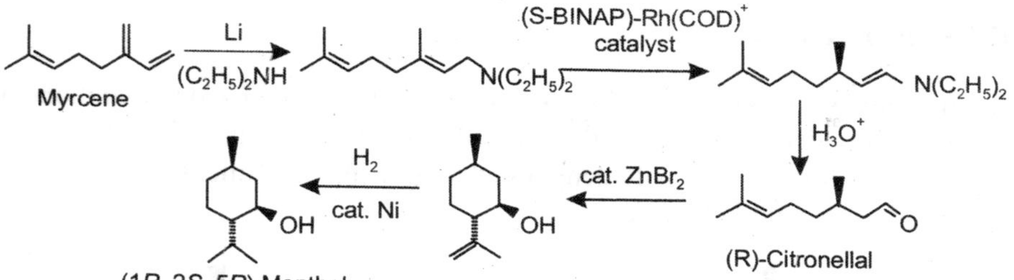

Myrcene → (Li, $(C_2H_5)_2NH$) → allylic amine ($N(C_2H_5)_2$) → ((S-BINAP)-Rh(COD)$^+$ catalyst) → enamine ($N(C_2H_5)_2$) → ($H_3O^+$) → (R)-Citronellal → (cat. $ZnBr_2$) → isopulegol → ($H_2$, cat. Ni) → (1*R*, 2*S*, 5*R*)-Menthol

The process begins by forming an allylic amine from myrcene, which undergoes asymmetric isomerisation in the presence of a BINAP rhodium complex to give (after hydrolysis) enantiomerically pure *R*-citronellal. This is cyclised by a carbonyl-ene-reaction initiated by zinc bromide to isopulegol which is then hydrogenated to give pure (*1R,2S,5R*)-menthol.

Racemic menthol can be prepared simply by hydrogenation of thymol, and menthol is also formed by hydrogenation of pulegone. For preparation of other isomers such as neomenthol.

### Natural Menthol

*Mentha arvensis* is the primary species of mint used to make natural menthol crystals and natural menthol flakes. This species is primarily grown in the Uttar Pradesh region in India.

## Physcial Properties

| | |
|---|---|
| Molecular Formula | $= C_{10}H_{20}O$ |
| Formula Weight | = 156.2652 |
| Composition | = C(76.86%) H(12.90%) O(10.24%) |
| Melting Point | = 36–38 °C, racemic 42–45 °C 35-33-31 °C, (-)-isomer |
| Boiling point | = 212 °C |
| Molar Refractivity | $= 47.83cm^3$ |
| Molar Volume | $= 175.5\ cm^3$ |
| Parachor | $= 409.8\ cm^3$ |
| Index of Refraction | = 1.457 |
| Surface Tension | = 29.7 dyne/cm |
| Density | $= 0.890g/cm^3$ |
| Polarizability | $= 18.96\ cm^3$ |
| Monoisotopic Mass | = 156.151415 Da |
| Nominal Mass | = 156 Da |
| Average Mass | = 156.2652 Da |
| Log P | = 3.20 |

## Chemical Properties

Menthol reacts in many ways like a normal secondary alcohol. It is oxidized to menthone by oxidizing agents such as chromic acid or dichromate, though under some conditions the oxidation can go further and break open the ring. Menthol is easily dehydrated to give mainly 3-menthene, by the action of 2% sulfuric acid. $PCl_5$ gives menthyl chloride.

Menthone $CrO_3$ Dil.$H_2SO_4$ > 90% 3-Menthene OH $CrO_3$ $CH_3COOH$ (–)-Menthol $PCl_5$ COOH Menthyl chloride Cl

## Biological properties

Menthol's ability to chemically trigger cold-sensitive receptors in the skin is responsible for the well known cooling sensation that it provokes when inhaled, eaten, or applied to the skin. Menthol does not cause an actual drop in temperature.[8] In this sense it is similar to capsaicin, the chemical responsible for the spiciness of hot peppers (which stimulates heat sensors, also without causing actual temperature rise).

## Applications

Menthol is included in many products for a variety of reasons. These include:

- In non-prescription products for short-term relief of minor sore throat and minor mouth or throat irritation

  * Examples: lip balms and cough medicines
- As an antipruritic to reduce itching
- As a topical analgesic to relieve minor aches and pains such as muscle cramps, sprains, headaches and similar conditions, alone or combined with products like Camphor or Capsaicin. In Europe it tends to appear as a gel or a cream, while in the US patches and body sleeves are very frequently used

  * Examples: Tiger Balm, or IcyHot patches or knee/elbow sleeves
- In decongestants for chest and sinuses (cream, patch or nose inhaler)

  * Examples: Vicks Vaporub
- In certain medications used to treat sunburns, as it provides a cooling sensation (then often associated with Aloe)
- As an additive in certain cigarette brands, for flavour, to reduce the throat and sinus irritation caused by smoking and arguably to reduce the bad-breath smokers experience and possibly improve the smell of second-hand smoke.
- Commonly used in oral hygiene products and bad-breath remedies like mouthwash, toothpaste, mouth and tongue-spray, and more generally as a food flavour agent; e.g. in chewing-gum, candy
- In a soda as well as in a syrup to be mixed with water to obtain a very low alcohol drink or (brand Riclès in France). The syrup is/was also used to alleviate nausea, in particular motion sickness, by pouring a few drops on a lump of sugar.
- As a pesticide against tracheal mites of honeybees
- In perfumery, menthol is used to prepare menthyl esters to emphasise floral notes (especially rose)
- In first aid products such as "mineral ice" to produce a cooling effect as a substitute for real ice in the absence of water or electricity (Pouch, Body patch/sleeve or cream)
- In various patches ranging from fever-reducing patches applied to children's foreheads to "foot patches" to relieve numerous ailments (the latter being much more frequent and elaborate

in Asia, especially Japan: some varieties use "functional protrusions", or small bumps to massage ones feet as well as soothing them and cooling them down).

- In some beauty products such as hair-conditioners, based on natural ingredients (e.g. St. Ives).

Some supporters of homeopathic alternative medicine believe that menthol interferes with the effects of homeopathic remedies. Members of the homeopathic community discourage its use for those seeking homeopathic cures, to the point of prohibiting the use of mint-flavoured toothpaste. (see this page). Currently no other reported nutrient or herb interactions involve menthol. It is used in Eastern medicine to treat indigestion, nausea, sore throat, diarrhea, colds, and headaches. (-)-Menthol has low toxicity: Oral (rat) $LD_{50}$: 3300 $mg{\cdot}kg^{-1}$; Skin (rabbit) $LD_{50}$: 15800$mg{\cdot}kg^{-1}$).

In organic chemistry, menthol is used as a chiral auxiliary in asymmetric synthesis. For example, sulphinate esters made from sulphinyl chlorides and menthol can be used to make enantiomerically pure sulphoxides by reaction with organolithium reagents or Grignard reagents. Menthol is also used for classical resolution of chiral carboxylic acids, via the menthyl esters.

# TERPIN

## Introduction

Terpene also known as 1:8 methane diol, exists in two possible stereo – isomeric state i.e. *cis* and *trans*. A crystalline hydrate is produced when pinene or dipentane is allowed to stand form sometime at the ordinary stand for sometime at the ordinary temperature in contact with dilute mineral acids. Terpin hydrate is also formed from geraniol by treatment with dilute sulphuric acid. On prolonged heating at 100°C it yields *cis* – terpin.

Geraniol $\xrightarrow{+ 2H_2O}$ Terpin hydrate → cis - terpin / trans -terpin

Perkin and Ray have synthesized terpin by treating cyclohexanone 4 – carboxylic ester with excess of methyl magnesium iodide, and have thus confirmed its constitution.

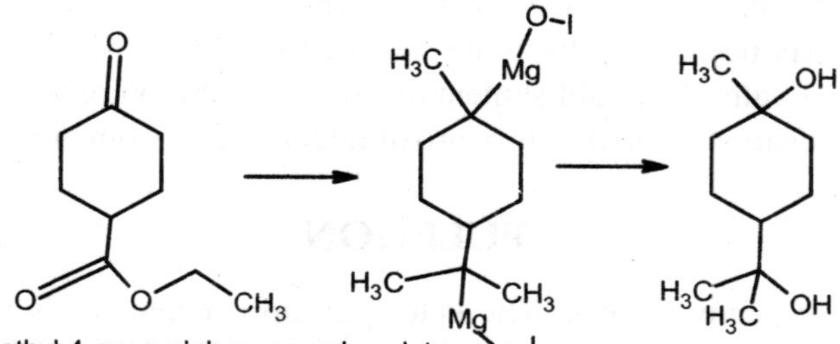

ethyl 4-oxocyclohexanecarboxylate → terpin

## CINEOL

Cineol of the annexed formula is an inner anhydride of terpin and occurs in many ethereal oils, such as oil of eucalyptus, wormed oil and rosemary oil. It is a liquid of boiling point 176°C with a smell of camphor, when treated with hydrochloric acid in glacial acetic acid solution; it is converted into dipentane hydrochloride.

## TERPINEOL

This compound is also called Δ methane 8 ol. It is obtained from terpin hydrate by treating it under certain conditions with dilute sulphuric acid, when two molecules of water are removed as follows:

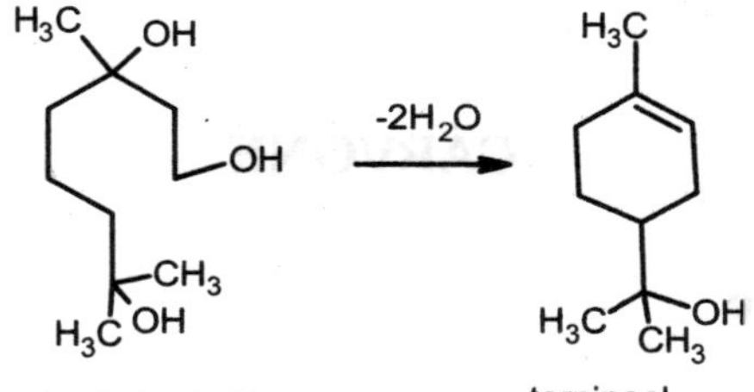

terpin hydrate $\xrightarrow{-2H_2O}$ terpineol

Terpineol is present in a number of essential oils and has an odour of lilac. Hence it is used in the manufacture of perfumes. When heated with potassium bisulphate it yields dipentene and with oxalic acid terpinolene. Carvacrol and carvone (see later) may be prepared from the nitrosochloride of terpineol.

Among the ketones of this group are menthone, piperitone, pulegone and carvone. Buchu – camphor is an example of a ketonic alcohol.

## PIPERITONE

Piperitone also known as Δ methane – 3 – one is an unsaturated ketone found in the 1 – form in a number of eucalyptus oils, especially in that of the Broad – leaved Peppermint. More recently a d –

piperitone has been isolated by Simson from the essential oil of a Himalayan grass Andropogon Jwarancusa. Piperitone has been carefully is investigated by Read and his co – workers and in his hands has proved of great value on establishment of the configuration of the methones, menthols and metylamines. On hydrogenation it yields a mixture of methone and isomenthone

## PULEGON

It is present in oil of penny – royal. The keto – group occupies a similar position to that in methone into which compound pulegone may be converted by hydrogenation, when superheated with formic acid or water pulegone is hydrolyzed to 3 – methyl cyclohexanone and acetone, from which its constitution is deduced. On the otherhand, by the condensation of methyl – cyclohexanone with acetone, an isomeride of pulegone is also obtained.

Pulegon also reacts with hydroxyl amine in the normal name to form an oxime, it also yields an addition product in which hydrooylamine attached to the unsaturated linking, the group C = C being converted into CH – C – NHOH.

methenone ← pulegone → 3-methylcyclohexanone + acetone

## CARVONE

### History and Occurrence

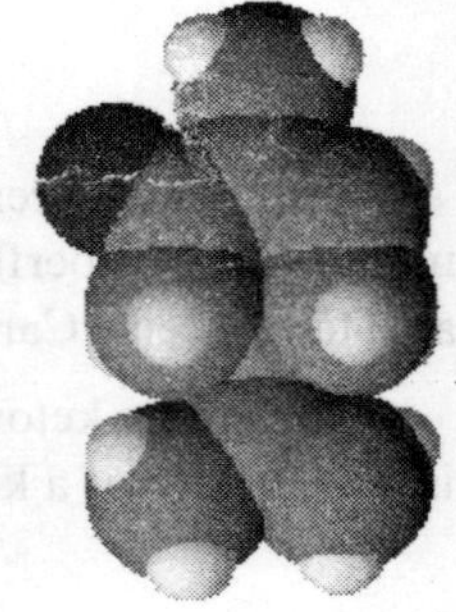

Caraway was used for medicinal purposes by the ancient Romans,[2] but carvone was probably not isolated as a pure compound until Varrentrapp obtained it in 1841.[1] It was originally called *carvol* by Schweizer. Goldschmidt and Zûrrer identified it as a ketone related to limonene, and the structure was finally elucidated by Wagner in 1894. *S*-(+)-Carvone is the principal constituent (50-70%) of the oil from caraway seeds (*Carum carvi*),[5], which is produced on a scale of about 10 tonnes per year.[2] It also occurs to the extent of about 40-60% in dill seed oil (from *Anethum graveolens*), and also in mandarin orange peel oil. *R*-(–)-Carvone is present at levels greater than 51% in spearmint oil (*Mentha spicata*), which is produced on a scale of around 1500 tonnes annually. This isomer also occurs in kuromoji oil. Some oils, like gingergrass oil, contain a mixture of both enantiomers. Many other natural oils, for example peppermint oil, contain lower concentrations of carvones.

## Preparation

The dextro-form is obtained practically pure by the fractional distillation of caraway oil; the laevo-form from the oils containing it, by first forming its addition compound with hydrogen sulfide, decomposing this by potassium hydroxide in ethanol, and distilling the product in a current of steam. It may be synthetically prepared from limonene nitrosochloride, alcoholic converting this compound into 1-carvoxime, which on boiling with dilute sulfuric acid yields l-carvone.

The biosynthesis of carvone is by oxidation of limonene.

### Organic Synthesis

Carvone is available inexpensively in both enantiomerically pure forms, making it an attractive starting material for the asymmetric total synthesis of natural products. For example, (*S*)-(+)-carvone was used to begin a 1998 synthesis of the terpenoid quassin[7]:

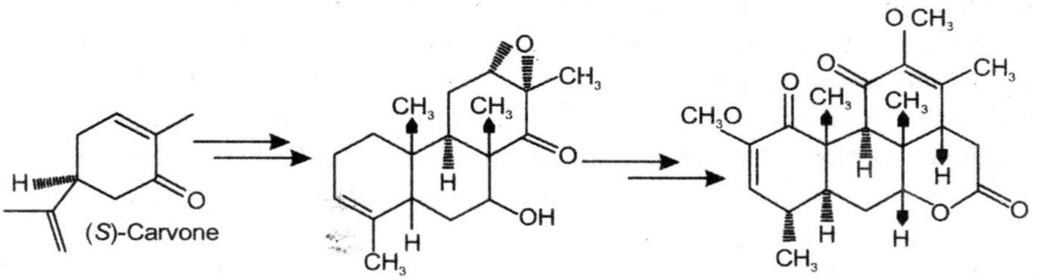

(*S*)-Carvone

(+)-Quassin
2.6% overall yield
in 28 steps

## Physcial Properties

|  |  |
|---|---|
| Molecular Formula | $= C_{10}H_{14}O$ |
| Formula Weight | = 150.21756 |
| Composition | = C(79.96%) H(9.39%) O(10.65%) |
| Melting point | = 89 °C |
| Boiling point | = 231 °C |
| Molar Refractivity | = 45.48 $cm^3$ |
| Molar Volume | = 159.7 $cm^3$ |
| Parachor | = 373.3 $cm^3$ |
| Index of Refraction | = 1.481 |
| Surface Tension | = 29.8 dyne/cm |

| | |
|---|---|
| Density | = 0.940 g/cm$^3$ |
| Polarizability | = 18.03 10$^{-24}$cm$^3$ |
| Monoisotopic Mass | = 150.104465 Da |
| Nominal Mass | = 150 Da |
| Average Mass | = 150.2176 Da |
| Log P | = 2.26 |

## Chemical Properties

### Reduction

There are three double bonds in carvone capable of reduction; the product of reduction depends on the reagents and conditions used. Catalytic hydrogenation of carvone (**1**) can give either carvomenthol (**2**) or carvomenthone (**3**). Zinc and acetic acid reduce carvone to give dihydrocarvone (**4**). MPV reduction using propan-2-ol and aluminium isopropoxide effects reduction of the carbonyl group only to provide carveol (**5**); a combination of sodium borohydride and $CeCl_3$ (Luche reduction) is also effective. Hydrazine and potassium hydroxide give limonene (**6**) via a Wolff-Kishner reduction.

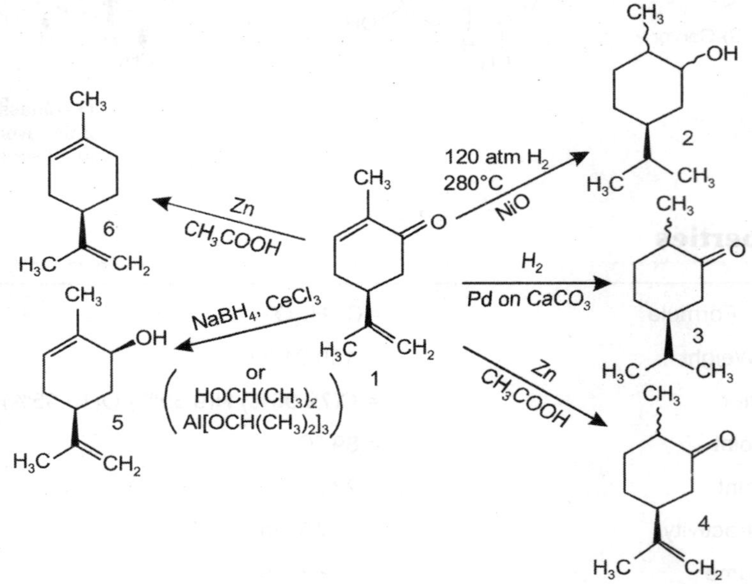

### Oxidation

Oxidation of carvone can also lead to a variety of products. In the presence of an alkali such as $Ba(OH)_2$, carvone is oxidised by air or oxygen to give the diketone **7**. With hydrogen peroxide the epoxide **8** is formed. Carvone may be cleaved using ozone followed by steam, giving dilactone **9**, while $KMnO_4$ gives **10**.

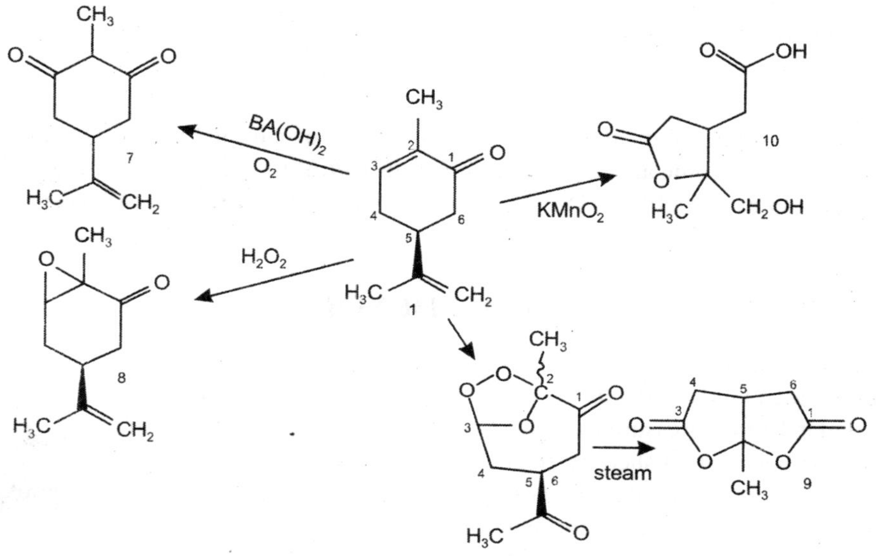

## Conjugate Additions

As an α,beta;-unsaturated ketone, carvone undergoes conjugate additions of nucleophiles. For example, carvone reacts with lithium dimethylcuprate to place a methyl group *trans* to the isopropenyl group with good stereoselectivity. The resulting enolate can then be allylated using allyl bromide to give ketone **11**

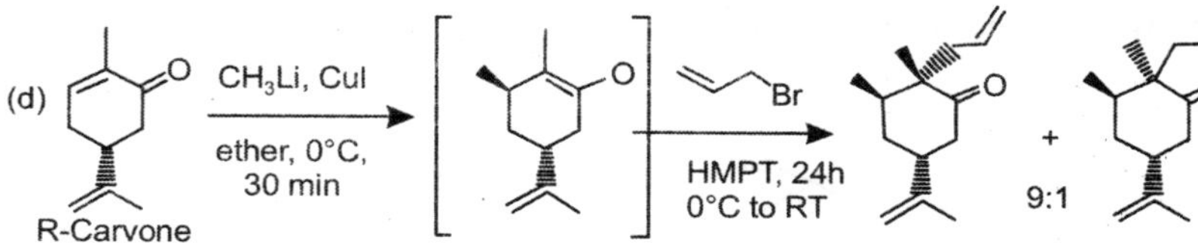

## Uses

Both carvones are used in the food and flavor industry. *R*-(-)-Carvone is also used for air freshening products and, like many essential oils, oils containing carvones are used in aromatherapy and alternative medicine.

## Food Applications

As the compound most responsible for the flavor of caraway, dill and spearmint, carvone has been used for millennia in food. Wrigley's Spearmint Gum is gum soaked in *R*-(–)-carvone and powdered with sugar.

### Agriculture

*S*-(+)-Carvone is also used to prevent premature sprouting of potatoes during storage, being marketed in the Netherlands for this purpose under the name *Talent*.

### Metabolism

In the body, *in vivo* studies indicate that both enantiomers of carvone are mainly metabolized into dihydrocarvonic acid, carvonic acid and uroterpenolone. (4*R*,6*S*)-(–)-carveol is also formed as a minor product via reduction by NADPH. (4*S*)-(+)-carvone is likewise converted to (4*S*,6*S*)-(+)-carveol. This mainly occurs in the liver and involves cytochrome P450 oxidase and (+)-trans-carveol dehydrogenase.

## DIOSPHENOL

This is also called Buchu – camphor, and is prepared from an essential oil obtained from various kinds of genus Baronisa found in South Africa. It boils at q09 – 110 under 10 mm pressure melting point is 83°C. The constitution of this substance has been established by Semmler and Mc Renzie. One of the two oxygen atoms is contained in an alcoholic yields acetate and benzoate. The second is a ketonic oxygen atom as diosphenol forms a normal oxime. Hence, the substance been shown to be monocyclic and unsaturated. Oxidation with ozone leads to the production of $\alpha$ - isopropyl $\gamma$ - acetyl - - butyric acid, thus proving the structure of that part of the molecule shown. The arrangement of the remaining disosphenol to the glycol, $C_{10}H_{20}O_2$ which on oxidation yields $\alpha$ - isopropyl methyl – adiphic acid, thus showing the presence of a six – membered ring with methyl and isopropyl groups in the para – position to one another. The disposition of the Ketonic and hydroxyl it follows that the double bond must be attached to the carbon atom linked to the methyl group the position 2:3 being impossible. Diosphenol therefore possess the structure II.

I II

$CH_3$ $CH_3$ OH O

$H_3C$—CH-$CH_3$ $H_3C$—CH $CH_3$

The synthesis of diosphenol has been affected by Semmler and Mc Renzie starting with

HC $CH_3$ CH—OH O $H_3C$—CH $CH_3$ → HC $CH_3$ O O $H_3C$—CH $CH_3$ → $CH_3$ OH O

Hydroxymethylene methanone Diketone Diosphenol

hydroxylmethylene – methone and oxidizing this is to the diketone, $C_{10}H_{16}O_2$. The latter may then be transformed in various ways e.g. with acid or alkalies into diosphenol.

## Dicyclic Terpenes and Camphors

The most important of this class is ordinary or Japanese Camphor. Before discussing this compound in detail, however, the parent hydrocarbon menthane, the dicyclic terpenes and their derivatives may be referred according to the type of "bridged ring" present, to one of the three saturated hydrocarbons, carane, pinane and caniphane.

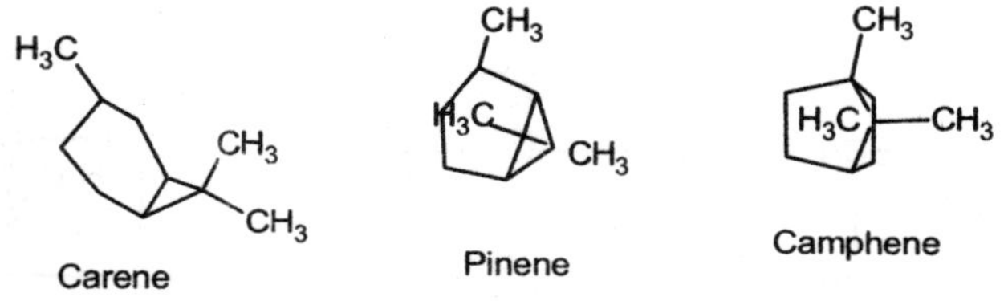

Carene Pinene Camphene

## Dicyclic Terpenes

Carenes propably occur more widely in nature than was formerly supposed, but during the purification of the essential oil with hydrogen chloride they become transformed into dipentane dihydrochloride and sylvestrene dihydrochloride by the opening of the cylcopropane ring.

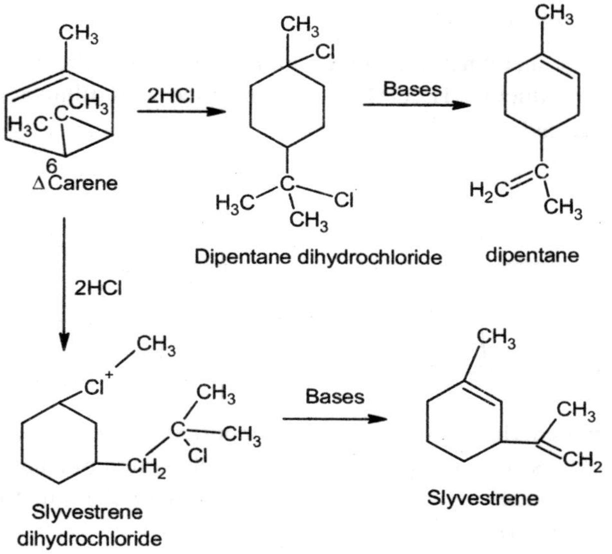

$\Delta^6$ Carene → (2HCl) Dipentane dihydrochloride → (Bases) dipentane

$\Delta^6$ Carene → (2HCl) Slyvestrene dihydrochloride → (Bases) Slyvestrene

**α - Pinene** is common constituent of many ethereal oils, and is especially abundant in the turpentine oils, obtained by distilling the resious exudations of the pines and firs. Associated with it is found an isomeride, β - pinine or norpinene (II). Although these compounds differ from them by means of simple addition reactions, because in many cases they yield the same products. With hydrogen chloride for example, each pinene gives the hydrochloride (III) and in the presence of dilute sulphuric acid each is converted into the same terpin. Indication of a difference in structure is furnished by their behaviour on oxidation with potassium permanganate, when α - pinine yields pinonic acid and noipinene gives *nopinic acid.* Under certain conditions α - pinene is oxidized much more readily than non – pinene by mercuric acetate.

**β - Pinene** also forms mercuric additions compounds which are readily isolated.

(I) (II) (III)

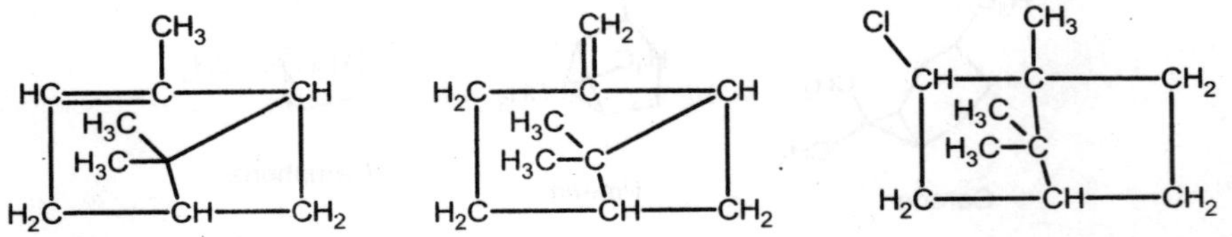

Pinene combines with two atoms of chlorine or bromine to from compounds which on being heated, breaks up into hydrogen halide and p – cymene. When dry hydrogen chloride gas is led into well – cooled pinene hydrochloride first formed isomerises with great rapidity into bornyl chloride (III) which separates out as a crystalline substance of m.p 131°C. Owing to its close resemblence to camphor in smell and a appearance this substance is sometimes known as "artificial camphor". In its formation not only has addition of hydrochloric acid taken place but the "pinane bridge" has simultaneously been changed into the "camphene bridge".

This is confirmed by the fact when the magnesium compound of the hydrochloride is treated with oxygen, and the resulting product $C_{10}H_{17}OMgCl$, decomposed with dilute acids, an almost theoretical yield of borneol (V) is obtained,

(IV) (V)

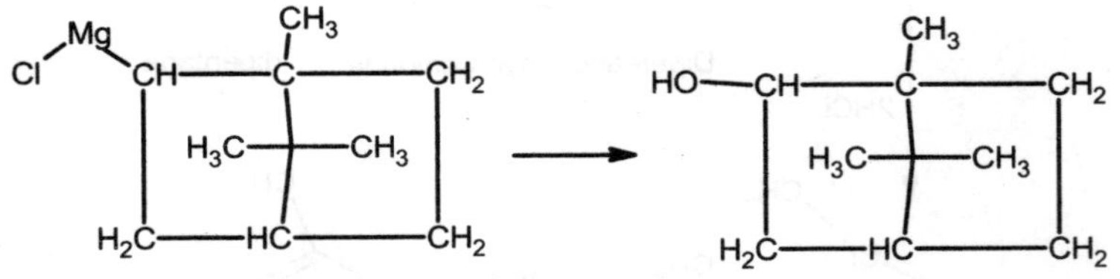

The presence of a tetramethylene ring in pinene is supported by constitution of its oxidation products pinonic acid and pinic acid, as well as by its synthesis, when heated pinene yields dipentine (together with isoprene, etc). Other changes are shown in the table showing the transformation and degradation of the terpenes and camphor series.

Pinene exists in optically active forms, δ - Pinene can be isolated by the fractional distillation of American oil of turpentine and *l* – pinene from French turpentine. An inactive pinene is obtained by the action of aniline on active pinene nitroso – chloride (obtained from pinene and nitrosyl chloride) when the nitrosyl chloride is again removed.

**Camphene, $C_{10}H_{16}$** : This is solid terpene found as the d – form in ginger, rosemary and spike oils, and as the oil – forms in citronella and valerian oils. It is produced from bornyl chloride by removing hydrogen chloride with sodium acetate and glacial acetic acid at 200°C or by heating with potassium cresolate. The constitution of camphene has given rise to much discussion, but may now be regarded as settled in accordance with formula (VI)

Camphene is a solid, melting point at about 50°C, with a smell of turpentine and camphor, when oxidized with chromic acid it yields camphor. If hydrogen chloride is passed into an ethereal solution of camphene, its hydrochloride is formed (VII). This very readily isomerises into isobornyl chloride. Recently it has been shown that camphene hydrochloride, bornyl chloride exists in a state of equilibrium with one another, when in the fused state or in solution. The isomerism of bornyl and isobornyl chloride and of the corresponding alcohols borneol and isoborneol is more readily explained as arising from *cis* or *trans* position of Cl or OH with respect to the $(CH_3)C<$ bridge across the cyclohexane ring (see formulæ VIII and IX in which the bracketed chlorine or hydroxyl is supposed to lie behind the plane of that part of the ring structure to which is attached. Although this problem has not been solved beyond dispute it appears portable that bornyl chloride and borneol are represented by exo – structures in which Cl or OH is in the cis position to the bridge, and that isobornyl chloride and isoborneol have the alternative endo – structures with a *trans* arrangement. Corresponding to bornyl and isobornyl chlorides are the two stereoisomeric secondary alcohols.

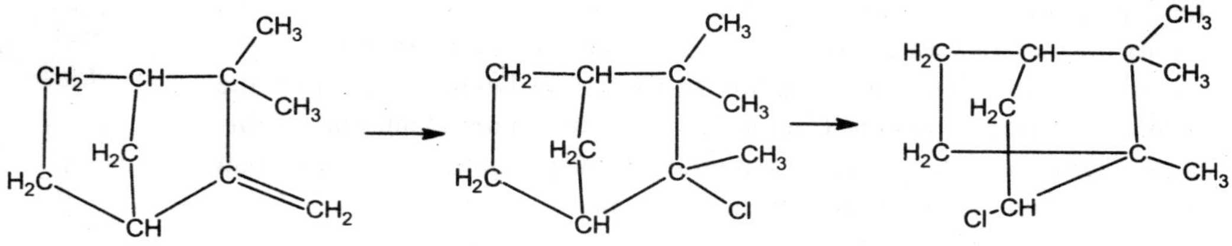

Camphene → Camphene hydrochloride → isobornyl chloride

VIIIa

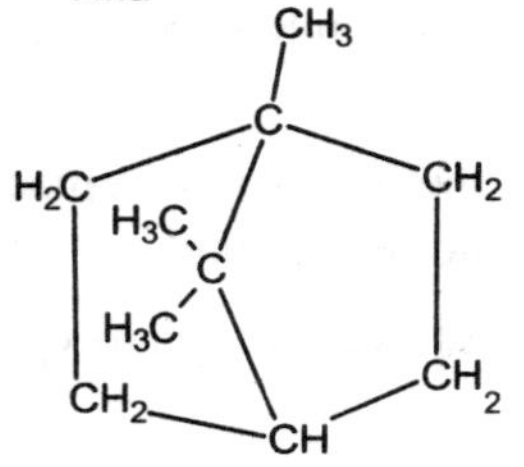

bornyl chloride

isobornyl chloride

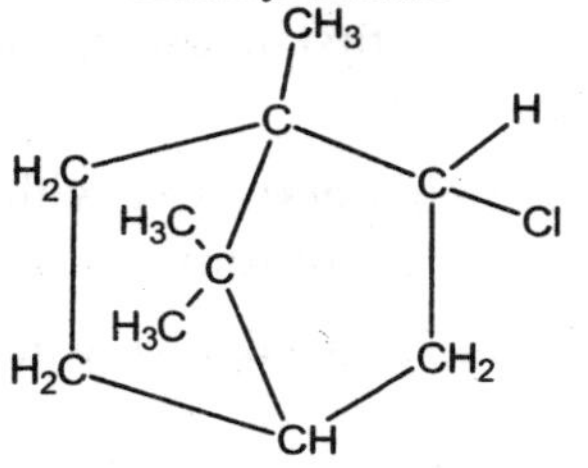

Chlorides are the two stereoisomeric secondary alcohols borneol and isoborneol. Borneol can be obtained in considerable portion from both of the above chlorides by way of the magnesium compounds, but camphene hydrochloride yields a layer proportion of isoborneol.

As many been seen from the formula, camphene is not related directly to camphene. An unsaturated derivative of camphene is bornylene (see below) which may be prepared by treating bornyl iodide with alkalies. The structure assigned to camphene as bornylene are confirmed by their reactions with diazo acetic ester. With unsaturated compounds this ester normally yields nitrogen and a cylcopropane derivative, formed by union of the resedue > CH.COOEt with the two C – atoms of the double bond Camphene should thus give a structure.

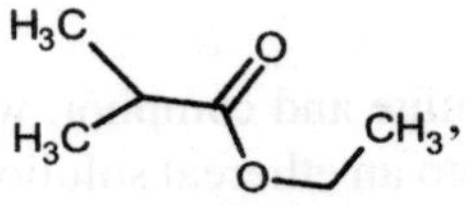

$H_3C$ O $H_3C$ O $CH_3$, which on oxidation should yield 1:1:2 tricaboxylic acid. Both of these changes have been found to take place.

# ALCOHOLS AND KETONES
# *BORNEOL, BORNEO CAMPHOR*

## History

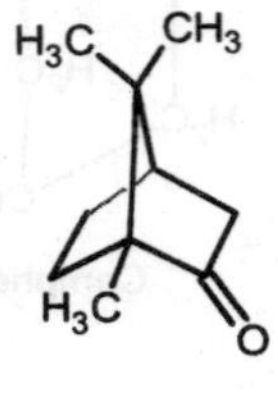

The word camphor derives from the French word *camphre*, itself from Medieval Latin *camphora*, from Arabic *kafur*, from Malay *kapur Barus* meaning "Barus chalk". In fact Malay traders from whom Indian and Middle East merchants would buy camphor called it *kapur*, "chalk" because of its white colour. Barus was the port on the western coast of the Indonesian island of Sumatra where foreign traders would call to buy camphor. In the Indian language Sanskrit, the word 'karpoor' is used to denote Camphore. A south-indian adaptation of this word, 'karpooram' has been used for camphor in many south-indian/dravidian languages (like Telugu, Tamil, Kannada and Malayalam).

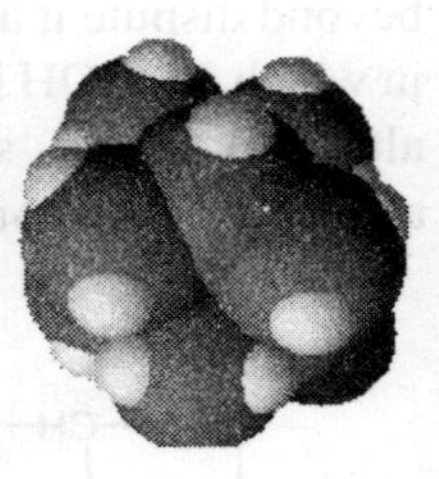

$H_3C$ $CH_3$ $H_3C$ O

Camphor was first synthesized by Gustaf Komppa in 1903. Previously, some organic compounds (such as urea) had been synthesized in the laboratory as a proof of concept, but camphor was a scarce natural product with a worldwide demand. The synthesis was the first industrial total synthesis, when Komppa began industrial production in Tainionkoski, Finland, in 1907.

**Norcamphor** is a camphor derivative with the three methyl groups replaced by hydrogen. Other substances deriving from trees are sometimes wrongly sold as camphor.

Camphor Trees are widely found in very deep jungles of Western Ghats of Tamil Nadu and Kerala states in South India.

## Occurrence and Synthesis

It exists in nature in the d, l and r forms, d – borneol is found in a tree Dryobalanops camphora, growing in Sumatra and Borneo and also in rosemary and spite oils, l – borneol and the inactive of borneol form pinene and camphene hydrochlorides has already been discussed. It is related to ordinary camphor in smell and burning taste. When warmed with potassium bisulphate it parts with water and yields camphene. It has already been mentioned that bornyl chloride is identical with "pinene hydrochloride" Isoborneol is a stereo isomeride of borneol which does not occur naturally, it may be obtained in the form of its acetate by warming camphene with glacial acetic acid and concentrated sulphuric acid at 50 to 60°C. If isoborneol dissolved in xylene is treated with sodium it is transformed into borneol, with oxidizing agents such as potassium permanganate, ozone chloride or oxides of nitrogen it is readily converted into borneol and isoborneol has been dealt with previously,

## Biosynthesis

In biosynthesis camphor is produced from geranyl pyrophosphate, via cyclisation of linaloyl pyrophosphate to bornyl pyrophosphate, followed by hydrolysis to borneol and oxidation to camphor.



## Physcial Porperties

| | |
|---|---|
| Molecular Formula | $= C_{10}H_{16}O$ |
| Formula Weight | = 152.23344 |
| Composition | = C(78.90%) H(10.59%) O(10.51%) |
| Melting point | =179.75 °C |
| Boiling Point | = 204 °C |
| Molar Refractivity | = 44.39 $cm^3$ |
| Molar Volume | = 154.8 $cm^3$ |
| Parachor | = 367.1 $cm^3$ |
| Index of Refraction | = 1.485 |
| Surface Tension | = 31.5 dyne/cm |
| Density | = 0.982 $g/cm^3$ |
| Polarizability | = 17.59 $cm^3$ |
| Monoisotopic Mass | = 152.120115 Da |
| Nominal Mass | = 152 Da |
| Average Mass | = 152.2334 Da |
| Log P | = 2.13 |

## Chemical Properties

Typical camphor reactions are:

- bromination

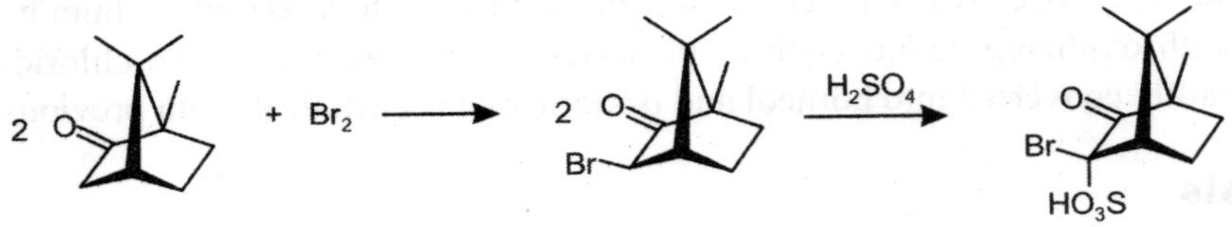

- conversion to isonitrosocamphor

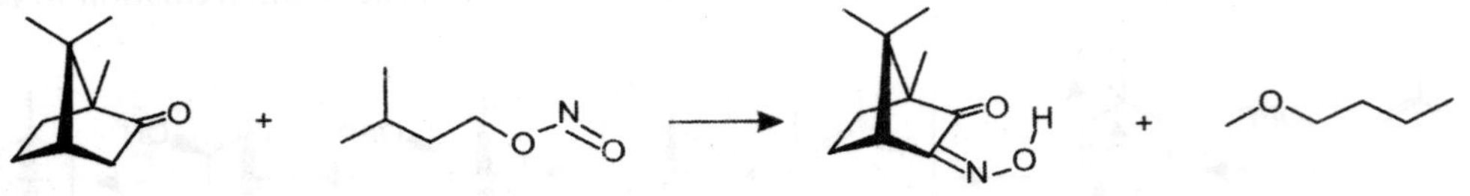

Camphor can also be reduced to isoborneol using sodium borohydride.

## Uses

Modern uses include as a plasticizer for cellulose nitrate, as a moth repellent, as an antimicrobial substance, in embalming, and in fireworks. Camphor crystals are also used to prevent damage to insect collections by other small insects. A form of anti-itch gel currently on the market uses camphor as its active ingredient. It is also used in medicine. Camphor is readily absorbed through the skin and produces a feeling of cooling similar to that of menthol and acts as slight local anesthetic and antimicrobial substance. Camphor is an active ingredient (along with menthol) in vapor-steam products, such as Vicks VapoRub, and it is effective as a cough suppressant. It may also be administered orally in small quantities (50 mg) for minor heart symptoms and fatigue.

In the 17th Century, it was used by Auenbrugger in the treatment of mania.

It is also believed that camphor will deter snakes and other reptiles due to its strong odor. Similarly, camphor is believed to be toxic to insects and is thus sometimes used as a repellent.

Camphor is also used in the Mahashiva ratri celebrations of Shiva, the Hindu god of destruction of evil. Its natural pitch substance burns cool without leaving an ash residue, which symbolizes the consciousness. Recently, carbon nanotubes were successfully synthesized using camphor in chemical vapor deposition process.

Currently, camphor is mostly used as a flavouring for sweets in Asia. In ancient and medieval Europe it was widely used as ingredient for sweets but it is now mainly used for medicinal purposes. Camphor was used as a flavouring in confections resembling ice cream in China during the Tang dynasty (A.D. 618-907). Camphor is widely used in cooking (mainly for desert dishes) in India where it is known as *Pachha Karpooram* (literally meaning "green camphor" though "Pachha" in Tamil can also be translated to mean "raw" which is "Pachha Karpooram's" intended meaning). It is widely available at Indian grocery stores and is labeled as "Edible Camphor." In Hindu poojas and ceremonies, camphor is burned in a ceremonial spoon for performing aarti. This type of camphor is also sold at Indian grocery stores but it is not suitable for cooking. The only type that should be used for food are those which are labelled as "Edible Camphor."

## Japanese Camphor

This camphor was until recently obtained exclusively from the camphor tree, Laurus Camphora (Bay camphor), growing in Japan (particularly in Formosa) and china

## Extraction and Preparation

The camphor wood is heated with water when camphor and camphor oil pass over with steam. The vapours are condensed in a suitable manner and the camphor is removed and purified by sublimation. It forms a colourless transparent mass of characteristic smell burning tastes melting point is 175°C and boiling point 209°C.

The most convincing prove of Bredt's formula for camphor is provided by Komppa's Synthesis of camphoric acid from which camphor itself had previously be obtained. The method employed was, the dimethyl ester of ββ - dimethyl glutaric acid was condensed with oxalic ester to give diketo apocamphoric ester. From this, by treatment with metallic sodium and methylation with methyl iodide, was obtained diketo camphoric ester which by way of various intermediate compounds was reduced to r – camphoric acid.

dimethyl oxalate + β,β' dimethyl glutaric acid → apocamphoric ester → diketocamphoric ester

Camphoric anhydride, when treated with sodium amalgam, can also be converted into the campholide, which with potassium cyanide yields the calcium salt of this acid, on distillation finally gives the corresponding ketone camphor.

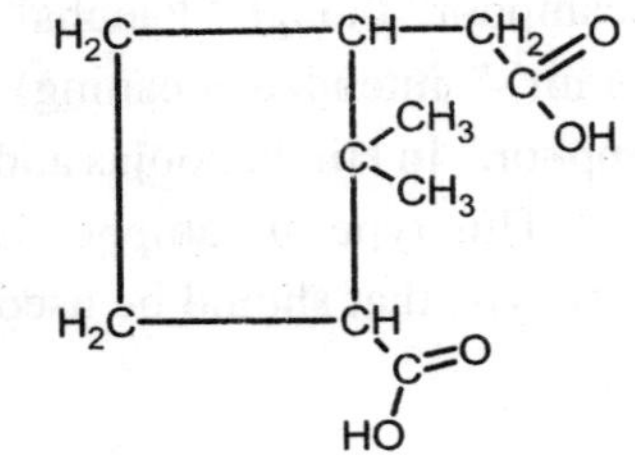

r - camphoric acid

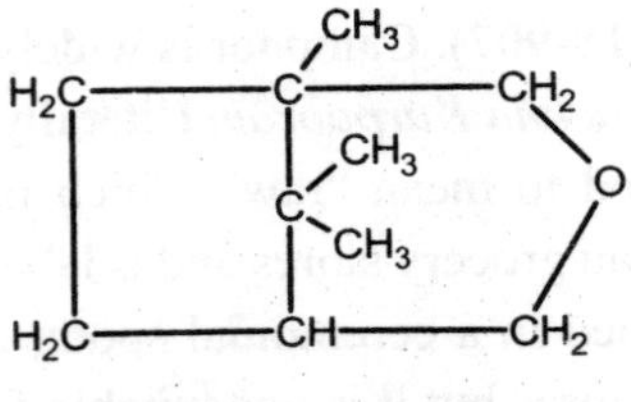

campholide

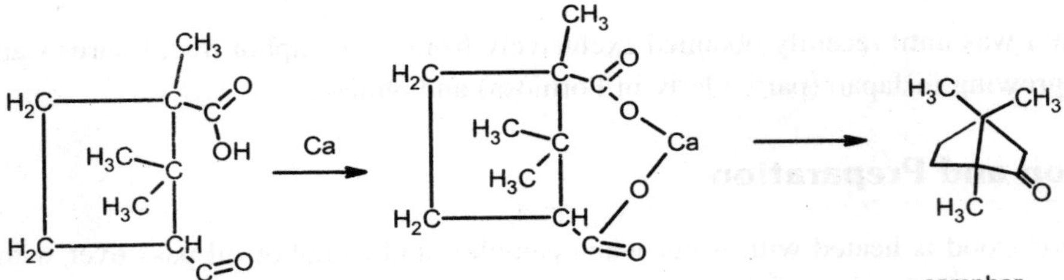

Homocamphoric acid

calcium homocamphorate

camphor

## Industrial Preparation

Camphor, now a days is being prepared in industry, from pinene here pinene is converted into isoborneol and it is oxidized to camphor

Pinene → Bornyl Chloride → isobornyl acetate → isoborneol → camphor

***l* – Camphor**: The camphor antipode of *d* – camphor, occurs in the oil of Martricaria pertenium, and apart from the sign of its rotation, shows the same properties as ordinary camphor. By mixing the two antipodes r – camphor, m.p 178°C, obtained identical with that produced by the oxidation of r – borneol or r – camphene

## Constitution of Camphor

The first suggestion as to the constitution of camphor was advanced in 1859 by Berthelot, and for the next forty years numerous workers were engaged on the problem. Information has been gained chiefly by the degradation of the camphor module by oxidation, and a series of detailed investigations on this line finally led Bredt to put forward a constitution (formula X) which is now generally accepted as the correct.

The oxidation of camphor with nitric acid leads to the formation of camphoric acid, $C_{10}H_{16}O_4$, camphanic acid $C_{10}H_{14}O_4$ and camphoronic acid $C_9H_{14}O_6$ as chief products. These three acids represent different stages of oxidation and camphoric acid. It is a tribasic acid which resembles tricarballylic

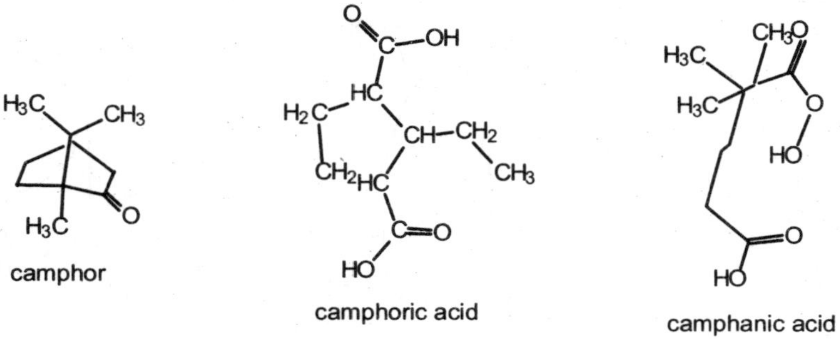

camphor

camphoric acid

camphanic acid

acid, in its properties on slow oxidation it breads up mainly to forms carbon dioxide isobutyric acid and trimethyl – succcinic acid. From this behaviour, Bredt concluded that camphoronic acid was trimethyl tricarballylic acid, a view confirmed later by its synthesis by Perkin and Thorpe.

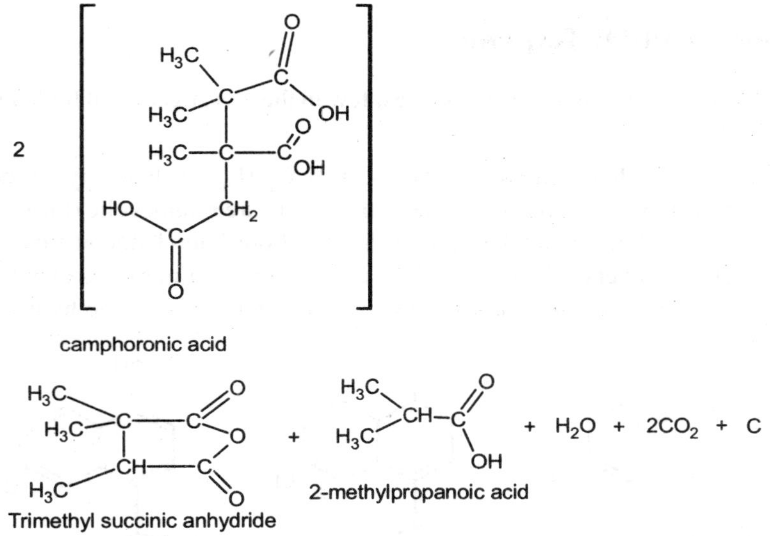

2 [camphoronic acid] → Trimethyl succinic anhydride + 2-methylpropanoic acid + $H_2O$ + $2CO_2$ + C

camphoronic acid

Trimethyl succinic anhydride

2-methylpropanoic acid

Since camphoronic acid is an oxidation product of camphanic acid camphoric acids and camphor, it may be concluded that the carbon framework of the camphoronic acid is present in each of these compounds, thus leading to the formula shown above. Trimethyl – succcinic acid has also been prepred directly from camphoric acid by oxidation with chromic acid. The constitution of camphoric acid has in addition been confirmed synthetically. The Ketonic character of camphor is proved by the conversion of camphor into caracrol $C_{10}H_{14}O$ on boiling with iodine. In the latter compound the hydroxyl is an the ortho – position to a methyl group. The presence of the group. $CH_2$ – CO is camphor follows from the treatment with amyl nitrate and sodium alcoholate, when this substance is boiled with dilute sulphuric acid it yields camphor quinone.

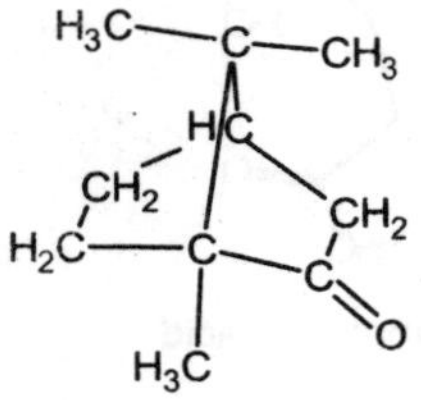

camphor 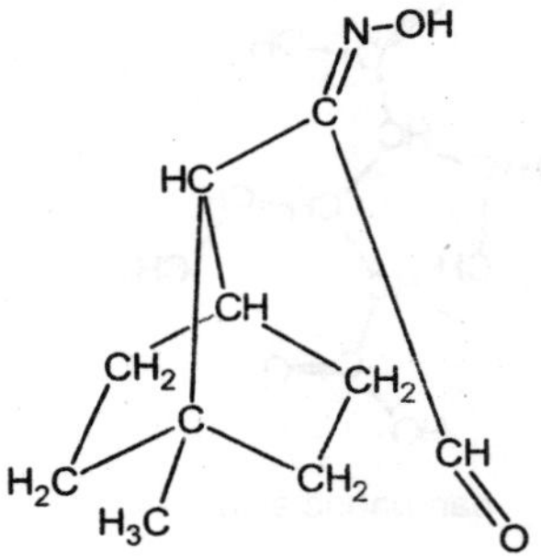 iso nitroso camphor 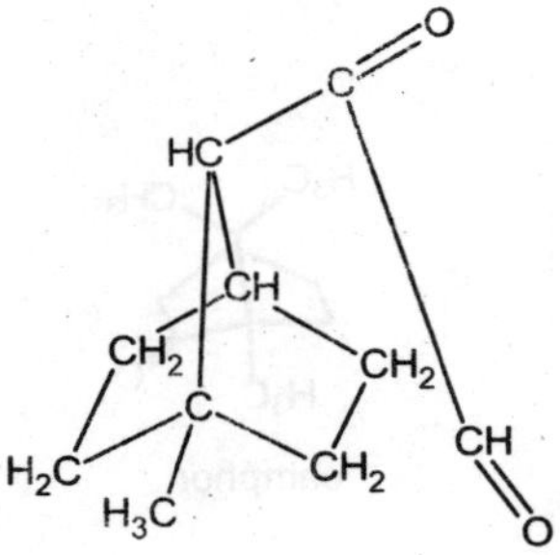 Camphor quinone

Among other reactions camphor yields p – cymene by loss of water when it is heated with phosphorous pentoxide

## Sesquiterpenes and Di Terpenes

Considerable progress has been made during the last few in the investigation of higher terpenes especially the sesquiterrpenes.

Sesquiterpenes include hydrocarbon of the formula $C_{15}H_{24}$ ad their oxygen derivatives and are found widely distributed in essential oils. Some of the others are monocyclic (bisaboline) Zingiberene, and still other dicyclic (cadenene and deudesmol) Much of our knowledge of these compounds is due to Ruzicka Simonsen and others. The relationships among the sesquiterpenes may be illustrated by the conversion of the open chain alcohol farnesol by loss of water into dicyclic hydrocarbon codinene

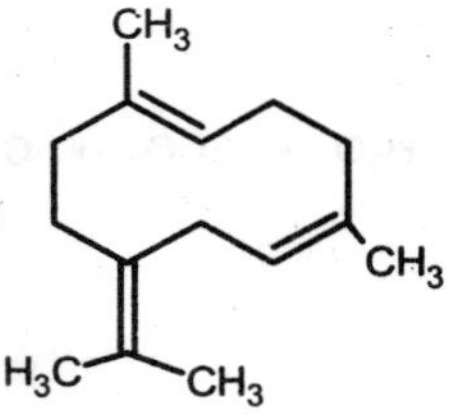

Farnesol 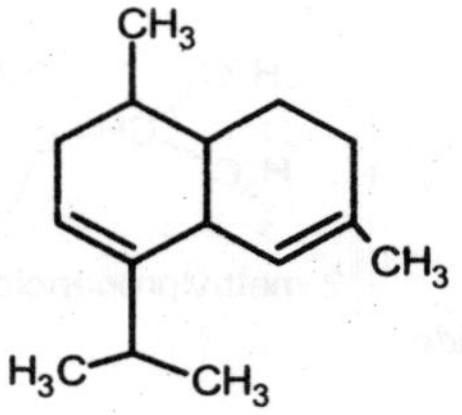 Cardenene 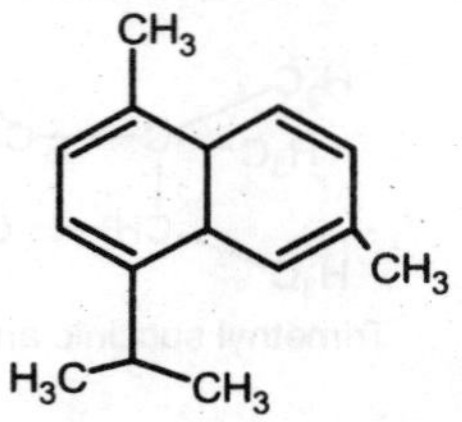 Cadalene

By fusion with sulphur (Veresterberg's hydrogenation method) the latter has been converted into cadalene, 1:6 dimethyl – 4 – isopropyl naphthalene, thus giving valuable information as to its structure.

Among the monocylic sesquiterpenes may be mentioned bisaboline (from the oil of lemons) and its isomerides Zingiberene (in the oil of ginger) . The former may be obtained from farnesol by gently warming with strong

The dicyclic sesquiterpenes are related to naphthalene in much the related to naphthalene in much the same manner as the terpenes are to benzene, some (e.g. cadenene, in oil of cubebs) being derived from cadalene other (e,g. eudesmol) in eucalyptus oil, from eudalene, I – methyl – 7 – iso propyi naphthalene, and similar types.

Bisabolene

Zingiberene

## SANTONIN

**Santonin** is a drug which was widely used in the past as an anthelminthic, a drug that expels parasitic worms (helminths) from the body, by either killing or stunning them. Santonin was formerly listed in U.S. and British pharmacopoeia.

### Chemical composition and derivation

Santonin ($C_{15}H_{18}O_3$) is the anhydride of **santonic acid** ($C_{15}H_{20}O_4$), which is a chemical derivative of dimethylnaphtalene ($C_{10}H_6[CH_3]_2$). Santonin dissolves in alkalies with formation of salts of this carboxylic acid.

Santonin, in acetic acid solution, when exposed to sunlight for about a month, is converted into (colourless) **photosantonic acid** ($C_{15}H_{22}O_5$) which is generally regarded as less toxic. The ethyl ester of the latter is obtained when an alcoholic solution of santonin is exposed to sunlight (Sestini). A yellow colouration is developed upon exposure of santonin to light. Santonin is optically levorotatory.

The full chemical name of santonin is (3*S*, 3a*S*, 5a*S*, 9b*S*)-3a, 5, 5a, 9b-tetrahydro-3, 5a, 9-trimethylnaptho[1,2-*b*]furan-2,8(3*H*,4*H*)-dione. Its molecular weight is 246.30 santonin is an organic chemical consisting of colourless flat prisms, turning slightly yellow from the action of light and soluble in alcohol, chloroform and boiling water. It is derived from **santonica** which is the unexpanded flower-heads of *Artemisia maritima*.

According to the US Pharmacopoeia Santonin occurs "in colourless, shining, flattened, prismatic crystals, odorless and nearly tasteless when first put in the mouth, but afterward developing a bitter taste; not altered by exposure to air, but turning yellow on exposure to light. Nearly insoluble in cold water; soluble in 40 parts of alcohol at 15°C. (59°F.), in 250 parts of boiling water, and in 8 parts of boiling alcohol; also soluble in 140 parts of ether, in 4 parts of chloroform, and in solutions of caustic alkalies. When heated to 170°C. (338°F.), santonin melts, and forms, if rapidly cooled, an amorphous mass, which instantly crystallizes oil coming in contact with a minute quantity of one of its solvents.

At a higher temperature, it sublimes partly unchanged, and, when ignited, it is consumed, leaving no residue. Santonin is neutral to litmus paper moistened with alcohol. Santonin yields, with an alcoholic solution of potassium hydrate, a bright pinkish-red liquid, which gradually becomes colourless. From its solution in caustic alkalies, santonin is completely precipitated by supersaturation with an acid".

## Historical Pharmacological Use

As noted above, santonin was formerly used as an anthelminthic, typically administered with a purgative. Santonin was used in treatment of infestation by the roundworm *Ascaris lumbricoides* and in ascarid parasitoses in general (including threadworm parasitosis). It is ineffective in treatment of tapeworm infestation.

The *Encyclopedia Britannica* (1911) notes that the typical dose was 2 to 5 grams. (It should be noted this was a *total* dose; many regimens called for 3 doses daily over 3 days, and the "3 teaspoons 3 times a day for 3 days" regimen was typical around the 50's when use of santonin was starting to wane; actual doses per *dose* were closer to 20-30 milligrams per adult dose in a typical "50's regimen", but "one-shot" doses of santonin (especially via suppository) were common in the late 1800s-early 1900s.) The only formerly registered British preparation (as of 1911) was the "trochiscus santonini" (santonin lozenge), but the preparation "sodii santoninas" (soda of santonin) was also formerly listed as an official preparation in the U.S. Pharmacopeia. Commercial preparations containing santonin (usually containing a purgative laxative as well) also appeared in US drug formularies as late as the 50's; the *Modern Drug Encyclopedia and Therapeutic Index* of 1955 listed Lumbricide (produced by Massengill) and a generic santonin prepraration made by Winthrop-Stearns (now Winthrop-Sanofi).

Santonin also was used in a lesser extent in treatment of atony of the bladder. This usage largely dropped off after the early 1900s.

Dosage forms varied for santonin; in the 1800s-1900s, santonin lozenges or suppositories designed for single-dosage treatment of ascarid infestation were the typical form of treatment, whilst in the 50's the two remaining santonin preparations on the market in the United States were liquid medications.

## Hazards and Difficulty of Use of Santonin

Santonin was an agent which (compared to more modern anthelminthic drugs) was very complicated to use and entailed rather serious risk to the patient. Nearly every formulary and herbal which lists santonin or santonin-containing plants lists the real risk of yellow vision and of fatal reactions; even small doses of santonin cause disturbances of vision, usually yellow vision or perhaps green (xanthopsia or chromatopsia). Even the *Encyclopedia Britannica* noted:

...These effects usually pass off in a few days. Large doses, however, produce toxic effects, aphasia, muscular tremors and epileptiform convulsions, and the disturbances of vision may go on to total blindness.

More typical is the warning given regarding side effects of santonin in *King's American Dispensatory*:

Santonin is an active agent, and, in improper doses, is capable of producing serious symptoms, and even death. As small a dose as 2 grains is said to have killed a weakly child of 5 years, and 5 grains produced death in about 1/2 hour in a child of the same age. Among the toxic effects may be mentioned gastric pain, pallor and coldness of the surface, followed by heat and injection of the head, tremors, dizziness, pupillary dilatation, twitching of the eyes, stertor, copious sweating, hematuria, convulsive movements, tetanic cramps stupor, and insensibility. Occasionally symptoms resembling cholera morbus have been produced, and in all cases the urine presents a characteristic yellowish or greenish-yellow hue. We have observed convulsions caused by the administration of "worm lozenges." Death from santonin is due to respiratory paralysis, and post-mortem examination revealed in one instance a contracted and empty right ventricle, and about an ounce of liquid, black blood in the left heart, an inflamed duodenum, and inflamed patches in the stomach (Kilner). Santonin often produces a singular effect upon the vision, causing surrounding objects to appear discolored, as if they were yellow or green, and occasionally blue or red; it also imparts a yellow or green colour to the urine, and a reddish-purple colour if that fluid be alkaline. Prof. Giovanni was led to believe that the apparent yellow colour of objects observed by the eye, when under the influence of santonin, did not depend upon an elective action on the optic nerves, but rather to the yellow color which the drug itself takes when exposed to the air. Santonin coloured by the air does not produce this effect, which only follows the white article. The air gives the yellow colour to santonin, to passed urine containing it, and to the serum of the blood when drawn from a vein, and, according to Giovanni, it is owing to its direct action upon the aqueous humor, where it is carried by absorption, that objects present this colour. The view now held, however, is that of Rose, that the alkaline serum dissolves the santonin, which then acts upon the perspective centers of the brain, producing the chromatopsia or xanthopsia.

At least one modern herbal has also noted these same severe side effects of santonin.

Even were it not for the fact that santonin is among the most toxic of herbal anthelminthic drugs, deworming using santonin is complicated in comparison to more modern anthelminthics. Typically, santonin must be taken whilst fasting completely (both before and after taking the drug) for "single dose" regimens or on a full stomach with all fats and oils in the diet being avoided for 2-3 days before treatment as well as during treatment and 2-3 days afterwards (due to santonin being fat soluble and having an increased risk of side effects); after a course of santonin, a purgative must be given to cleanse the body of the dead worms. (The two remaining registered santonin preparations in the United States as of 1955 were in fact santonin/purgative combinations; Lumbricide contained santonin and senna (among other ingredients) and the Winthrop-Stearns generic preparation was a santonin/cascara sagrada combination drug.)

Due to the severe side effects (even when used as directed), the need for a purgative, and the development of many safer deworming drugs, santonin has largely fallen out of use. Typically mebendazole and pyrantel pamoate are used in modern pharmacopoeia practice where santonin was formerly used; even guides on holistic medicine strongly recommend avoiding the use of santonin due to its severe and occasionally fatal side effects and the availability of far safer anthelminthics. The Council Directive 65/65 European Economic Community (EEC) (in regards to pharmaceuticals and naturopathic preparations) has officially ruled santonin preparations to have an "unacceptable" risk-benefit ratio and preparations containing santonin are no longer eligible for registration in EU countries.

## Santonin and absinthe

**Whilst absinthe is certainly more infamous for its content of thujone, the liquor does also contain small amounts of santonin. It has been speculated by some parties that Impressionist art—in particular, Van Gogh's artwork—may have been inspired *not* by thujone and its presumed psychotropic effects, but on the "yellow vision" or xanthopsia which is a known side effect of santonin. This has been disputed, however, most notably by Arnold and Loftus (1991) who have noted the santonin content would have been insufficient to cause xanthopsia.**

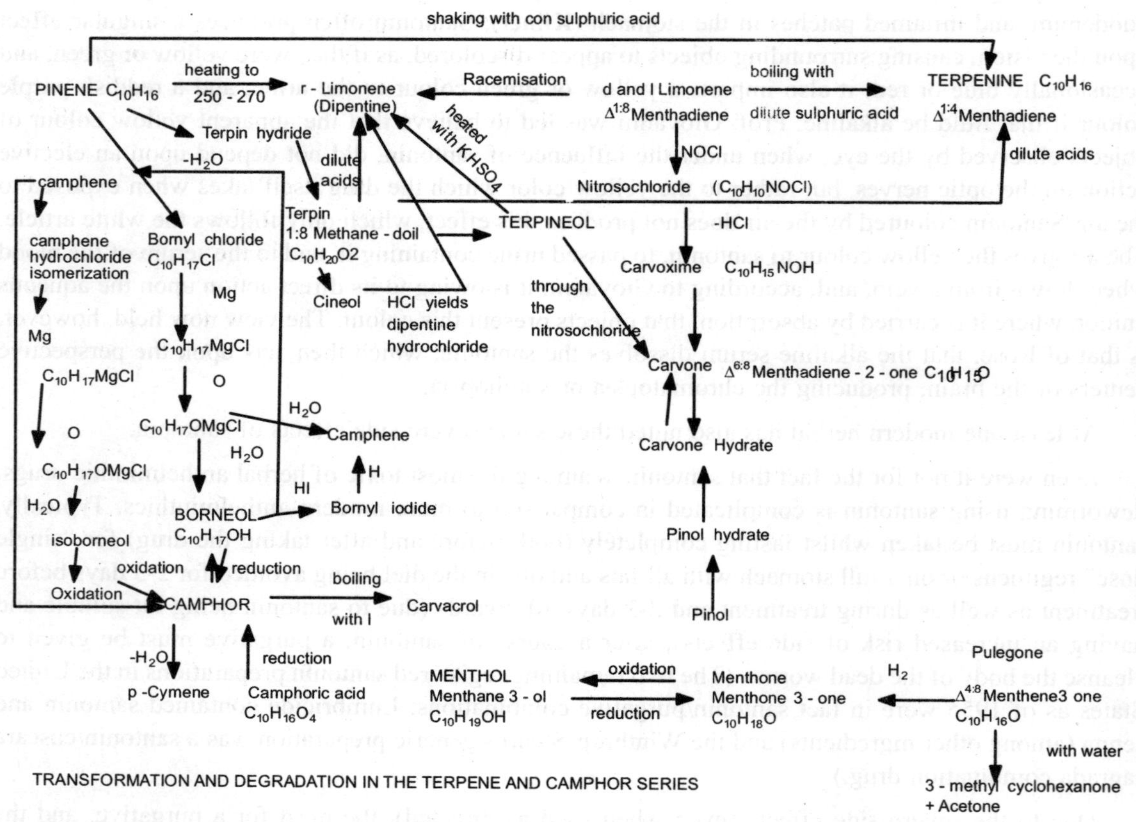


TRANSFORMATION AND DEGRADATION IN THE TERPENE AND CAMPHOR SERIES

# RUBBER

## History

In its native Central America and South America, rubber has been collected for a long time. The Mesoamerican civilizations used rubber mostly from *Castilla elastica*. The Ancient Mesoamericans had a ball game using rubber balls (*see: Mesoamerican ballgame*), and a few Pre-Columbian rubber balls have been found (always in sites that were flooded under fresh water), the earliest dating to about

1600 BC. According to Bernal Díaz del Castillo, the Spanish Conquistadores were so astounded by the vigorous bouncing of the rubber balls of the Aztecs that they wondered if the balls were enchanted by evil spirits. The Maya also made a type of temporary rubber shoe by dipping their feet into a latex mixture. Rubber was used in various other contexts, such as strips to hold stone and metal tools to wooden handles, and padding for the tool handles. While the ancient Mesoamericans did not have vulcanization, they developed organic methods of processing the rubber with similar results, mixing the raw latex with various saps and juices of other vines, particularly *Ipomoea alba*, a species of Morning glory. In Brazil the indigenous population understood the use of rubber to make water-resistant cloth. A story says that the first European to return to Portugal from Brazil with samples of such water-repellent rubberized cloth so shocked people that he was brought to court on the charge of witchcraft.

When samples of rubber first arrived in England, it was observed by Joseph Priestley, in 1770, that a piece of the material was extremely good for rubbing out pencil marks on paper, hence the name **rubber**.

The para rubber tree initially grew in South America, where it was the main source of what limited amount of latex rubber was consumed during much of the 19th century. About 100 years ago, the Congo Free State in Africa was a significant source of natural rubber latex, mostly gathered by forced labour. The Congo Free State was forged and ruled as a personal colony by the Belgian King Leopold II. After repeated efforts (see Henry Wickham) rubber was successfully cultivated in Southeast Asia, where it is now widely grown.

In India commercial cultivation of natural rubber was introduced by the British Planters, although the experimental efforts to grow rubber on a commercial scale in India were initiated as early as 1873 at the Botanical Gardens, Kolkata. The first commercial Hevea plantations in India were established at Thattekadu in Kerala in 1902.

The Rubber Board is a statutory body constituted by the Government of India, under the Rubber Act 1947, for the overall development of the rubber industry in the country.The head office of the rubber board is situated in Kottayam, Kerala, where the main production of natural rubber.

**Natural rubber** is an elastic hydrocarbon polymer that naturally occurs as a milky colloidal suspension, or *latex*, in the sap of some plants. It can also be synthesized. The entropy model of rubber was developed in 1934 by Werner Kuhn. The scientific name for the rubber tree is *Hevea brasiliensis*.

Collection of rubber from tree

The major commercial source of natural rubber latex is the Para rubber tree, *Hevea brasiliensis* (Euphorbiaceae). This is largely because it responds to wounding by producing more latex. Henry Wickham gathered thousands of seeds from Brazil in 1876 and they were germinated in Kew Gardens, England. The seedlings were sent to Colombo, Indonesia, Singapore and British Malaya. Malaya was later to become the biggest producer of rubber. Liberia and Nigeria are examples of African rubber-producing countries.

Other plants containing latex include figs (*Ficus elastica*), *Castilla*, euphorbias, and the common dandelion. Although these have not been major sources of rubber, Germany attempted to use such sources during World War II when it was cut off from rubber supplies. These attempts were later supplanted by the development of synthetic rubber.

Synthetic rubbers are made by the polymerization of a single monomer or a mixture of monomers to produce polymers. These form part of a broad range of products extensively studied by polymer science and rubber technology. Examples are SBR, or styrene-butadiene rubber, BR or butadiene rubber CR or chloroprene rubber and EPDM (ethylene-propylene-diene rubber)

# SYNTHETIC RUBBER

## History

The expanded use of motor vehicles, and particularly motor vehicle tires, starting in the 1890s, created increased demand for rubber. Political problems that resulted from great fluctuations in the cost of natural rubber led to enactment of the Stevenson Act in 1921. This act essentially created a cartel which supported rubber prices by regulating production (see OPEC), but insufficient supply, especially due to wartime shortages, also led to a search for alternative forms of synthetic rubber.

In 1879, Bouchardt created one form of synthetic rubber, producing a polymer of isoprene in a laboratory. Scientists in England and Germany developed alternate methods for creating isoprene polymers from 1910-1912.

The first large-scale commercial production occurred in Germany during World War I, as a result of shortages of natural rubber. However, it used a different form of synthetic rubber based on a polymer of butadiene, building on the laboratory work of the Russian scientist Sergei Lebedev. This early form of synthetic rubber was again replaced with natural rubber after the war ended, but investigations of synthetic rubber continued, leading to the 1933 invention which German scientists designated "Buna S". This type of synthetic rubber, a copolymer of butadiene and styrene, still represents about one-half of total world production. Dr. Waldo Semon of the B.F. Goodrich Company developed *Koroseal* in 1935, and *Ameripol* (from American Polymer) in 1940, while Russian researchers created *Sovprene*

Synthetic rubber is any type of artificially made polymer material which acts as an elastomer. An elastomer is a material with the mechanical (or material) property that it can undergo much more elastic deformation under stress than most materials and still return to its previous size without permanent deformation. Synthetic rubber serves as a substitute for natural rubber in many cases, especially when improved material properties are needed.

Natural rubber coming from latex is mostly polymerized isoprene with a small percentage of impurities in it. This will limit the range of properties available to it. Also, there are limitations on the proportions of *cis* and *trans* double bonds resulting from methods of polymerizing natural latex. This also limits the range of properties available to natural rubber, although addition of sulfur and vulcanization are used to improve the properties.

However, synthetic rubber can be made from the polymerization of a variety of monomers including isoprene (2-methyl-1,3-butadiene), 1,3-butadiene, chloroprene (2-chloro-1,3-butadiene), and isobutylene (methylpropene) with a small percentage of isoprene for cross-linking. Furthermore, these and other monomers can be mixed in various desirable proportions to be copolymerized for a wide range of physical, mechanical, and chemical properties. The monomers can be produced pure and addition of impurities or additives can be controlled by design to give optimal properties. Polymerization of pure monomers can be better controlled to give a desired proportion of *cis* and *trans* double bonds.

## Manufacture of Synthetic Rubber

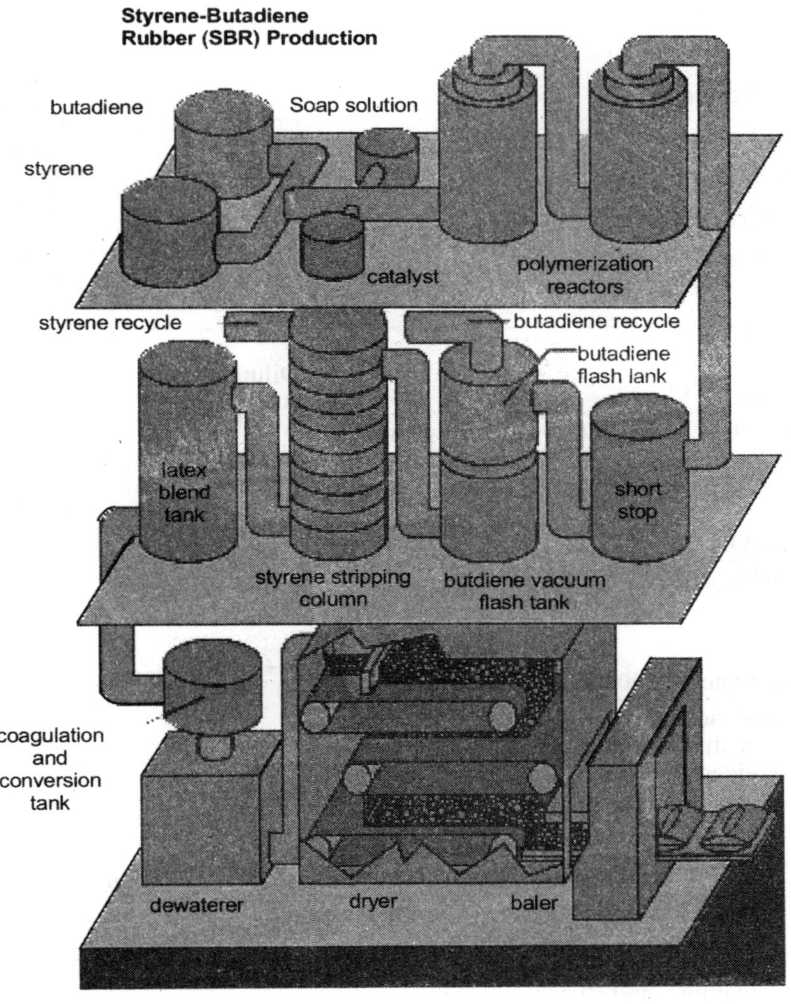


**Fig. 39.1. Showing the Industrial Process for the Manufacture of Artificial Rubber (SBR)**

## Production of Artificial Styrene—Butadiene Rubber (SBR)

In this process an initiator catalyst potassium peroxy di sulphate or potassium persulphate ($K_2S_2O_8$) is used initially for catalyzing the reaction, later on the reaction proceeds in the soap solution which acts as the catalyst. During the reaction "Some chain modifier (D. D. M.) reaction arrester (hydroquinone) and an antioxidant (N-phenyl 2 naphthyl amine) are also added. Following table gives the different materials and their pro-portions used.

| | |
|---|---|
| Butadine | 75% |
| Styrene | 25% |
| Total water | 180% |
| Soap | 5% |
| Sodium Hydroxide | 0.03% |
| Trisodium phosphate ($Na_3PO_4$) | 0.5% |
| potassium peroxy di sulphate | 0.3% |
| D.D.M (tertiary mecapten) | 0.3% |
| Hydroquinone | 0.1% |
| Antioxidant | 1.25% |

6 p.c. potash soap solution —diluted with rest of water in wooden vat at slightly elevated temp.→ Diluted soap solution in wooden vat (1)

(1) —NaOH and $Na_3PO_4$added.→ Soap solution with NaOH and $Na_2PO_4$ (2)

(2) —Passed into a 2500 gallon glass lined jacketted autoclave fitted with stirrer, and initiator catalyst added→ Initiatorcatalyst + soap solution (3)

to (3) is added ←Styrene and D.D.M. added to auto-clave from which air is evacuated to guard against foaming, and stirred (300 r.p.m.)→ Styrene + D.D.M. + I. Catalyst + Soap solution in autoclave (4)

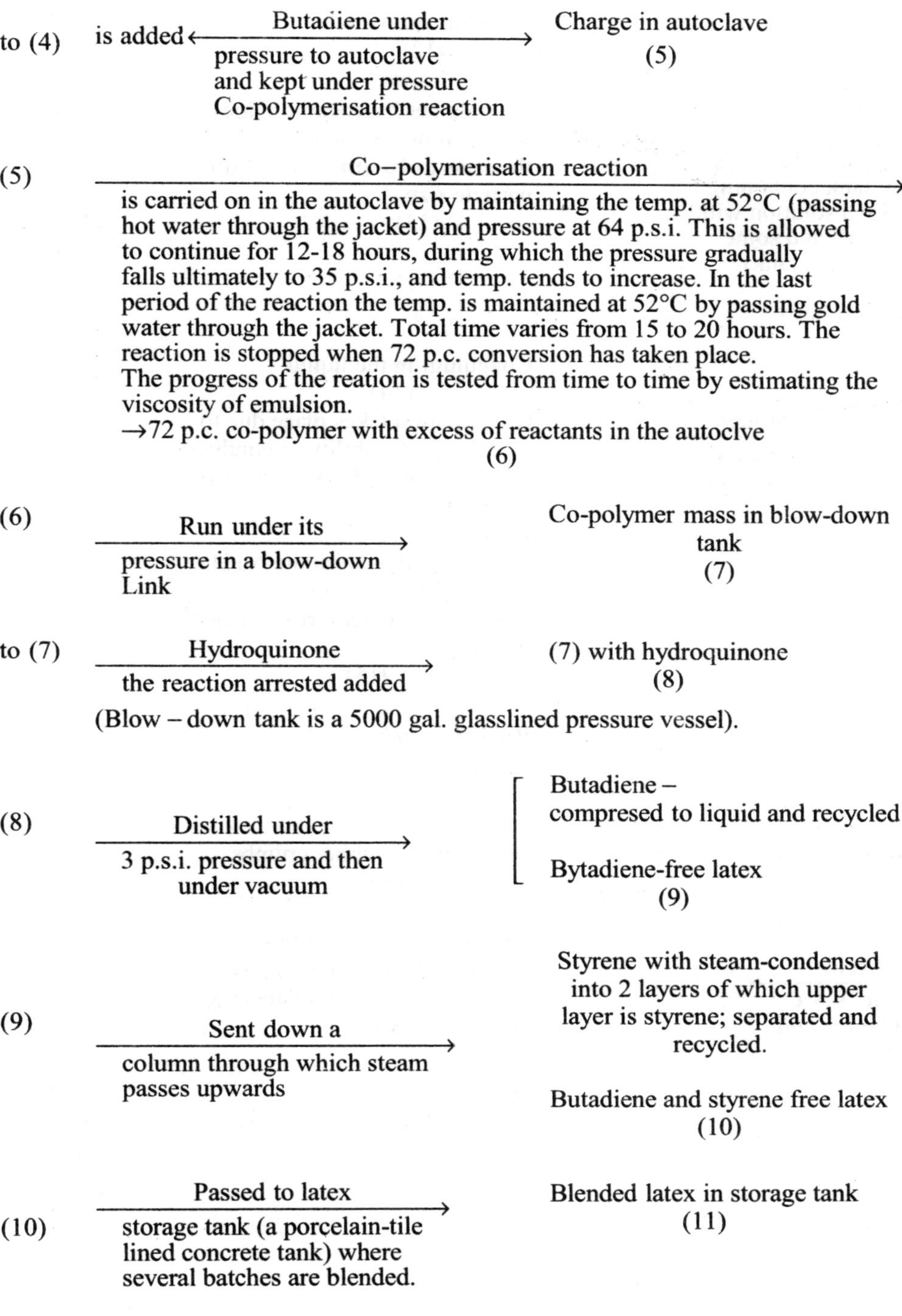

to (4)
is added ← Butadiene under → Charge in autoclave (5)
pressure to autoclave
and kept under pressure
Co-polymerisation reaction
(5)
Co–polymerisation reaction →
is carried on in the autoclave by maintaining the temp. at 52°C (passing hot water through the jacket) and pressure at 64 p.s.i. This is allowed to continue for 12-18 hours, during which the pressure gradually falls ultimately to 35 p.s.i., and temp. tends to increase. In the last period of the reaction the temp. is maintained at 52°C by passing gold water through the jacket. Total time varies from 15 to 20 hours. The reaction is stopped when 72 p.c. conversion has taken place.
The progress of the reation is tested from time to time by estimating the viscosity of emulsion.
→72 p.c. co-polymer with excess of reactants in the autoclve
(6)
(6)
Run under its →
pressure in a blow-down Link
Co-polymer mass in blow-down tank
(7)
to (7)
Hydroquinone →
the reaction arrested added
(7) with hydroquinone
(8)
(Blow – down tank is a 5000 gal. glasslined pressure vessel).
(8)
Distilled under →
3 p.s.i. pressure and then under vacuum
Butadiene – compresed to liquid and recycled
Bytadiene-free latex
(9)
(9)
Sent down a →
column through which steam passes upwards
Styrene with steam-condensed into 2 layers of which upper layer is styrene; separated and recycled.
Butadiene and styrene free latex
(10)
(10)
Passed to latex →
storage tank (a porcelain-tile lined concrete tank) where several batches are blended.
Blended latex in storage tank
(11)

to (11) $\xrightarrow[\text{added}]{\text{Antioxidant}}$ (11) with antioxidant (12)

(12) $\xrightarrow[\text{wooden tank containing 4 p.c. NaCl solution with a little alkali and some glue (creamery tank)}]{\text{Passed into}}$ Co-polymer hydrocarbon does not separate but creams or thickens to 10 times the originl particle size. (13)

(13) is the creamed latex.

(13) $\xrightarrow[\text{wooden tanks containing dilute sulphuric acid−coaguation takes place.}]{\text{Passed through}}$ Coagulum in the liquid (14) (Coagulation takes place due to vanishing of stability of emulision by the action of $H_2SO_4$ on soap solution).

(14) $\xrightarrow[\text{screen}]{\text{Vibrating}}$ Liquid – serves to dilute NaCl soln. and $H_2SO_4$ soln. Crumbs of rubber – re – slurried inwater (15)

(15) $\xrightarrow[\text{Oliver suction filter and washed.}]{\text{Filtered in}}$ Rubber crumbs (16)

(16) $\xrightarrow{\text{Disintegrated}}$ Disintegrated rubber crumbs (17)

(17) $\xrightarrow[\text{hot air at 79.5°C (Seventynine and half degree C)}]{\text{Dried in}}$ Dried rubber is baled. It is dusted with soap stone powder and slipped into craft paper bags 70 lbs. in a bag.

CHAPTER

# 40

# The Compounds of Biological and Biochemical Importance

## Introduction

The compounds of biological and biochemical importance; this chapter channelizes the direction of the students towards the biochemistry and medicine. This chapter includes those chemical compounds which are directly related to the human metabolism. Though in the introductory level but an endeavour has been made to introduce the following chemical compounds, such as protein, lipids, vitamins, hormones steroids, enzymes, etc this will become a successful introduction if the students feel comfortable to study biochemistry after reading this chapter.

## PROTEIN

The word *protein* comes from the Greek word πρωτωσ ("protos"), meaning of primary importance Proteins are large organic compounds made of amino acids arranged in a linear chain and joined together by peptide bonds between the carboxyl and amino groups of adjacent amino acid residues. The sequence of amino acids in a protein is defined by a gene and encoded in the genetic code. Although this genetic code specifies 20 "standard" amino acids plus selenocysteine and - in certain archaea - pyrrolysine, the residues in a protein are sometimes chemically altered in post-translational modification: either before the protein can function in the cell, or as part of control mechanisms. Proteins can also work together to achieve a particular function, and they often associate to form stable complexes.

Like other biological macromolecules such as polysaccharides and nucleic acids, proteins are essential parts of organisms and participate in every process within cells. Many proteins are enzymes that catalyze biochemical reactions and are vital to metabolism. Proteins also have structural or mechanical functions, such as actin and myosin in muscle and the proteins in the cytoskeleton, which form a system of scaffolding that maintains cell shape. Other proteins are important in cell signaling, immune responses, cell adhesion, and the cell cycle. Proteins are also necessary in animals' diets, since animals cannot synthesize all the amino acids they need and must obtain essential amino acids from food. Through the process of digestion, animals break down ingested protein into free amino acids that are then used in metabolism.

## History

Proteins were first described and named by the Swedish chemist Jöns Jakob Berzelius in 1838. However, central role of proteins in living organisms was not fully appreciated until 1926, when James B. Sumner showed that the enzyme urease was a protein. The first protein to be sequenced was insulin, by Frederick Sanger, who won the Nobel Prize for this achievement in 1958. The first protein structures to be solved included hemoglobin and myoglobin, by Max Perutz and Sir John Cowdery Kendrew, respectively, in 1958. The three-dimensional structures of both proteins were first determined by x-ray diffraction analysis; the structures of myoglobin and hemoglobin won the 1962 Nobel Prize in Chemistry for their discoveries.

## Biochemistry

Resonance structures of the peptide bond that links individual amino acids to form a protein polymer.

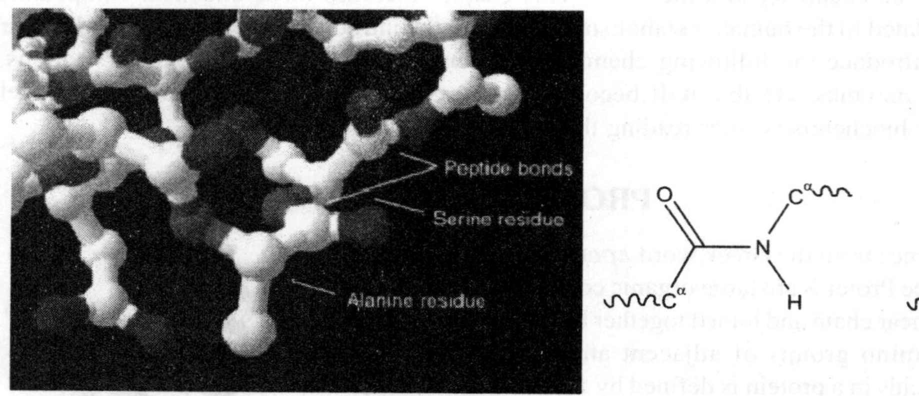


**Fig. 40.1. Section of a protein structure showing serine and alanine residues linked together by peptide bonds. Carbons are shown in white and hydrogens are omitted for clarity**

Proteins are linear polymers built from 20 different L-α-amino acids. All amino acids possess common structural features, including an á carbon to which an amino group, a carboxyl group, and a variable side chain are bonded. Only proline differs from this basic structure as it contains an unusual ring to the N-end amine group, which forces the CO–NH amide moiety into a fixed conformation. The side chains of the standard amino acids, detailed in the list of standard amino acids, have different chemical properties that produce three-dimensional protein structure and are therefore critical to protein function. The amino acids in a polypeptide chain are linked by peptide bonds formed in a dehydration reaction. Once linked in the protein chain, an individual amino acid is called a *residue,* and the linked series of carbon, nitrogen, and oxygen atoms are known as the *main chain* or *protein backbone.* The peptide bond has two resonance forms that contribute some double-bond character and inhibit rotation around its axis, so that the alpha carbons are roughly coplanar. The other two dihedral angles in the peptide bond determine the local shape assumed by the protein backbone.

Due to the chemical structure of the individual amino acids, the protein chain has directionality. The end of the protein with a free carboxyl group is known as the C-terminus or carboxy terminus, whereas the end with a free amino group is known as the N-terminus or amino terminus.

The words *protein, polypeptide,* and *peptide* are a little ambiguous and can overlap in meaning. *Protein* is generally used to refer to the complete biological molecule in a stable conformation, whereas *peptide* is generally reserved for a short amino acid oligomers often lacking a stable three-dimensional structure. However, the boundary between the two is not well defined and usually lies near 20–30 residues.[5] *Polypeptide* can refer to any single linear chain of amino acids, usually regardless of length, but often implies an absence of a defined conformation.

## Synthesis

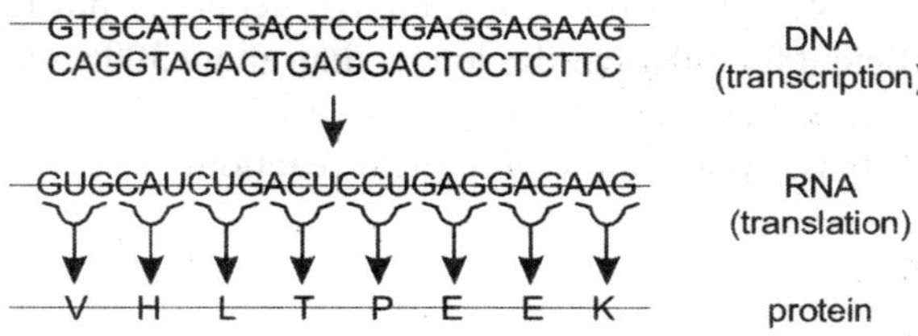


**Fig. 40.2. The DNA sequence of a gene encodes the amino acid sequence of a protein.**

Proteins are assembled from amino acids using information encoded in genes. Each protein has its own unique amino acid sequence that is specified by the nucleotide sequence of the gene encoding this protein. The genetic code is a set of three-nucleotide sets called codons and each three-nucleotide combination stands for an amino acid, for example AUG stands for methionine. Because DNA contains four nucleotides, the total number of possible codons is 64; hence, there is some redundancy in the genetic code, with some amino acids specified by more than one codon. Genes encoded in DNA are first transcribed into pre-messenger RNA (mRNA) by proteins such as RNA polymerase. Most organisms then process the pre-mRNA (also known as a *primary transcript*) using various forms of post-transcriptional modification to form the mature mRNA, which is then used as a template for protein synthesis by the ribosome. In prokaryotes the mRNA may either be used as soon as it is produced, or be bound by a ribosome after having moved away from the nucleoid. In contrast, eukaryotes make mRNA in the cell nucleus and then translocate it across the nuclear membrane into the cytoplasm, where protein synthesis then takes place. The rate of protein synthesis is higher in prokaryotes than eukaryotes and can reach up to 20 amino acids per second.

The process of synthesizing a protein from an mRNA template is known as translation. The mRNA is loaded onto the ribosome and is read three nucleotides at a time by matching each codon to its base pairing anticodon located on a transfer RNA molecule, which carries the amino acid corresponding to the codon it recognizes. The enzyme aminoacyl tRNA synthetase "charges" the tRNA molecules with the correct amino acids. The growing polypeptide is often termed the *nascent chain*. Proteins are always biosynthesized from N-terminus to C-terminus.

The size of a synthesized protein can be measured by the number of amino acids it contains and by its total molecular mass, which is normally reported in units of *daltons* (synonymous with atomic mass units), or the derivative unit kilodalton (kDa). Yeast proteins are on average 466 amino acids long and 53 kDa in mass.[5] The largest known proteins are the titins, a component of the muscle sarcomere, with a molecular mass of almost 3,000 kDa and a total length of almost 27,000 amino acids.

## Chemical Synthesis

Short proteins can also be synthesized chemically by a family of methods known as peptide synthesis, which rely on organic synthesis techniques such as chemical ligation to produce peptides in high yield.[8] Chemical synthesis allows for the introduction of non-natural amino acids into polypeptide chains, such as attachment of fluorescent probes to amino acid side chains.[9] These methods are useful in laboratory biochemistry and cell biology, though generally not for commercial applications. Chemical synthesis is inefficient for polypeptides longer than about 300 amino acids, and the synthesized proteins may not readily assume their native tertiary structure. Most chemical synthesis methods proceed from C-terminus to N-terminus, opposite the biological reaction.

## Structure of Proteins

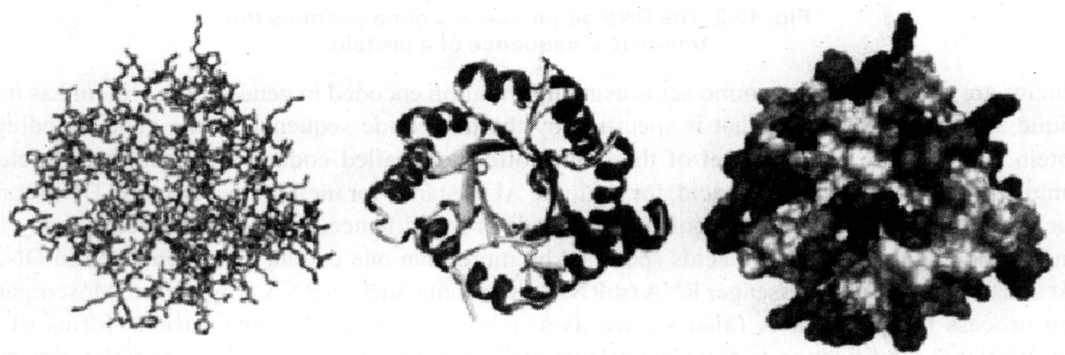

**Fig. 40.3. Three possible representations of the three-dimensional structure of the protein triose phosphate isomerase. Left: all-atom representation colored by atom type. Middle: simplified representation illustrating the backbone conformation, coloured by secondary structure. Right: Solvent-accessible surface representation colored by residue type (acidic residues red, basic residues blue, polar residues green, nonpolar residues white)**

Most proteins fold into unique 3-dimensional structures. The shape into which a protein naturally folds is known as its native state. Although many proteins can fold unassisted, simply through the chemical properties of their amino acids, others require the aid of molecular chaperones to fold into their native states. Biochemists often refer to four distinct aspects of a protein's structure:

- *Primary structure*: the amino acid sequence.
- *Secondary structure*: regularly repeating local structures stabilized by hydrogen bonds. The most common examples are the alpha helix and beta sheet.[10] Because secondary structures are local, many regions of different secondary structure can be present in the same protein molecule.
- *Tertiary structure*: the overall shape of a single protein molecule; the spatial relationship of the secondary structures to one another. Tertiary structure is generally stabilized by nonlocal interactions, most commonly the formation of a hydrophobic core, but also through salt bridges, hydrogen bonds, disulfide bonds, and even post-translational modifications. The term "tertiary structure" is often used as synonymous with the term *fold*.
- *Quaternary structure*: the shape or structure that results from the interaction of more than one protein molecule, usually called *protein subunits* in this context, which function as part of the larger assembly or protein complex.

NMR structures of the protein cytochrome c in solution show the constantly shifting dynamic structure of the protein. Larger version.

Proteins are not entirely rigid molecules. In addition to these levels of structure, proteins may shift between several related structures while they perform their biological function. In the context of these functional rearrangements, these tertiary or quaternary structures are usually referred to as "conformations," and transitions between them are called *conformational changes*. Such changes are often induced by the binding of a substrate molecule to an enzyme's active site, or the physical region of the protein that participates in chemical catalysis. In solution all proteins also undergo variation in structure through thermal vibration and the collision with other molecules, see the animation on the right.

**Fig. 40.4. Molecular surface of several proteins showing their comparative sizes. From left to right are: Antibody (IgG), Hemoglobin, Insulin (a hormone), Adenylate kinase (an enzyme), and Glutamine synthetase (an enzyme).**

Proteins can be informally divided into three main classes, which correlate with typical tertiary structures: globular proteins, fibrous proteins, and membrane proteins. Almost all globular proteins are soluble and many are enzymes. Fibrous proteins are often structural; membrane proteins often serve as receptors or provide channels for polar or charged molecules to pass through the cell membrane.

A special case of intramolecular hydrogen bonds within proteins, poorly shielded from water attack and hence promoting their own dehydration, are called dehydrons.

## Structure Determination

Discovering the tertiary structure of a protein, or the quaternary structure of its complexes, can provide important clues about how the protein performs its function. Common experimental methods of structure determination include X-ray crystallography and NMR spectroscopy, both of which can produce information at atomic resolution. Cryoelectron microscopy is used to produce lower-resolution structural information about very large protein complexes, including assembled viruses;[10] a variant known as electron crystallography can also produce high-resolution information in some cases, especially for two-dimensional crystals of membrane proteins. Solved structures are usually deposited in the Protein Data Bank (PDB), a freely available resource from which structural data about thousands of proteins can be obtained in the form of Cartesian coordinates for each atom in the protein.

Many more gene sequences are known than protein structures. Further, the set of solved structures is biased toward proteins that can be easily subjected to the conditions required in X-ray crystallography, one of the major structure determination methods. In particular, globular proteins are comparatively easy to crystallize in preparation for X-ray crystallography. Membrane proteins, by contrast, are difficult to crystallize and are underrepresented in the PDB.[12] Structural genomics initiatives have attempted to remedy these deficiencies by systematically solving representative structures of major fold classes. Protein structure prediction methods attempt to provide a means of generating a plausible structure for proteins whose structures have not been experimentally determined.

## Cellular Functions

Proteins are the chief actors within the cell, said to be carrying out the duties specified by the information encoded in genes.[5] With the exception of certain types of RNA, most other biological molecules are relatively inert elements upon which proteins act. Proteins make up half the dry weight of an *Escherichia coli* cell, whereas other macromolecules such as DNA and RNA make up only 3% and 20%, respectively. The set of proteins expressed in a particular cell or cell type is known as its proteome.

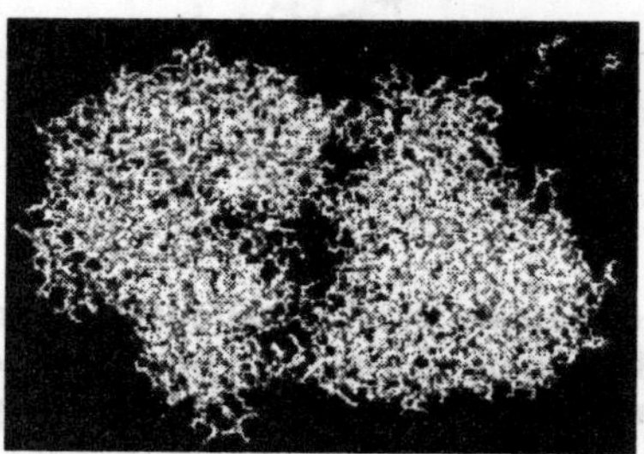

**Fig. 40.5. The enzyme hexokinase is shown as a simple ball-and-stick molecular model. To scale in the top right-hand corner are two of its substrates, ATP and glucose.**

The chief characteristic of proteins that allows their diverse set of functions is their ability to bind other molecules specifically and tightly. The region of the protein responsible for binding another molecule is known as the binding site and is often a depression or "pocket" on the molecular surface. This binding ability is mediated by the tertiary structure of the protein, which defines the binding site pocket, and by the chemical properties of the surrounding amino acids' side chains. Protein binding can be extraordinarily tight and specific; for example, the ribonuclease inhibitor protein binds to human angiogenin with a sub-femtomolar dissociation constant ($<10^{-15}$ M) but does not bind at all to its amphibian homolog onconase (>1 M). Extremely minor chemical changes such as the addition of a single methyl group to a binding partner can sometimes suffice to nearly eliminate binding; for example, the aminoacyl tRNA synthetase specific to the amino acid valine discriminates against the very similar side chain of the amino acid isoleucine.

Proteins can bind to other proteins as well as to small-molecule substrates. When proteins bind specifically to other copies of the same molecule, they can oligomerize to form fibrils; this process occurs often in structural proteins that consist of globular monomers that self-associate to form rigid fibers. Protein-protein interactions also regulate enzymatic activity, control progression through the cell cycle, and allow the assembly of large protein complexes that carry out many closely related reactions with a common biological function. Proteins can also bind to, or even be integrated into, cell membranes. The ability of binding partners to induce conformational changes in proteins allows the construction of enormously complex signaling networks.

# ENZYMES

## Etymology and History

As early as the late 1700s and early 1800s, the digestion of meat by stomach secretions and the conversion of starch to sugars by plant extracts and saliva were known. However, the mechanism by which this occurred had not been identified.

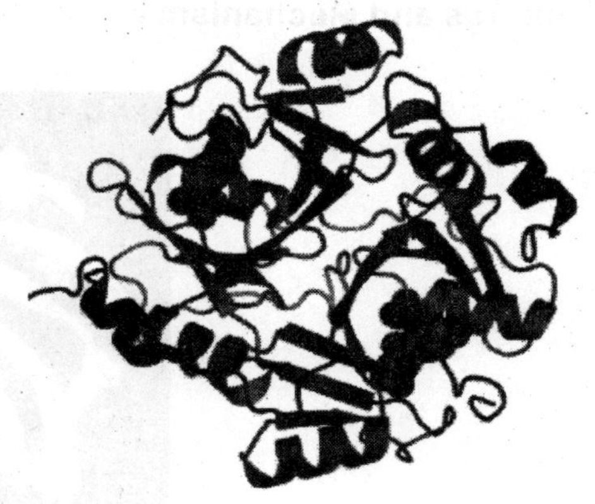

In the 19th century, when studying the fermentation of sugar to alcohol by yeast, Louis Pasteur came to the conclusion that this fermentation was catalyzed by a vital force contained within the yeast cells called "ferments", which were thought to function only within living organisms. He wrote that "alcoholic fermentation is an act correlated with the life and organization of the yeast cells, not with the death or putrefaction of the cells."

In 1878 German physiologist Wilhelm Kühne (1837–1900) first used the term *enzyme*, which comes from Greek ενζυμοον "in leaven", to describe this process. The word *enzyme* was used later to refer to nonliving substances such as pepsin, and the word *ferment* used to refer to chemical activity produced by living organisms.

In 1897 Eduard Buchner began to study the ability of yeast extracts that lacked any living yeast cells to ferment sugar. In a series of experiments at the University of Berlin, he found that the sugar was fermented even when there were no living yeast cells in the mixture. He named the enzyme that brought about the fermentation of sucrose "zymase". In 1907 he received the Nobel Prize in Chemistry "for his biochemical research and his discovery of cell-free fermentation". Following Buchner's example; enzymes are usually named according to the reaction they carry out. Typically the suffix *-ase* is added to the name of the substrate (*e.g.*, lactase is the enzyme that cleaves lactose) or the type of reaction (*e.g.*, DNA polymerase forms DNA polymers).

Having shown that enzymes could function outside a living cell, the next step was to determine their biochemical nature. Many early workers noted that enzymatic activity was associated with proteins, but several scientists (such as Nobel laureate Richard Willstätter) argued that proteins were merely carriers for the true enzymes and that proteins *per se* were incapable of catalysis. However, in 1926, James B. Sumner showed that the enzyme urease was a pure protein and crystallized it; Sumner did likewise for the enzyme catalase in 1937. The conclusion that pure proteins can be enzymes was definitively proved by Northrop and Stanley, who worked on the digestive enzymes pepsin (1930), trypsin and chymotrypsin. These three scientists were awarded the 1946 Nobel Prize in Chemistry.

This discovery that enzymes could be crystallized eventually allowed their structures to be solved by x-ray crystallography. This was first done for lysozyme, an enzyme found in tears, saliva and egg whites that digests the coating of some bacteria; the structure was solved by a group led by David Chilton Phillips and published in 1965. This high-resolution structure of lysozyme marked the beginning of the field of structural biology and the effort to understand how enzymes work at an atomic level of detail.

## Structures and Mechanisms

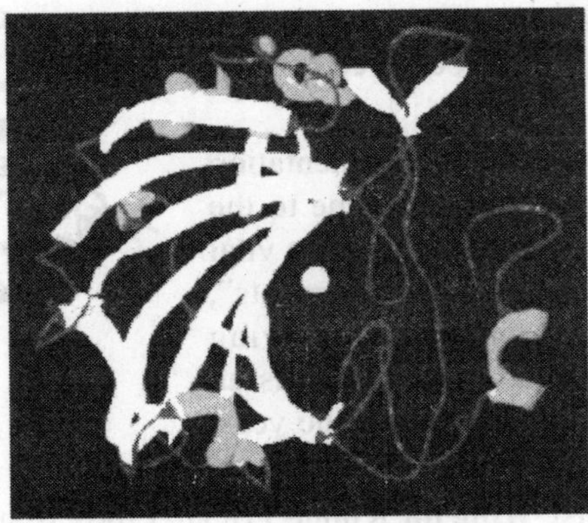

**Fig. 40.6. Ribbon-diagram showing carbonic anhydrase II. The grey sphere is the zinc cofactor in the active site. Diagram drawn from PDB 1MOO.**

Enzymes are generally globular proteins and range from just 62 amino acid residues in size, for the monomer of 4-oxalocrotonate tautomerase, to over 2,500 residues in the animal fatty acid synthase. A small number of RNA-based biological catalysts exist, with the most common being the ribosome, these are either referred to as *RNA-enzymes*, or ribozymes. The activities of enzymes are determined by their three-dimensional structure. Most enzymes are much larger than the substrates they act on, and only a small portion of the enzyme (around 3–4 amino acids) is directly involved in catalysis. The region that contains these catalytic residues, binds the substrate, and then carries out the reaction is known as the active site. Enzymes can also contain sites that bind cofactors, which are needed for catalysis. Some enzymes also have binding sites for small molecules, which are often direct or indirect products or substrates of the reaction catalyzed. This binding can serve to increase or decrease the enzyme's activity, providing a means for feedback regulation.

Like all proteins, enzymes are made as long, linear chains of amino acids that fold to produce a three-dimensional product. Each unique amino acid sequence produces a specific structure, which has unique properties. Individual protein chains may sometimes group together to form a protein complex. Most enzymes can be denatured—that is, unfolded and inactivated—by heating, which destroys the three-dimensional structure of the protein. Depending on the enzyme, denaturation may be reversible or irreversible.

## Specificity

Enzymes are usually very specific as to which reactions they catalyze and the substrates that are involved in these reactions. Complementary shape, charge and hydrophilic/hydrophobic characteristics of enzymes and substrates are responsible for this specificity. Enzymes can also show impressive levels of stereospecificity, regioselectivity and chemoselectivity.[18]

Some of the enzymes showing the highest specificity and accuracy are involved in the copying and expression of the genome. These enzymes have "proof-reading" mechanisms. Here, an enzyme such as DNA polymerase catalyzes a reaction in a first step and then checks that the product is correct in a second step.[19] This two-step process results in average error rates of less than 1 error in 100 million reactions in high-fidelity mammalian polymerases. Similar proofreading mechanisms are also found in RNA polymerase, aminoacyl tRNA synthetases and ribosomes.

Some enzymes that produce secondary metabolites are described as promiscuous, as they can act on a relatively broad range of different substrates. It has been suggested that this broad substrate specificity is important for the evolution of new biosynthetic pathways.

## "Lock and key" Model

Enzymes are very specific, and it was suggested by Emil Fischer in 1894 that this was because both the enzyme and the substrate possess specific complementary geometric shapes that fit exactly into one another. This is often referred to as "the lock and key" model. However, while this model explains enzyme specificity, it fails to explain the stabilization of the transition state that enzymes achieve. The "lock and key" model has proven inaccurate and the induced fit model is the most currently accepted enzyme-substrate-coenzyme figure.

## Induced Fit Model

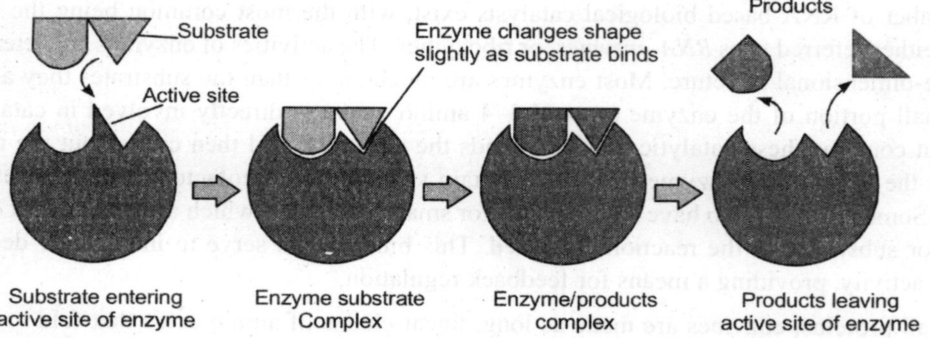


**Fig. 40.7. Diagrams to show the induced fit hypothesis of enzyme action.**

In 1958 Daniel Koshland suggested a modification to the lock and key model: since enzymes are rather flexible structures, the active site is continually reshaped by interactions with the substrate as the substrate interacts with the enzyme. As a result, the substrate does not simply bind to a rigid active site, the amino acid side chains which make up the active site are moulded into the precise positions that enable the enzyme to perform its catalytic function. In some cases, such as glycosidases, the substrate molecule also changes shape slightly as it enters the active site. The active site continues to change until the substrate is completely bound, at which point the final shape and charge is determined.

## Mechanisms

Enzymes can act in several ways, all of which lower $\Delta G^{\ddagger}$:

- Lowering the activation energy by creating an environment in which the transition state is stabilized (e.g. straining the shape of a substrate - by binding the transition-state conformation of the substrate/product molecules, the enzyme distorts the bound substrate(s) into their transition state form, thereby reducing the amount of energy required to complete the transition).
- Lowering the energy of the transition state, but without distorting the substrate, by creating an environment with the opposite charge distribution to that of the transition state.
- Providing an alternative pathway. For example, temporarily reacting with the substrate to form an intermediate ES complex, which would be impossible in the absence of the enzyme.
- Reducing the reaction entropy change by bringing substrates together in the correct orientation to react. Considering $\Delta H^{\ddagger}$ alone overlooks this effect.

Interestingly, this entropic effect involves destabilization of the ground state, and its contribution to catalysis is relatively small.

## Transition State Stabilization

The understanding of the origin of the reduction of $\Delta G^{\ddagger}$ requires one to find out how the enzymes can stabilize its transition state more than the transition state of the uncatalyzed reaction. Apparently, the most effective way for reaching large stabilization is the use of electrostatic effects, in particular, by having a relatively fixed polar environment that is oriented toward the charge distribution of the transition state. Such an environment does not exist in the uncatalyzed reaction in water.

### Dynamics and Function

Recent investigations have provided new insights into the connection between internal dynamics of enzymes and their mechanism of catalysis. An enzyme's internal dynamics are described as the movement of internal parts (*e.g.* amino acids, a group of amino acids, a loop region, an alpha helix, neighbouring beta-sheets or even entire domain) of these biomolecules, which can occur at various time-scales ranging from femtoseconds to seconds. Networks of protein residues throughout an enzyme's structure can contribute to catalysis through dynamic motions. Protein motions are vital to many enzymes, but whether small and fast vibrations or larger and slower conformational movements are more important depends on the type of reaction involved. These new insights also have implications in understanding allosteric effects and developing new drugs.

It should be clarified, however, that the dynamical time-dependent processes are not likely to help to accelerate enzymatic reactions, since such motions randomize and the rate constant is determined by the probability (P) of reaching the transition state, ($P = \exp\{\Delta G^{\ddagger}/RT\}$). Furthermore, the reduction of $\Delta G^{\ddagger}$ requires having relatively smaller motions (in relation to the corresponding motions in solution reactions) for the transition between the reactant and the product states. Thus, it is not clear that motional or dynamical effects contribute to the catalysis of the chemical step.

### Allosteric Modulation

Allosteric enzymes change their structure in response to binding of effectors. Modulation can be direct, where the effector binds directly to binding sites in the enzyme, or indirect, where the effector binds to other proteins or protein subunits that interact with the allosteric enzyme and thus influence catalytic activity.

## Cofactors and Coenzymes

### Cofactors

Some enzymes do not need any additional components to show full activity. However, others require non-protein molecules called cofactors to be bound for activity. Cofactors can be either inorganic (*e.g.*, metal ions and iron-sulfur clusters) or organic compounds, (e.g., flavin and heme). Organic cofactors can be either prosthetic groups, which are tightly bound to an enzyme, or coenzymes, which are released from the enzyme's active site during the reaction. Coenzymes include NADH, NADPH and adenosine triphosphate. These molecules act to transfer chemical groups between enzymes.

An example of an enzyme that contains a cofactor is carbonic anhydrase, and is shown in the ribbon diagram above with a zinc cofactor bound as part of its active site. These tightly-bound molecules are usually found in the active site and are involved in catalysis. For example, flavin and heme cofactors are often involved in redox reactions.

Enzymes that require a cofactor but do not have one bound are called *apoenzymes*. An apoenzyme together with its cofactor(s) is called a *holoenzyme* (this is the active form). Most cofactors are not covalently attached to an enzyme, but are very tightly bound. However, organic prosthetic groups can be covalently bound (*e.g.*, thiamine pyrophosphate in the enzyme pyruvate dehydrogenase).

## Coenzymes

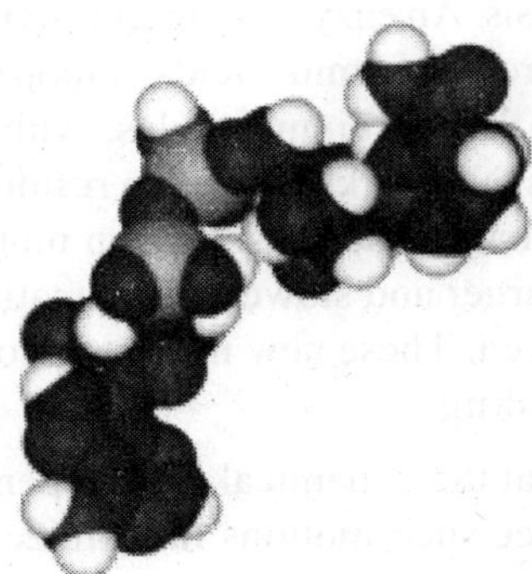

**Fig. 40.8. Space-filling model of the coenzyme NADH**

Coenzymes are small organic molecules that transport chemical groups from one enzyme to another. Some of these chemicals such as riboflavin, thiamine and folic acid are vitamins, this is when these compounds cannot be made in the body and must be acquired from the diet. The chemical groups carried include the hydride ion ($H^-$) carried by NAD or $NADP^+$, the acetyl group carried by coenzyme A, formyl, methenyl or methyl groups carried by folic acid and the methyl group carried by S-adenosylmethionine.

Since coenzymes are chemically changed as a consequence of enzyme action, it is useful to consider coenzymes to be a special class of substrates, or second substrates, which are common to many different enzymes. For example, about 700 enzymes are known to use the coenzyme NADH.

Coenzymes are usually regenerated and their concentrations maintained at a steady level inside the cell: for example, NADPH is regenerated through the pentose phosphate pathway and *S*-adenosylmethionine by methionine adenosyltransferase.

Coenzymes are small organic molecules that transport chemical groups from one enzyme to another. Some of these chemicals such as riboflavin, thiamine and folic acid are vitamins, this is when these compounds cannot be made in the body and must be acquired from the diet. The chemical groups carried include the hydride ion ($H^-$) carried by NAD or $NADP^+$, the acetyl group carried by coenzyme A, formyl, methenyl or methyl groups carried by folic acid and the methyl group carried by S-adenosylmethionine.

Since coenzymes are chemically changed as a consequence of enzyme action, it is useful to consider coenzymes to be a special class of substrates, or second substrates, which are common to many different enzymes. For example, about 700 enzymes are known to use the coenzyme NADH.

Coenzymes are usually regenerated and their concentrations maintained at a steady level inside the cell: for example, NADPH is regenerated through the pentose phosphate pathway and *S*-adenosylmethionine by methionine adenosyltransferase.

## Thermodynamics

The energies of various stages of a chemical reaction. Substrates need a large amount of energy to reach a transition state, which then decays into products. The enzyme stabilizes the transition state, reducing the energy needed to form products.

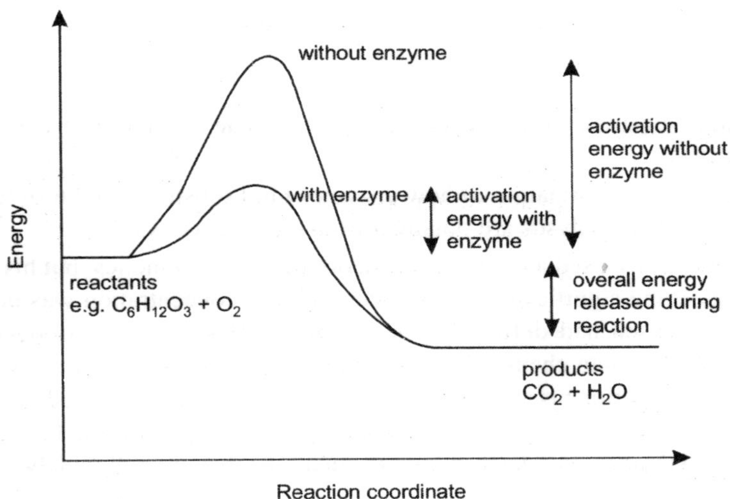


**Fig. 40.9. Showing the Graphical representation of the energy and reaction coordination in the presence and absence of enzymes**

As all catalysts, enzymes do not alter the position of the chemical equilibrium of the reaction. Usually, in the presence of an enzyme, the reaction runs in the same direction as it would without the enzyme, just more quickly. However, in the absence of the enzyme, other possible uncatalyzed, "spontaneous" reactions might lead to different products, because in those conditions this different product is formed faster.

Furthermore, enzymes can couple two or more reactions, so that a thermodynamically favourable reaction can be used to "drive" a thermodynamically unfavourable one. For example, the hydrolysis of ATP is often used to drive other chemical reactions.

Enzymes catalyze the forward and backward reactions equally. They do not alter the equilibrium itself, but only the speed at which it is reached. For example, carbonic anhydrase catalyzes its reaction in either direction depending on the concentration of its reactants.

$$CO_2 + H_2O \xrightarrow{\text{Carbonic anhydrase}} H_2CO_3 \text{ (in tissues; high } CO_2 \text{ concentration)}$$

$$H_2CO_3 \xrightarrow{\text{Carbonic anhydrase}} CO_2 + H_2O \text{ (in lungs; low } CO_2 \text{ concentration)}$$

Nevertheless, if the equilibrium is greatly displaced in one direction, that is, in a very exergonic reaction, the reaction is *effectively* irreversible. Under these conditions the enzyme will, in fact, only catalyze the reaction in the thermodynamically allowed direction.

## Kinetics

Catalytic step

$$E + S \rightleftharpoons ES \longrightarrow E + P$$

Substrate binding

Mechanism for a single substrate enzyme catalyzed reaction. The enzyme (E) binds a substrate (S) and produces a product (P).

Enzyme kinetics is the investigation of how enzymes bind substrates and turn them into products. The rate data used in kinetic analyses are obtained from enzyme assays.

In 1902 Victor Henri proposed a quantitative theory of enzyme kinetics, but his experimental data were not useful because the significance of the hydrogen ion concentration was not yet appreciated. After Peter Lauritz Sørensen had defined the logarithmic pH-scale and introduced the concept of buffering in 1909[47] the German chemist Leonor Michaelis and his Canadian postdoc Maud Leonora Menten repeated Henri's experiments and confirmed his equation which is referred to as Henri-Michaelis-Menten kinetics (sometimes also Michaelis-Menten kinetics). Their work was further developed by G. E. Briggs and J. B. S. Haldane, who derived kinetic equations that are still widely used today.

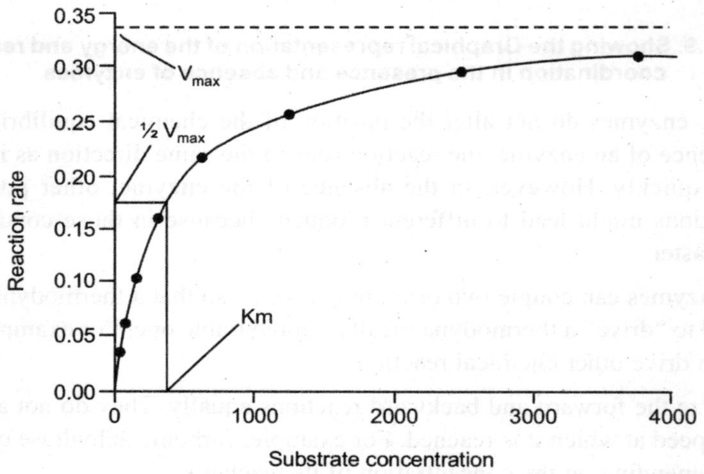


**Fig. 40.10. Saturation curve for an enzyme reaction showing the relation between the substrate concentration (S) and rate (*v*).**

The major contribution of Henri was to think of enzyme reactions in two stages. In the first, the substrate binds reversibly to the enzyme, forming the enzyme-substrate complex. This is sometimes called the Michaelis complex. The enzyme then catalyzes the chemical step in the reaction and releases.

Enzymes can catalyze up to several million reactions per second. For example, the reaction catalyzed by orotidine 5'-phosphate decarboxylase will consume half of its substrate in 78 million years if no enzyme is present. However, when the decarboxylase is added, the same process takes just 25 milliseconds.[50] Enzyme rates depend on solution conditions and substrate concentration. Conditions that denature the protein abolish enzyme activity, such as high temperatures, extremes of pH or high salt concentrations, while raising substrate concentration tends to increase activity. To find the maximum speed of an enzymatic reaction, the substrate concentration is increased until a constant rate of product formation is seen. This is shown in the saturation curve on the right. Saturation happens because, as substrate concentration increases, more and more of the free enzyme is converted into the substrate-bound ES form. At the maximum velocity ($V_{max}$) of the enzyme, all the enzyme active sites are bound to substrate, and the amount of ES complex is the same as the total amount of enzyme. However, $V_{max}$ is only one kinetic constant of enzymes. The amount of substrate needed to achieve a given rate of reaction is also important. This is given by the Michaelis-Menten constant ($K_m$), which is the substrate concentration required for an enzyme to reach one-half its maximum velocity. Each enzyme has a characteristic $K_m$ for a given substrate, and this can show how tight the binding of the substrate is to the enzyme. Another useful constant is $k_{cat}$, which is the number of substrate molecules handled by one active site per second.

The efficiency of an enzyme can be expressed in terms of $k_{cat}/K_m$. This is also called the specificity constant and incorporates the rate constants for all steps in the reaction. Because the specificity constant reflects both affinity and catalytic ability, it is useful for comparing different enzymes against each other, or the same enzyme with different substrates. The theoretical maximum for the specificity constant is called the diffusion limit and is about $10^8$ to $10^9$ ($M^{-1}\ s^{-1}$). At this point every collision of the enzyme with its substrate will result in catalysis, and the rate of product formation is not limited by the reaction rate but by the diffusion rate. Enzymes with this property are called *catalytically perfect* or *kinetically perfect*. Example of such enzymes are triose-phosphate isomerase, carbonic anhydrase, acetylcholinesterase, catalase, fumarase, β-lactamase, and superoxide dismutase.

Michaelis-Menten kinetics relies on the law of mass action, which is derived from the assumptions of free diffusion and thermodynamically-driven random collision. However, many biochemical or cellular processes deviate significantly from these conditions, because of very high concentrations, phase-separation of the enzyme/substrate/product, or one or two-dimensional molecular movement. In these situations, a fractal Michaelis-Menten kinetics may be applied.

Some enzymes operate with kinetics which are faster than diffusion rates, which would seem to be impossible. Several mechanisms have been invoked to explain this phenomenon. Some proteins are believed to accelerate catalysis by drawing their substrate in and pre-orienting them by using dipolar electric fields. Other models invoke a quantum-mechanical tunneling explanation, whereby a proton or an electron can tunnel through activation barriers, although for proton tunneling this model remains somewhat controversial. Quantum tunneling for protons has been observed in tryptamine. This suggests that enzyme catalysis may be more accurately characterized as "through the barrier" rather than the traditional model, which requires substrates to go "over" a lowered energy barrier.

## Inhibition

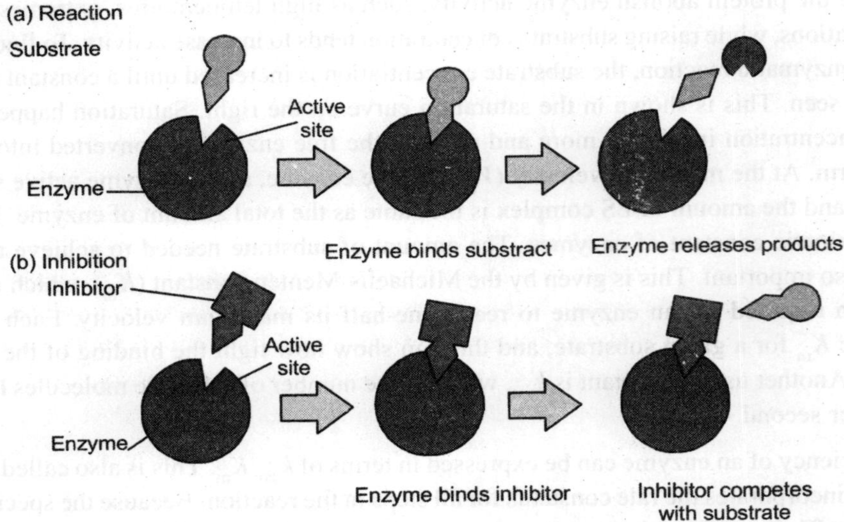


**Fig. 40.11. Competitive inhibitors bind reversibly to the enzyme, preventing the binding of substrate. On the other hand, binding of substrate prevents binding of the inhibitor. Substrate and inhibitor compete for the enzyme**

$$E + S \underset{K_s}{\rightleftharpoons} ES \xrightarrow{K_{cat}} E + P \quad \text{competitive inhibition}$$

$$E + I \underset{K_I}{\rightleftharpoons} EI$$

$$E + S \underset{K_s}{\rightleftharpoons} ES \xrightarrow{K_{cat}} E + P \quad \text{uncompetitive inhibition}$$

$$ES + I \underset{K_a}{\rightleftharpoons} EIS$$

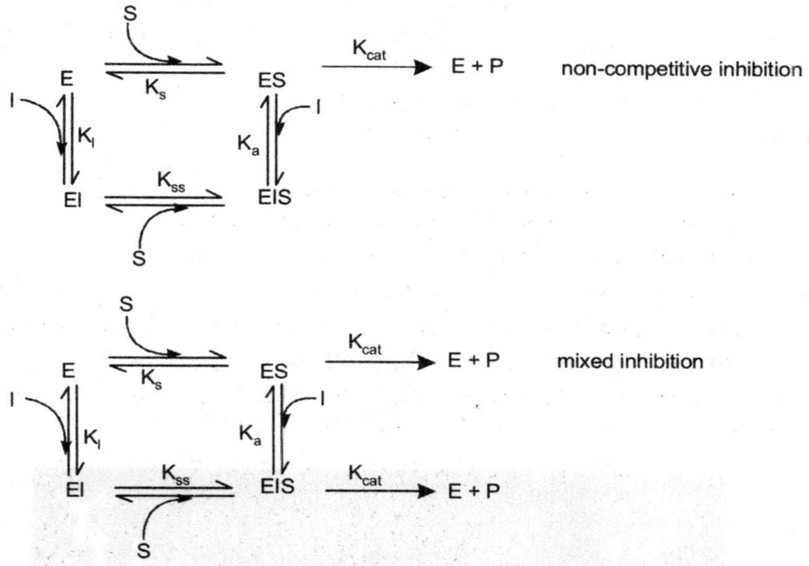

**Fig. 40.12. Types of inhibition. This classification was introduced by W.W. Cleland**

Enzyme reaction rates can be decreased by various types of enzyme inhibitors.

## Competitive Inhibition

In competitive inhibition, the inhibitor and substrate compete for the enzyme (i.e., they can not bind at the same time). Often competitive inhibitors strongly resemble the real substrate of the enzyme. For example, methotrexate is a competitive inhibitor of the enzyme dihydrofolate reductase, which catalyzes the reduction of dihydrofolate to tetrahydrofolate. The similarity between the structures of folic acid and this drug are shown in the figure to the *right* bottom. Note that binding of the inhibitor need *not* be to the substrate binding site (as frequently stated), if binding of the inhibitor changes the conformation of the enzyme to prevent substrate binding and *vice versa*. In competitive inhibition the maximal velocity of the reaction is not changed, but higher substrate concentrations are required to reach a given velocity, increasing the apparent $K_m$.

## Uncompetitive Inhibition

In uncompetitive inhibition the inhibitor can not bind to the free enzyme, but only to the ES-complex. The EIS-complex thus formed is enzymatically inactive. This type of inhibition is rare, but may occur in multimeric enzymes.

## Non-competitive Inhibition

Non-competitive inhibitors can bind to the enzyme at the same time as the substrate, i.e. they *never* bind to the active site. Both the EI and EIS complexes are enzymatically inactive. Because the inhibitor cannot be driven from the enzyme by higher substrate concentration (in contrast to competitive inhibition), the apparent $V_{max}$ changes. But because the substrate can still bind to the enzyme, the $K_m$ stays the same.

### Mixed Inhibition

This type of inhibition resembles the non-competitive, except that the EIS-complex has residual enzymatic activity.

In many organisms inhibitors may act as part of a feedback mechanism. If an enzyme produces too much of one substance in the organism, that substance may act as an inhibitor for the enzyme at the beginning of the pathway that produces it, causing production of the substance to slow down or stop when there is sufficient amount. This is a form of negative feedback. Enzymes which are subject to this form of regulation are often multimeric and have allosteric binding sites for regulatory substances. Their substrate/velocity plots are not hyperbolar, but sigmoidal (S-shaped).

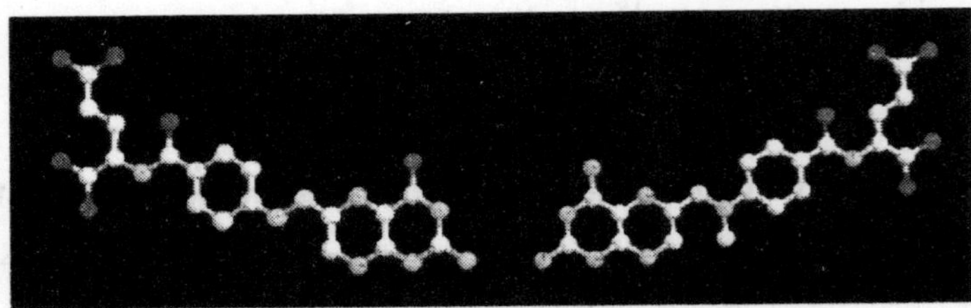

**Fig. 40.13. The coenzyme folic acid (left) and the anti-cancer drug methotrexate (right) are very similar in structure**

As a result, methotrexate is a competitive inhibitor of many enzymes that use folates.

Irreversible inhibitors react with the enzyme and form a covalent adduct with the protein. The inactivation is irreversible. These compounds include eflornithine a drug used to treat the parasitic disease sleeping sickness. Penicillin and Aspirin also act in this manner. With these drugs, the compound is bound in the active site and the enzyme then converts the inhibitor into an activated form that reacts irreversibly with one or more amino acid residues.

## Uses of Inactivators

Inhibitors are often used as drugs, but they can also act as poisons. However, the difference between a drug and a poison is usually only a matter of amount, since most drugs are toxic at some level, as Paracelsus wrote, "*In all things there is a poison, and there is nothing without a poison.*"[61] Equally, antibiotics and other anti-infective drugs are just specific poisons that kill a pathogen but not its host.

An example of an inactivator being used as a drug is aspirin, which inhibits the COX-1 and COX-2 enzymes that produce the inflammation messenger prostaglandin, thus suppressing pain and inflammation. The poison cyanide is an irreversible enzyme inactivator that combines with the copper and iron in the active site of the enzyme cytochrome c oxidase and blocks cellular respiration.

## Biological Function

Enzymes serve a wide variety of functions inside living organisms. They are indispensable for signal transduction and cell regulation, often via kinases and phosphatases. They also generate movement, with myosin hydrolysing ATP to generate muscle contraction and also moving cargo around the cell as part of the cytoskeleton. Other ATPases in the cell membrane are ion pumps involved in active transport. Enzymes are also involved in more exotic functions, such as luciferase generating light in fireflies.[65] Viruses can also contain enzymes for infecting cells, such as the HIV integrase and reverse transcriptase, or for viral release from cells, like the influenza virus neuraminidase.

An important function of enzymes is in the digestive systems of animals. Enzymes such as amylases and proteases break down large molecules (starch or proteins, respectively) into smaller ones, so they can be absorbed by the intestines. Starch molecules, for example, are too large to be absorbed from the intestine, but enzymes hydrolyse the starch chains into smaller molecules such as maltose and eventually glucose, which can then be absorbed. Different enzymes digest different food substances. In ruminants which have a herbivorous diets, microorganisms in the gut produce another enzyme, cellulase to break down the cellulose cell walls of plant fiber.

Several enzymes can work together in a specific order, creating metabolic pathways. In a metabolic pathway, one enzyme takes the product of another enzyme as a substrate. After the catalytic reaction, the product is then passed on to another enzyme. Sometimes more than one enzyme can catalyze the same reaction in parallel, this can allow more complex regulation: with for example a low constant activity being provided by one enzyme but an inducible high activity from a second enzyme.

Enzymes determine what steps occur in these pathways. Without enzymes, metabolism would neither progress through the same steps, nor be fast enough to serve the needs of the cell. Indeed, a metabolic pathway such as glycolysis could not exist independently of enzymes. Glucose, for example, can react directly with ATP to become phosphorylated at one or more of its carbons. In the absence of enzymes, this occurs so slowly as to be insignificant. However, if hexokinase is added, these slow reactions continue to take place except that phosphorylation at carbon 6 occurs so rapidly that if the mixture is tested a short time later, glucose-6-phosphate is found to be the only significant product. Consequently, the network of metabolic pathways within each cell depends on the set of functional enzymes that are present.

### Control of Activity

There are five main ways that enzyme activity is controlled in the cell.

1. **Enzyme production** (transcription and translation of enzyme genes) can be enhanced or diminished by a cell in response to changes in the cell's environment. This form of gene regulation is called enzyme induction and inhibition. For example, bacteria may become resistant to antibiotics such as penicillin because enzymes called beta-lactamases are induced that hydrolyse the crucial beta-lactam ring within the penicillin molecule. Another example are enzymes in the liver called cytochrome P450 oxidases, which are important in drug metabolism. Induction or inhibition of these enzymes can cause drug interactions.

2. Enzymes can be **compartmentalized**, with different metabolic pathways occurring in different cellular compartments. For example, fatty acids are synthesized by one set of enzymes in the cytosol, endoplasmic reticulum and the Golgi apparatus and used by a different set of enzymes as a source of energy in the mitochondrion, through â-oxidation.[67]
3. Enzymes can be regulated by **inhibitors and activators**. For example, the end product(s) of a metabolic pathway are often inhibitors for one of the first enzymes of the pathway (usually the first irreversible step, called *committed step*), thus regulating the amount of end product made by the pathways. Such a regulatory mechanism is called a negative feedback mechanism, because the amount of the end product produced is regulated by its own concentration. Negative feedback mechanism can effectively adjust the rate of synthesis of intermediate metabolites according to the demands of the cells. This helps allocate materials and energy economically, and prevents the manufacture of excess end products. Like other homeostatic devices, the control of enzymatic action helps to maintain a stable internal environment in living organisms.
4. Enzymes can be regulated through **post-translational modification**. This can include phosphorylation, myristoylation and glycosylation. For example, in the response to insulin, the phosphorylation of multiple enzymes, including glycogen synthase, helps control the synthesis or degradation of glycogen and allows the cell to respond to changes in blood sugar.[68] Another example of post-translational modification is the cleavage of the polypeptide chain. Chymotrypsin, a digestive protease, is produced in inactive form as chymotrypsinogen in the pancreas and transported in this form to the stomach where it is activated. This stops the enzyme from digesting the pancreas or other tissues before it enters the gut. This type of inactive precursor to an enzyme is known as a zymogen.
5. Some enzymes may become **activated when localized to a different environment** (eg. from a reducing (cytoplasm) to an oxidising (periplasm) environment, high pH to low pH etc). For example, hemagglutinin of the influenza virus undergoes a conformational change once it encounters the acidic environment of the host cell vesicle causing its activation.[69]

## Involvement in Disease

Since the tight control of enzyme activity is essential for homeostasis, any malfunction (mutation, overproduction, underproduction or deletion) of a single critical enzyme can lead to a genetic disease. The importance of enzymes is shown by the fact that a lethal illness can be caused by the malfunction of just one type of enzyme out of the thousands of types present in our bodies.

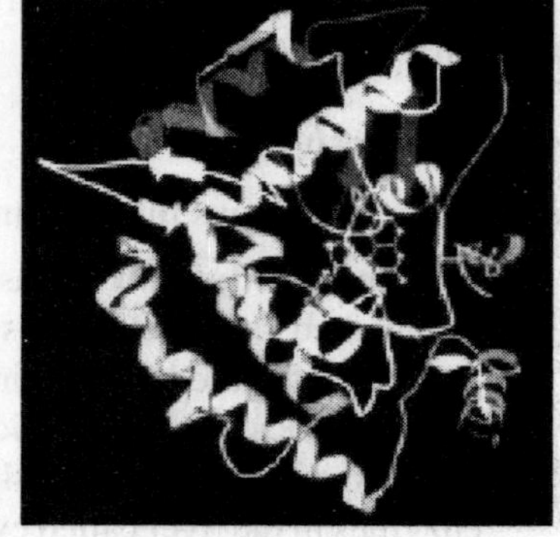

**Fig. 40.14. Phenylalanine hydroxylase**

One example is the most common type of phenylketonuria. A mutation of a single amino acid in the enzyme phenylalanine hydroxylase, which catalyzes the first step in the degradation of phenylalanine, results in build-up of phenylalanine and related products. This can lead to mental retardation if the disease is untreated.[70]

Another example is when germline mutations in genes coding for DNA repair enzymes cause hereditary cancer syndromes such as xeroderma pigmentosum. Defects in these enzymes cause cancer since the body is less able to repair mutations in the genome. This causes a slow accumulation of mutations and results in the development of many types of cancer in the sufferer.

## Naming Conventions

An enzyme's name is often derived from its substrate or the chemical reaction it catalyzes, with the word ending in ***-ase***. Examples are lactase, alcohol dehydrogenase and DNA polymerase. This may result in different enzymes, called isoenzymes, with the same function having the same basic name. Isoenzymes have a different amino acid sequence and might be distinguished by their optimal pH, kinetic properties or immunologically. Furthermore, the normal physiological reaction an enzyme catalyzes may not be the same as under artificial conditions. This can result in the same enzyme being identified with two different names. *E.g.* Glucose isomerase, used industrially to convert glucose into the sweetener fructose, is a xylose isomerase *in vivo*.

The International Union of Biochemistry and Molecular Biology have developed a nomenclature for enzymes, the **EC numbers**; each enzyme is described by a sequence of four numbers preceded by "EC". The first number broadly classifies the enzyme based on its mechanism:

The top-level classification is

- EC 1 *Oxidoreductases*: catalyze oxidation/reduction reactions
- EC 2 *Transferases*: transfer a functional group (*e.g.* a methyl or phosphate group)
- EC 3 *Hydrolases*: catalyze the hydrolysis of various bonds
- EC 4 *Lyases*: cleave various bonds by means other than hydrolysis and oxidation
- EC 5 *Isomerases*: catalyze isomerization changes within a single molecule
- EC 6 *Ligases*: join two molecules with covalent bonds

The complete nomenclature can be browsed at http://www.chem.qmul.ac.uk/iubmb/enzyme/.

## Industrial Applications

Enzymes are used in the chemical industry and other industrial applications when extremely specific catalysts are required. However, enzymes in general are limited in the number of reactions they have evolved to catalyze and also by their lack of stability in organic solvents and at high temperatures. Consequently, protein engineering is an active area of research and involves attempts to create new enzymes with novel properties, either through rational design or *in vitro* evolution.

| Application | Enzymes used | Uses |
|---|---|---|
| **Baking industry** alpha-amylase catalyzes the release of sugar monomers from starch | Fungal alpha-amylase enzymes are normally inactivated at about 50 degrees Celsius, but are destroyed during the baking process. | Catalyze breakdown of starch in the flour to sugar. Yeast action on sugar produces carbon dioxide. Used in production of white bread, buns, and rolls. |
| | Proteases | Biscuit manufacturers use them to lower the protein level of flour. |
| **Baby foods** | Trypsin | To predigest baby foods. |
| Brewing industry | Enzymes from barley are released during the mashing stage of beer production. | They degrade starch and proteins to produce simple sugar, amino acids and peptides that are used by yeast for fermentation. |
| | Industrially produced barley enzymes | Widely used in the brewing process to substitute for the natural enzymes found in barley. |
| | Amylase, glucanases, proteases | Split polysaccharides and proteins in the malt. |
| Germinating barley used for malt. | Betaglucanases and arabinoxylanases | Improve the wort and beer filtration characteristics. |
| | Amyloglucosidase and pullulanases | Low-calorie beer and adjustment of fermentability. |
| | Proteases | Remove cloudiness produced during storage of beers. |
| | Acetolactatedecarboxylase (ALDC) | Avoid the formation of diacetyl |
| **Fruit juices** | Cellulases, pectinases | Clarify fruit juices |
| **Dairy industry** | Rennin, derived from the stomachs of young ruminant animals (like calves and lambs). | Manufacture of cheese, used to hydrolyze protein. |
| | Microbially produced enzyme | Now finding increasing use in the dairy industry. |
| Roquefort cheese | Lipases | Is implemented during the production of Roquefort cheese to enhance the ripening of the blue-mould cheese. |
| | Lactases | Break down lactose to glucose and galactose. |
| **Meat tenderizers** | Papain | To soften meat for cooking. |
| **Starch industry** | Amylases, amyloglucosideases and glucoamylases | Converts starch into glucose and various syrups. |

| Application | Enzymes used | Uses |
|---|---|---|
| Glucose Fructose | Glucose isomerase | Converts glucose into fructose in production of high fructose syrups from starchy materials. These syrups have enhanced sweetening properties and lower calorific values than sucrose for the same level of sweetness. |
| **Paper industry**<br>A paper mill in South Carolina. | Amylases, Xylanases, Cellulases and ligninases | Degrade starch to lower viscosity, aiding sizing and coating paper. Xylanases reduce bleach required for decolorising; cellulases smooth fibers, enhance water drainage, and promote ink removal; lipases reduce pitch and lignin-degrading enzymes remove lignin to soften paper. |
| **Biofuel industry** | Cellulases | Used to break down cellulose into sugars that can be fermented (see cellulosic ethanol). |
| Cellulose in 3D | Ligninases | Use of lignin waste |
| **Biological detergent** | Frimarily proteases, produced in an extracellular form from bacteria | Used for presoak conditions and direct liquid applications helping with removal of protein stains from clothes. |
| Laundry soap | Amylases | Detergents for machine dish washing to remove resistant starch residues. |
| | Lipases | Used to assist in the removal of fatty and oily stains. |
| | Cellulases | Used in biological fabric conditioners. |
| **Contact lens cleaners** | Proteases | To remove proteins on contact lens to prevent infections. |
| **Rubber industry** | Catalase | To generate oxygen from peroxide to convert latex into foam rubber. |
| **Photographic industry** | Protease (ficin) | Dissolve gelatin off scrap film, allowing recovery of its silver content. |
| **Molecular biology**<br>Part of the DNA double helix. | Restriction enzymes, DNA ligase and polymerases | Used to manipulate DNA in genetic engineering, important in pharmacology, agriculture and medicine. Essential for restriction digestion and the polymerase chain reaction. Molecular biology is also important in fore |

# VITAMINS

## Introduction and History

The word vitamin comes from the two latin words *Vitæ misscere* meaning message of life.

A vitamin is an organic compound required as a nutrient in tiny amounts by an organism. A compound is called a vitamin when it cannot be synthesized in sufficient quantities by an organism, and must be obtained from the diet. Thus, the term is conditional both on the circumstances and the particular organism. For example, ascorbic acid functions as vitamin C for some animals but not others, and vitamins D and K are required in the human diet only in certain circumstances. Vitamins are defined by their biological activity, not their structure. Thus, each "vitamin" actually refers to a number of vitamer compounds, which form a set of distinct chemical compounds that show the biological activity of a particular vitamin. Such a set of chemicals are grouped under an alphabetized vitamin "generic descriptor" title, such as "vitamin A," which (for example) includes retinal, retinol, and many carotenoids.[3] Vitamers are often inter-convertible in the body. The term *vitamin* does not include other essential nutrients such as dietary minerals, essential fatty acids, or essential amino acids, nor does it encompass the large number of other nutrients that promote health but that are not essential for life.

Vitamins have diverse biochemical functions, including function as hormones (e.g. vitamin D), antioxidants (e.g. vitamin E), and mediators of cell signaling and regulators of cell and tissue growth and differentiation (e.g. vitamin A). The largest number of vitamins (e.g. B complex vitamins) function as precursors for enzyme cofactor bio-molecules (coenzymes), that help act as catalysts and substrates in metabolism. When acting as part of a catalyst, vitamins are bound to enzymes and are called prosthetic groups. For example, biotin is part of enzymes involved in making fatty acids. Vitamins also act as coenzymes to carry chemical groups between enzymes. For example, folic acid carries various forms of carbon group – methyl, formyl and methylene - in the cell. Although these roles in assisting enzyme reactions are vitamins' best-known function, the other vitamin functions are equally important.[5]

Until the 1800s, vitamins were obtained solely through food intake, and changes in diet (which, for example, could occur during a particular growing season) can alter the types and amounts of vitamins ingested. Vitamins have been produced as commodity chemicals and made widely available as inexpensive pills for several decades,[6] allowing supplementation of the dietary intake.

The Ancient Egyptians knew that feeding a patient liver (back, right) would help cure night blindness.

The value of eating certain foods to maintain health was recognized long before vitamins were identified. The ancient Egyptians knew that feeding a patient liver would help cure night blindness, an illness now known to be caused by a vitamin A deficiency. The advancement of ocean voyage during the Renaissance resulted in prolonged periods without access to fresh fruits and vegetables, and made illnesses from vitamin deficiency common among ship's crew.

In 1749, the Scottish surgeon James Lind discovered that citrus foods helped prevent scurvy, a particularly deadly disease in which collagen is not properly formed, causing poor wound healing, bleeding of the gums, severe pain, and death. In 1753, Lind published his *Treatise on the Scurvy*, which recommended using lemons and limes to avoid scurvy, which was adopted by the British Royal Navy. This led to the nickname Limey for sailors of that organization. Lind's discovery, however, was not widely accepted by individuals in the Royal Navy's Arctic expeditions in the 19th century, where it was widely believed that scurvy could be prevented by practicing good hygiene, regular exercise, and by maintaining the morale of the crew while on board, rather than by a diet of fresh food. As a result, Arctic expeditions continued to be plagued by scurvy and other deficiency diseases. In the early 20th century, when Robert Falcon Scott made his two expeditions to the Antarctic, the prevailing medical theory was that scurvy was caused by "tainted" canned food.

### Types of Vitamins and There Respective Source

| Year of discovery | Vitamin | Isolation from |
|---|---|---|
| 1909 | Vitamin A (Retinol) | Cod liver oil |
| 1912 | Vitamin $B_1$ (Thiamin) | Rice bran |
| 1912 | Vitamin C (Ascorbic acid) | Lemons |
| 1918 | Vitamin D (Calciferol) | Cod liver oil |
| 1920 | Vitamin $B_2$ (Riboflavin) | Eggs |
| 1922 | Vitamin E (Tocopherol) | Wheat germ oil |
| 1926 | Vitamin $B_{12}$ (Cobalamine) | Liver |
| 1929 | Vitamin K (Phyllochinone) | Luzerne |
| 1931 | Vitamin $B_5$ (Pantothenic acid) | Liver |
| 1931 | Vitamin $B_7$ (Biotin) | Liver |
| 1934 | Vitamin $B_6$ (Pyridoxine) | Rice bran |
| 1936 | Vitamin $B_3$ (Niacin) | Liver |
| 1941 | Vitamin $B_9$ (Folic acid) | Liver |

In 1881, Russian surgeon Nikolai Lunin studied the effects of scurvy while at the University of Tartu in present-day Estonia.[8] He fed mice an artificial mixture of all the separate constituents of milk known at that time, namely the proteins, fats, carbohydrates, and salts. The mice that received only the individual constituents died, while the mice fed by milk itself developed normally. He made a conclusion that "a natural food such as milk must therefore contain, besides these known principal ingredients, small quantities of unknown substances essential to life."[8] However, his conclusions were rejected by other researchers when they were unable to reproduce his results. One difference was that he had used table sugar (sucrose), while other researchers had used milk sugar (lactose) that still contained small amounts of vitamin B.

In the Orient where polished white rice was the common staple food of the middle class, beriberi resulting from lack of vitamin B was endemic. In 1884, Takaki Kanehiro, a British trained medical doctor of the Japanese Navy observed that beriberi was endemic among low ranking crew who often ate nothing but rice but not among crews of Western navies and officers who were entitled to a Western-style diet. Kanehiro initially believed that lack of protein was the chief cause of beriberi. With the support of Japanese navy, he experimented using crews of two battleships, one crew was fed only white rice, while the other was fed a diet of meat, fish, barley, rice, and beans. The group that ate only white rice documented 161 crew with beriberi and 25 deaths, while the latter group had only 14 cases of beriberi and no deaths. This convinced Kanehiro and the Japanese Navy that diet was the cause of beriberi. This was confirmed in 1897, when Christiaan Eijkman discovered that feeding unpolished rice instead of the polished variety to chickens helped to prevent beriberi in the chickens. The following year, Frederick Hopkins postulated that some foods contained "accessory factors"—in addition to proteins, carbohydrates, fats, et cetera—that were necessary for the functions of the human body. Hopkins was awarded the 1929 Nobel Prize for Physiology or Medicine with Christiaan Eijkman for their discovery of several vitamins.

In 1910, Japanese scientist Umetaro Suzuki succeeded in extracting a water-soluble complex of micronutrients from rice bran and named it aberic acid. He published this discovery in a Japanese scientific journal.[9] When the article was translated into German, the translation failed to state that it was a newly discovered nutrient, a claim made in the original Japanese article, and hence his discovery failed to gain publicity. Polish biochemist Kazimierz Funk isolated the same complex of micronutrients and proposed the complex be named "Vitamine" (a portmanteau of "vital amine") in 1912. The name soon became synonymous with Hopkins' "accessory factors", and by the time it was shown that not all vitamins were amines, the word was already ubiquitous. In 1920, Jack Cecil Drummond proposed that the final "e" be dropped to deemphasize the "amine" reference after the discovery that vitamin C had no amine component.

Throughout the early 1900s, the use of deprivation studies allowed scientists to isolate and identify a number of vitamins. Initially, lipid from fish oil was used to cure rickets in rats, and the fat-soluble nutrient was called "antirachitic A". The irony here is that the first "vitamin" bioactivity ever isolated, which cured rickets, was initially called "vitamin A", the bioactivity of which is now called vitamin D.[11] What we now call "vitamin A" was identified in fish oil because it was inactivated by ultraviolet light. In 1931, Albert Szent-Györgyi and a fellow researcher Joseph Svirbely determined that "hexuronic acid" was actually vitamin C and noted its anti-scorbutic activity. In 1937, Szent-Györgyi was awarded the Nobel Prize for his discovery. In 1943 Edward Adelbert Doisy and Henrik Dam were awarded the Nobel Prize for their discovery of vitamin K and its chemical structure.

## In Humans

Vitamins are classified as either water-soluble, meaning that they dissolve easily in water, or fat-soluble vitamins, which are absorbed through the intestinal tract with the help of lipids (fats). In general, water-soluble vitamins are readily excreted from the body. Each vitamin is typically used in multiple reactions and, therefore, most have multiple functions.

In humans there are 13 vitamins: 4 fat-soluble (A, D, E and K) and 9 water-soluble (8 B vitamins and vitamin C).

| Vitamin generic descriptor name | Vitamer chemical name(s) | Solubility | Recommended dietary allowances (male, age 19–70) | Deficiency disease | Upper Intake Level (UL/day) | Overdose disease |
|---|---|---|---|---|---|---|
| **Vitamin A** | Retinoids(retinol, retinoidsand carotenoids) | Water | **900 µg** | Night-blindness and Keratomalacia | 3,000 µg | Hypervitaminosis A |
| **Vitamin $B_1$** | Thiamine | Fat | **1.2 mg** | Beriberi | N/D | ? |
| **Vitamin $B_2$** | Riboflavin | Fat | **1.3 mg** | Ariboflavinosis | N/D | ? |
| **Vitamin $B_3$** | Niacin, niacinamide | Fat | **16.0 mg** | Pellagra | 35.0 mg | Liver damage |
| **Vitamin $B_5$** | Pantothenic acid | Fat | 5.0 mg | Paresthesia | N/D | ? |

| Vitamin generic descriptor name | Vitamer chemical name(s) | Solubility | Recommended dietary allowances (male, age 19–70) | Deficiency disease | Upper Intake Level (UL/day) | Overdose disease |
|---|---|---|---|---|---|---|
| **Vitamin $B_6$** | Pyridoxine, pyridoxamine, pyridoxal | Fat | **1.3-1.7 mg** | Anaemia | 100 mg | Impairment of proprioception, nerve damage |
| **Vitamin $B_7$** | Biotin | Fat | 30.0 μg | Dermatitis, enteritis | N/D | ? |
| **Vitamin $B_9$** | Folic acid, folinic acid | Fat | **400 μg** | Deficiency during pregnancy is associated with birth defects, such as neural tube defects | 1,000 μg | Refer to deficiency of Vitamin $B_6$ |
| **Vitamin $B_{12}$** | Cyanocobalamin, hydroxycobalamin, methylcobalamin | Fat | **2.4 μg** | Megaloblastic anaemia | N/D | ? |
| **Vitamin C** | Ascorbic acid | Fat | **90.0 mg** | Scurvy | 2,000 mg | Refer to Vitamin C megadosage |
| **Vitamin D** | Ergocalciferol, cholecalciferol | Water | 5.0 μg-10 μg | Rickets and Osteomalacia | 50 μg | Hypervitaminosis D |
| **Vitamin E** | Tocopherols, tocotrienols | Water | **15.0 mg** | Deficiency is very rare; mild hemolytic anemia in newborn infants. | 1,000 mg | ? |
| **Vitamin K** | phylloquinone, menaquinones | Water | 120 μg | Bleeding diathesis | N/D | ? |

## In Nutrition and Diseases

Vitamins are essential for the normal growth and development of a multicellular organism. Using the genetic blueprint inherited from its parents, a fetus begins to develop, at the moment of conception, from the nutrients it absorbs. It requires certain vitamins and minerals to be present at certain times. These nutrients facilitate the chemical reactions that produce among other things, skin, bone, and muscle. If there is serious deficiency in one or more of these nutrients, a child may develop a deficiency disease. Even minor deficiencies may cause permanent damage. For the most part, vitamins are obtained with food, but a few are obtained by other means. For example, microorganisms in the intestine—commonly known as "gut flora"—produce vitamin K and biotin, while one form of vitamin D is synthesized in the skin with the help of natural ultraviolet in sunlight. Humans can produce some vitamins from precursors they consume. Examples include vitamin A, produced from beta carotene, and niacin, from the amino acid tryptophan. Once growth and development are completed, vitamins remain essential nutrients for the healthy maintenance of the cells, tissues, and organs that make up a multicellular organism; they also enable a multicellular life form to efficiently use chemical energy provided by food it eats, and to help process the proteins, carbohydrates, and fats required for respiration.

## Deficiencies

Deficiencies of vitamins are classified as either primary or secondary. A **primary deficiency** occurs when an organism does not get enough of the vitamin in its food. A **secondary deficiency** may be due to an underlying disorder that prevents or limits the absorption or use of the vitamin, due to a "lifestyle factor", such as smoking, excessive alcohol consumption, or the use of medications that interfere with the absorption or use of the vitamin. People who eat a varied diet are unlikely to develop a severe primary vitamin deficiency. In contrast, restrictive diets have the potential to cause prolonged vitamin deficits, which may result in often painful and potentially deadly diseases.Because human bodies do not store most vitamins, humans must consume them regularly to avoid deficiency. Human bodily stores for different vitamins vary widely; vitamins A, D, and $B_{12}$ are stored in significant amounts in the human body, mainly in the liver, and an adult human's diet may be deficient in vitamins A and $B_{12}$ for many months before developing a deficiency condition. Vitamin $B_3$ is not stored in the human body in significant amounts, so stores may only last a couple of weeks. Well-known human vitamin deficiencies involve thiamine (beriberi), niacin (pellagra), vitamin C (scurvy) and vitamin D (rickets). In much of the developed world, such deficiencies are rare; this is due to (1) an adequate supply of food; and (2) the addition of vitamins and minerals to common foods, often called fortification. Recent lines of evidence also suggest a link between nutrition and mental disorders, as evidenced by Lakhan & Vieira (2008).

## Side Effects and Overdose

In large doses, some vitamins have documented side effects that tend to be more severe with a larger dosage. The likelihood of consuming too much of any vitamin from food is remote, but overdosing from vitamin supplementation does occur. At high enough dosages some vitamins cause side effects such as nausea, diarrhea, and vomiting,

**Overdosage via Vitamin Supplements can be a Problem**

| Vitamin | Amount | Problem |
|---|---|---|
| Vitamin A (Retinol) | | Hypervitaminosis A |
| Vitamin $B_1$ (Thiamin) | | |
| Vitamin $B_2$ (Riboflavin) | | |
| Vitamin $B_3$ (Niacin) | > 2 g/day | Liver damage[24] and other problems |
| Vitamin $B_5$ (Pantothenic acid) | | |
| Vitamin $B_6$ (Pyridoxine) | > 100 mg/day | Nerve damage |
| Vitamin $B_7$ (Biotin) | | |
| Vitamin $B_9$ (Folic acid) | | |
| Vitamin $B_{12}$ (Cobalamine) | | |
| Vitamin C (Ascorbic acid) | | |
| Vitamin D (Calciferol) bones, organs, etc. | | Hypervitaminosis DOver-calcification ofthe |
| Vitamin E (Tocopherol) | | Possible heart problems |
| Vitamin K (Phyllochinone) | | |

When side effects emerge, recovery is often accomplished by reducing the dosage. The concentrations of vitamins an individual can tolerate vary widely, and appear to be related to age and state of health.[25] In the United States, overdose exposure to all formulations of vitamins was reported by 62,562 individuals in 2004 (nearly 80% of these exposures were in children under the age of 6), leading to 53 "major" life-threatening outcomes and 3 deaths—a small number in comparison to the 19,250 people who died of unintentional poisoning of all kinds in the U.S. in the same year (2004).

## Supplements

Dietary supplements, often containing vitamins, are used to ensure that adequate amounts of nutrients are obtained on a daily basis, if optimal amounts of the nutrients cannot be obtained through a varied diet. Scientific evidence supporting the benefits of some dietary supplements is well established for certain health conditions, but others need further study.

In the India, advertising for dietary supplements is required to include a disclaimer that the product is not intended to treat, diagnose, mitigate, prevent, or cure disease, and that any health claims have not been evaluated by the Food and Drug Administration In some cases, dietary supplements may have unwanted effects, especially if taken before surgery, with other dietary supplements or medicines, or if the person taking them has certain health conditions. Vitamin supplements may also contain levels of vitamins many times higher, and in different forms, than one may ingest through food.

Intake of excessive quantities can cause vitamin poisoning, often due to overdose of Vitamin A and Vitamin D (The most common poisoning with multinutrient supplement pills does not involve a vitamin, but is rather due to the mineral iron). Due to toxicity, most common vitamins have recommended upper daily intake amounts.

We shall now study these vitamins individually

# VITAMIN A

## Discovery

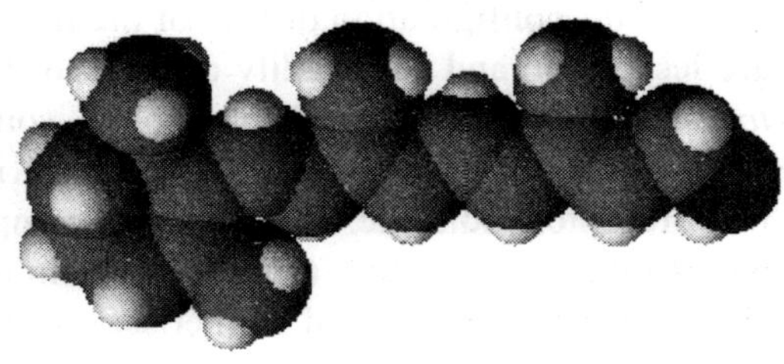

In 1913, Elmer McCollum, a biochemist at the University of Wisconsin-Madison, and colleague Marguerite Davis identified a fat-soluble nutrient in butterfat and cod liver oil. Their work confirmed that of Thomas Osborne and Lafayette Mendel, at Yale, which suggested a fat-soluble nutrient in butterfat, also in 1913. Vitamin A was first synthesized in 1947 by two Dutch chemists, David Adriaan van Dorp and Jozef Ferdinand Arens. Retinol (Afaxin), the animal form of vitamin A, is a fat-soluble vitamin important in vision and bone growth. It belongs to the family of chemical compounds known as retinoids. Retinol is ingested in a precursor form; animal sources (liver and eggs) contain retinyl esters, whereas plants (carrots, spinach) contain pro-vitamin A carotenoids.

Hydrolysis of retinyl esters results in retinol while pro-vitamin A carotenoids can be cleaved to produce retinal. Retinal, also known as retinaldehyde, can be reversibly reduced to produce retinol or it can be irreversibly oxidized to produce retinoic acid. The best described active retinoid metabolites are 11-*cis*-retinal and the all-trans and 9-cis-isomers of retinoic acid.

## Sources

All sources of vitamin A can provide retinol, but retinoids are found naturally in some foods of animal origin. Each of the following contains at least 0.15 mg of retinoids per 1.75-7 oz. (50-200 g):

- Cod Liver Oil
- Butter
- Liver (beef, pork, chicken, turkey, fish)
- Eggs

### Synthetic Sources

Synthetic retinol is marketed under the following trade names: Acon, Afaxin, Agiolan, Alphalin, Anatola, Aoral, Apexol, Apostavit, Atav, Avibon, Avita, Avitol, Axerol, Dohyfral A, Epiteliol, Nio-A-Let, Prepalin, Testavol, Vaflol, Vi-Alpha, Vitpex, Vogan, and Vogan-Neu.

## Units of Measurement

When referring to dietary allowances or nutritional science, retinol is usually measured in international units (IU). IU refers to biological activity and therefore is unique to each individual compound, however 1 IU of retinol is equivalent to approximately 0.3 micrograms (300 nanograms).

## Chemical Structure and Function

Many different geometric isomers of retinol, retinal and retinoic acid are possible as a result of either a *trans* or *cis* configuration of four of the five double bonds found in the polyene chain. The *cis* isomers are less stable and can readily convert to the all-*trans* configuration (as seen in the structure of all-*trans*-retinol shown here). Nevertheless, some *cis* isomers are found naturally and carry out essential functions. For example, the 11-*cis*-retinal isomer is the chromophore of rhodopsin, the vertebrate photoreceptor molecule. Rhodopsin is comprised of the 11-cis-retinal covalently linked via a Schiff base to the opsin protein (either rod opsin or blue, red or green cone opsins). The process of vision relies on the light-induced isomerisation of the chromophore from 11-*cis* to all-*trans* resulting in a change of the conformation and activation of the photoreceptor molecule. One of the earliest signs of vitamin A deficiency is night-blindness followed by decreased visual acuity. George Wald won the 1967 Nobel Prize in Physiology or Medicine for his work with retina pigments (also called visual pigments), which led to the understanding of the role of vitamin A in vision. Many of the non-visual functions of vitamin A are mediated by retinoic acid, which regulates gene expression by activating intracellular retinoic acid receptors. The non-visual functions of vitamin A are essential in the immunological function, reproduction and embryonic development of vertebrates as evidenced by the impaired growth, susceptibility to infection and birth defects observed in populations receiving suboptimal vitamin A in their diet.

## Role in Embryology

Retinoic acid via the retinoic acid receptor influences the process of cell differentiation, hence, the growth and development of embryos. During development there is a concentration gradient of retinoic acid along the anterior-posterior (head-tail) axis. Cells in the embryo respond differently to retinoic acid depending on the amount present. For example, in vertebrates the hindbrain transiently forms eight rhombomers and each rhombomere has a specific pattern of genes being expressed. If retinoic acid is not present the last four rhombomeres do not develop. Instead rhombomeres 1-4 grow to cover the same amount of space as all eight would normally occupy. Retinoic acid has its effects by turning on a differential pattern of Hox genes which encode different homeodomain transcription factors which in turn can turn on cell type specific genes. Deletion of the Hox-1 gene from rhombomere 4 makes the neurons growing in that region behave like neurons from rhombomere 2. The retina is also patterned by retinoic acid, with a concentration gradient that is high on the ventral side of the retina and low on the dorsal side.

### Vision

Vitamin A is required in the production of rhodopsin, the visual pigment used in low light levels. This is why eating foods rich in vitamin A is said to allow an individual to see in the dark.

### Epithelial Cells

Vitamin A is essential for the correct functioning of epithelial cells. In Vitamin A deficiency, mucus-secreting cells are replaced by keratin producing cells, leading to xerosis.

### Glycoprotein Synthesis

Glycoprotein synthesis requires adequate Vitamin A status. In severe Vitamin A deficiency, lack of glycoproteins may lead to corneal ulcers or liquefaction.

### Immune System

Vitamin A is essential to maintain intact epithelial tissues as a physical barrier to infection; it is also involved in maintaining a number of immune cell types from both the innate and acquired immune systems. These include the lymphocytes (B-cells, T-cells, and natural killer cells), as well as many myelocytes (neutrophils, macrophages, and myeloid dendritic cells).

### Formation of Red Blood Cells (Haematopoiesis)

Vitamin A may be needed for normal haematopoiesis; deficiency causes abnormalities in iron metabolism.

## Growth

Vitamin A affects the production of human growth hormone.

## Clinical Use

All retinoid forms of vitamin A are used in cosmetic and medical applications applied to the skin. Retinoic acid, termed Tretinoin in clinical usage, is used in the treatment of acne and keratosis pilaris in a topical cream. An isomer of tretinoin, isotretinoin is also used orally (under the trade names *Accutane* and *Roaccutane*), generally for severe or recalcitrant acne. In cosmetics, vitamin A derivatives are used as anti-aging chemicals- vitamin A is absorbed through the skin and increases the rate of skin turnover, and gives an increase in collagen giving a more youthful appearance. Tretinoin, under the alternative name of **all-trans retinoic acid** (ATRA), is used as chemotherapy for acute promyelocytic leukemia, a subtype of acute myelogenous leukemia. This is because cells of this subtype of leukemia are sensitive to agonists of the retinoic acid receptors (RARs).

# Nutrition

This vitamin plays an essential role in vision, particularly night vision, normal bone and tooth development, reproduction, and the health of skin and mucous membranes (the mucus-secreting layer that lines body regions such as the respiratory tract). Vitamin A also acts in the body as an antioxidant, a protective chemical that may reduce the risk of certain cancers.

| Vitamin properties | |
|---|---|
| Solubility | Fat |
| RDA (adult male) | 900 μg/day |
| RDA (adult female) | 700 μg/day |
| RDA upper limit (adult male) | 3,000 μg/day |
| RDA upper limit (adult female) | 3,000 μg/day |
| **Deficiency symptoms** | |
| • Night blindness | |
| • Keratomalacia | |
| • Pale, dry skin | |
| **Excess symptoms** | |
| • Liver toxicity | |
| • Dry skin | |
| • Hair loss | |
| • Teratological effects | |
| • Osteoporosis (suspected, long-term) | |
| **Common sources** | |
| • Liver | |
| • fortified Dairy products | |
| • Darkly colored fruits | |
| • Leafy vegetables | |

There are two sources of dietary vitamin A. Active forms, which are immediately available to the body are obtained from animal products. These are known as retinoids and include retinal and retinol. Precursors, also known as provitamins, which must be converted to active forms by the body, are obtained from fruits and vegetables containing yellow, orange and dark green pigments, known as carotenoids, the most well-known being beta-carotene. For this reason, amounts of vitamin A are measured in Retinal Equivalents (RE). One RE is equivalent to 0.001 mg of retinal, or 0.006 mg of beta-carotene, or 3.3 International Units of vitamin A.

In the intestine, vitamin A is protected from being chemically changed by vitamin E. Vitamin

A is fat-soluble and can be stored in the body. Most of the vitamin A you eat is stored in the liver. When required by a particular part of the body, the liver releases some vitamin A, which is carried by the blood and delivered to the target cells and tissues.

### Dietary Intake

The Dietary Reference Intake (DRI) Recommended Daily Amount (RDA) for Vitamin A for a 25-year old male is 900 micrograms/day, or 3000 IU.

The Food Standards Agency states that an average adult should not consume more than 1500 micrograms (5000 IU) per day, because this increases the chance of osteoporosis.

During the absorption process in the intestines, retinol is incorporated into chylomicrons as the ester form, and it is these particles that mediate transport to the liver. Liver cells (hepatocytes) store vitamin A as the ester, and when retinol is needed in other tissues, it is de-esterifed and released into the blood as the alcohol. Retinol then attaches to a serum carrier, retinol binding protein, for transport to target tissues. A binding protein inside cells, cellular retinoic acid binding protein, serves to store and move retinoic acid intracellularly. Carotenoid bioavailability ranges between 1/5 to 1/10 of retinol's. Carotenoids are better absorbed when ingested as part of a fatty meal. Also, the carotenoids in vegetables, especially those with tough cell walls (e.g. carrots), are better absorbed when these cell walls are broken up by cooking or mincing.

### Dietary Intake

The Dietary Reference Intake (DRI) Recommended Daily Amount (RDA) for Vitamin A for a 25-year old male is 900 micrograms/day, or 3000 IU. The Food Standards Agency states that an average adult should not consume more than 1500 micrograms (5000 IU) per day, because this increases the chance of osteoporosis. During the absorption process in the intestines, retinol is incorporated into chylomicrons as the ester form, and it is these particles that mediate transport to the liver. Liver cells (hepatocytes) store vitamin A as the ester, and when retinol is needed in other tissues, it is de-esterifed and released into the blood as the alcohol. Retinol then attaches to a serum carrier, retinol binding protein, for transport to target tissues. A binding protein inside cells, cellular retinoic acid binding protein, serves to store and move retinoic acid intracellularly. Carotenoid bioavailability ranges between 1/5 to 1/10 of retinol's. Carotenoids are better absorbed when ingested as part of a fatty meal. Also, the carotenoids in vegetables, especially those with tough cell walls (e.g. carrots), are better absorbed when these cell walls are broken up by cooking or mincing.

## VITAMIN $B_1$

### History

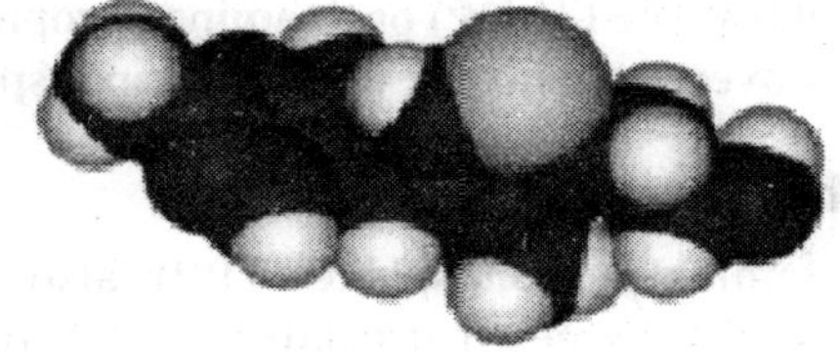

Thiamine was first discovered in 1910 by Umetaro Suzuki in Japan when researching how rice bran cured patients of beriberi. He named it **aberic acid** (later **oryzanin**). He did not determine its chemical composition, nor that it was an amine. It was first crystallized by Jansen and Donath in 1926 (they named it aneurin, for antineuritic vitamin). Its chemical composition and synthesis was finally reported by Robert R. Williams in 1935. He also coined the current name for it, **thiamine**. Thiamine or thiamin, also known as vitamin $B_1$ and aneurine hydrochloride, is one of the B vitamins. It is a colourless compound with chemical formula $C_{12}H_{17}N_4OS$. It is soluble in water and insoluble in alcohol. Thiamine decomposes if heated. Its chemical structure contains a pyrimidine ring and a thiazole ring.

## Good Sources

Thiamine is found naturally in the following foods, each of which contains at least 0.1 mg of the vitamin per 28-100 g (1-3.5 oz) green peas, spinach, liver, beef, pork, navy beans, nuts, pinto beans, bananas, soybeans, goji berries, whole-grains, breads, yeast, vegemite, the aleurone layer of unpolished rice, and legumes.

## Nutrition

Thiamine plays an important role in helping the body metabolize carbohydrates and fat to produce energy. It is essential for normal growth and development and helps to maintain proper functioning of the heart and the nervous and digestive systems. Thiamine is water-soluble and cannot be stored in the body; however, once absorbed, the vitamin is concentrated in muscle tissue.

## Deficiency

Systemic thiamine deficiency can lead to myriad problems including neurodegeneration, wasting and death. A lack of thiamine can be caused by malnutrition, alcoholism, a diet high in thiaminase-rich foods (raw freshwater fish, raw shellfish, ferns) and/or foods high in anti-thiamine factors (tea, coffee, betel nuts). Well-known syndromes caused by thiamine deficiency include Wernicke-Korsakoff syndrome and beriberi, diseases also common with chronic alcoholism. It is thought that many people with diabetes have a deficiency of thiamine and that this may be linked to some of the complications that can occur.

## Diagnostic Testing

A positive diagnosis test for thiamine deficiency can be ascertained by measuring the activity of the enzyme transketolase in erythrocytes. Thiamine can also be seen directly in whole blood following the conversion of thiamine to a fluorescent thiochrome derivative. However, this test may not reveal the deficiency in diabetic patients.

## Thiamine Phosphate Derivatives

There are four known natural thiamine phosphate derivatives: thiamine monophosphate (ThMP), thiamine diphosphate (ThDP) or thiamine pyrophosphate (TPP), thiamine triphosphate (ThTP), and the recently discovered adenosine thiamine triphosphate (AThTP).

## Thiamine Pyrophosphate

Thiamine pyrophosphate (TPP), also known as *thiamine diphosphate* (ThDP), is a coenzyme for several enzymes that catalyze the dehydrogenation (decarboxylation and subsequent conjugation to

Coenzyme A) of alpha-keto acids. Examples include:

- In mammals:
  - pyruvate dehydrogenase and α-ketoglutarate dehydrogenase (metabolism of carbohydrates)
  - branched-chain alpha-keto acid dehydrogenase
  - 2-hydroxyphytanoyl-CoA lyase
  - transketolase (functions in the pentose phosphate pathway to synthesize NADPH and the pentose sugars deoxyribose and ribose)
- In other species:
  - pyruvate decarboxylase (in yeast)
  - several additional bacterial enzymes

TPP is synthesized by the enzyme thiamine pyrophosphokinase, which requires free thiamine, magnesium, and adenosine triphosphate.

### Thiamine Triphosphate

Thiamine triphosphate (ThTP) was long considered a specific neuroactive form of thiamine. However, recently it was shown that ThTP exists in bacteria, fungi, plants and animals suggesting a much more general cellular role. In particular in *E. coli* it seems to play a role in response to amino acid starvation.

### Adenosine Thiamine Triphosphate

Adenosine thiamine triphosphate (AThTP) or thiaminylated adenosine triphosphate has recently been discovered in *Escherichia coli* where it accumulates as a result of carbon starvation. In *E. coli*, AThTP may account for up to 20 % of total thiamine. It also exists in lesser amounts in yeast, roots of higher plants and animal tissues.

### Genetic Diseases

Genetic diseases of thiamine transport are rare but serious. Thiamine Responsive Megaloblastic Anemia with diabetes mellitus and sensorineural deafness (TRMA)[5] is an autosomal recessive disorder caused by mutations in the gene SLC19A2,[6] a high affinity thiamine transporter. TRMA patients do not show signs of systemic thiamine deficiency, suggesting redundancy in the thiamine transport system. This has led to the discovery of a second high affinity thiamine transporter, SLC19A3.

## Research

### High Doses

The RDA in most countries is set at about 1.4 mg. However, tests on volunteers at daily doses of about 50 mg have claimed an increase in mental acuity.

### Thiamine as an Insect Repellent

Some studies suggest that taking thiamine 25 to 50 mg three times per day is effective in reducing mosquito bites. A large intake of thiamine produces a skin odor that is not detectable by humans, but is disagreeable to female mosquitoes. Thiamine takes more than 2 weeks before the odor fully saturates the skin. With the advances in topical preparations there is an increasing number of thiamine based repellent products. There is anecdotal evidence of thiamine products being effective in the field (Australia, US and Canada), but one study found thiamine had no effect.

### Autism

A 2002 pilot study administered thiamine tetrahydrofurfuryl disulphide (TTFD) rectally to ten autism spectrum children, and found beneficial clinical effect in eight. This study has not been replicated and a 2006 review of thiamine by the same author did not mention thiamine's possible effect on autism.

## VITAMIN $B_2$

**Riboflavin (E101)**, also known as **vitamin $B_2$**, is an easily absorbed micronutrient with a key role in maintaining health in animals. It is the central component of the cofactors FAD and FMN, and is therefore required by all flavoproteins. As such, vitamin $B_2$ is required for a wide variety of cellular processes. Like the other B vitamins, it plays a key role in energy metabolism, and is required for the metabolism of fats, carbohydrates, and proteins. Milk, cheese, leafy green vegetables, liver, legumes such as mature soybeans, yeast and almonds are good sources of vitamin $B_2$, but exposure to light destroys riboflavin.

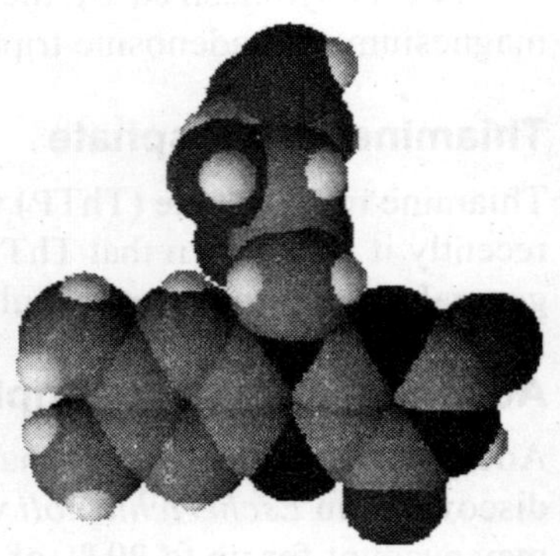

### Good Sources

Riboflavin is found naturally in asparagus, bananas, okra, chard, cottage cheese, milk, yogurt, meat, eggs and fish, each of which contain at least 0.1 mg of the vitamin per 3-10.5 oz (85-300 g) serving. Marmite is also a good source of riboflavin, containing 6.4 mg per 4 g serving.

**Properties** Molecular formula $C_{17}H_{20}N_4O_6$ Molar mass 376.36 g/mol Melting point 290 °C (dec.)

### Industrial Synthesis

Various biotechnological processes have been developed for industrial scale riboflavin biosynthesis using different microorganisms, including filamentous fungi such as *Ashbya gossypii*, *Candida famata* and *Candida flaveri* as well as the bacteria *Corynebacterium ammoniagenes* and *Bacillus subtilis*. The latter organism has been genetically modified to both increase the bacteria's production of riboflavin and to introduce an antibiotic (ampicillin) resistance marker, and is now successfully employed at a commercial scale to produce riboflavin for feed and food fortification purposes. The chemical company BASF has installed a plant in South Korea, which is specialized on riboflavin production using *Ashbya gossypii*. The concentrations of riboflavin in their modified strain are so high, that the mycelium has a reddish / brownish color and accumulates riboflavin crystals in the vacuoles, which will eventually burst the mycelium.

### Riboflavin in Food

Riboflavin is yellow or yellow-orange in colour and in addition to being used as a food colouring it is also used to fortify some foods. It is used in baby foods, breakfast cereals, pastas, sauces, processed cheese, fruit drinks, vitamin-enriched milk products, some energy drinks, and is widely used in vitamin supplements.

Large quantities of riboflavin are often included in multi-vitamins; often, the dose is far more than a normal human can use in a day. The excess is excreted in the urine, causing the urine to be colored bright yellow within a few hours of ingestion of the vitamin.

It is difficult to incorporate riboflavin into many liquid products because it has poor solubility in water. Hence the requirement for riboflavin-5'-phosphate (E101a), a more expensive but more soluble form of riboflavin.

## Nutrition

### Riboflavin Deficiency

Riboflavin is continuously excreted in the urine of healthy individuals[1], making deficiency relatively common when dietary intake is insufficient. However, riboflavin deficiency is always accompanied by deficiency of other vitamins· A deficiency of riboflavin can be primary - poor vitamin sources in one's daily diet - or secondary, which may be a result of conditions that affect absorption in the intestine, the body not being able to use the vitamin, or an increase in the excretion of the vitamin from the body.

In humans, signs and symptoms of riboflavin deficiency (ariboflavinosis) include cracked and red lips, inflammation of the lining of mouth and tongue, mouth ulcers, cracks at the corners of the mouth (angular cheilitis), and a sore throat. A deficiency may also cause dry and scaling skin, fluid in the mucous membranes, and iron-deficiency anemia. The eyes may also become bloodshot, itchy, watery and sensitive to bright light. Riboflavin deficiency is classically associated with the oral-ocular-genital syndrome. Angular cheilitis, photophobia, and scrotal dermatitis are the classic remembered signs.

In animals, riboflavin deficiency results in lack of growth, failure to thrive, and eventual death. Experimental riboflavin deficiency in dogs results in growth failure, weakness, ataxia, and inability to stand. The animals collapse, become comatose, and die. During the deficiency state, dermatitis develops together with hair-loss. Other signs include corneal opacity, lenticular cataracts, hemorrhagic adrenals, fatty degeneration of the kidney and liver, and inflammation of the mucus membrane of the gastrointestinal tract. Post-mortem studies in rhesus monkeys fed a riboflavin-deficient diet revealed that about one-third the normal amount of riboflavin was present in the liver, which is the main storage organ for riboflavin in mammals. These overt clinical signs of riboflavin deficiency are rarely seen among inhabitants of the developed countries. However, about 28 million Americans exhibit a common 'sub-clinical' stage, characterized by a change in biochemical indices (e.g. reduced plasma erythrocyte glutathione reductase levels). Although the effects of long-term sub-clinical riboflavin deficiency are unknown, in children this deficiency results in reduced growth. Subclinical riboflavin deficiency has also been observed in women taking oral contraceptives, in the elderly, in people with eating disorders, and in disease states such as HIV, inflammatory bowel disease, diabetes and chronic heart disease. The fact that riboflavin deficiency does not immediately lead to gross clinical manifestations indicates that the systemic levels of this essential vitamin are tightly regulated.

## Diagnostic Testing of $B_2$ Deficiency

A positive diagnostic test for measuring levels of riboflavin in serum is ascertained by measuring erythrocyte levels of glutathione reductase.

## Clinical Uses

Riboflavin has been used in several clinical and therapeutic situations. For over 30 years, riboflavin supplements have been used as part of the phototherapy treatment of neonatal jaundice. The light used to irradiate the infants breaks down not only the toxin causing the jaundice, but the naturally occurring riboflavin within the infant's blood as well.

More recently there has been growing evidence that supplemental riboflavin may be a useful additive along with beta-blockers in the treatment of migraine headaches.

Development is underway to use riboflavin to improve the safety of transfused blood by reducing pathogens found in collected blood. Riboflavin attaches itself to the nucleic acids (DNA and RNA) in cells, and when light is applied, the nucleic acids are broken, effectively killing those cells. The technology has been shown to be effective for inactivating pathogens in all three major blood components: (platelets, red blood cells, and plasma). It has been shown to inactivate a broad spectrum of pathogens, including known and emerging viruses, bacteria, and parasites.

Recently riboflavin has been used in a new treatment to slow or stop the progression of the corneal disorder keratoconus. This is called corneal collagen crosslinking (CXL). In corneal crosslinking, riboflavin drops are applied to the patient's corneal surface. Once the riboflavin has penetrated through the cornea, Ultraviolet A light therapy is applied. This induces collagen crosslinking, which increases the tensile strength of the cornea. The treatment has been shown in several studies to stabilise keratoconus.

## Industrial Uses

Because riboflavin is fluorescent under UV light, dilute solutions (0.015-0.025% w/w) are often used to detect leaks or to demonstrate cleanability in an industrial system such a chemical blend tank or bioreactor. (See the ASME BPE section on Testing and Inspection for additional details.)

# VITAMIN $B_{12}$

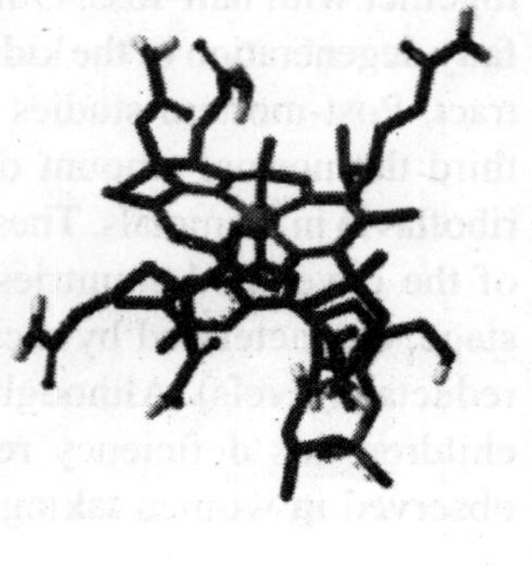

**Vitamin B-12**, is important for the normal functioning of the brain and nervous system and for the formation of blood. It is normally involved in the metabolism of every cell of the body, especially affecting the DNA synthesis and regulation but also fatty acid synthesis and energy production.

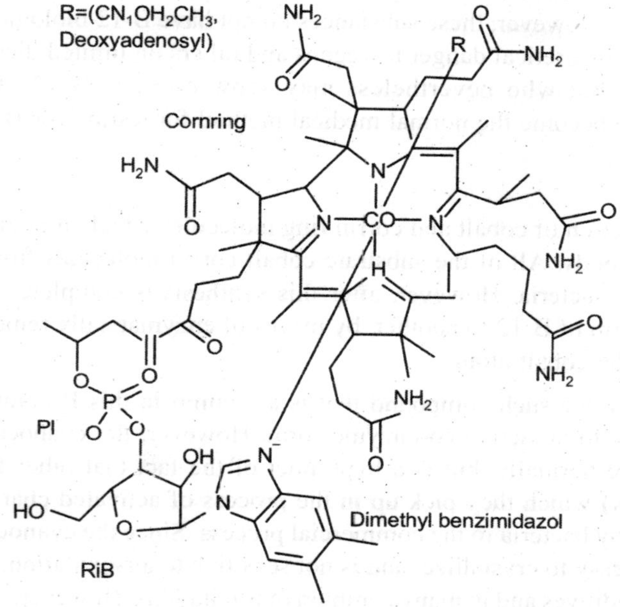

## Terminology

The name **vitamin B-12**, known also as **vitamin $B_{12}$** (or commonly **$B_{12}$** or **B-12** for short) generally refers to all forms of the vitamin. Some medical practitioners have suggested that its use be split into two different categories, however.

- In a broad sense B-12 still refers to a group of cobalt-containing vitamer compounds known as cobalamins: these include cyanocobalamin (an artifact formed as a result of the use of cyanide in the purification procedures), hydroxocobalamin (another medicinal form), and finally, the two naturally occurring cofactor forms of B-12: 5-deoxyadenosylcobalamin (adenosylcobalamin—AdoB-12), the cofactor of Methylmalonyl Coenzyme A mutase (MUT), and methylcobalamin (MeB-12), the cofactor of 5-methyltetrahydrofolate-homocysteine methyltransferase (MTR).
- The term B-12 may be properly used to refer to cyanocobalamin, the principal B-12 form used for foods and in nutritional supplements. This ordinarily creates no problem, except perhaps in rare cases of eye nerve damage, where the body is only marginally able to use this form due to high cyanide levels in the blood due to cigarette smoking, and thus requires cessation of smoking, or else B-12 given in another form, for the optic symptoms to abate.[1] However, tobacco amblyopia is a rare enough condition that debate continues about whether or not it represents a peculiar B-12 deficiency which is resistant to treatment with cyanocobalamin.

Finally, so-called **Pseudo-B-12** refers to B-12-like substances which are found in certain organisms, including spirulina (a cyanobacterium) and some algae. These substances are active in tests of B-12 activity by highly sensitive antibody-binding serum assay tests, which measure levels of B-12 and B-12 like compounds in blood. However, these substances do not have B-12 biological activity for humans, a fact which may pose a theoretical danger to vegans and others on limited diets who do not ingest B-12 producing bacteria, but who nevertheless may show normal "B-12" levels in the standard immunoassay which has become the normal medical method for testing for B-12 deficiency.

## Structure

**Vitamin B-12** is a collection of cobalt and corrin ring molecules which are defined by their particular vitamin function in the body. All of the substrate cobalt-corrin molecules from which B-12 is made must be synthesized by bacteria. However, after this synthesis is complete, the body has a limited power to convert any form of B-12 to another, by means of enzymatically removing certain prosthetic chemical groups from the cobalt atom.

**Cyanocobalamin** is one such compound that is a vitamin in this B complex, because it can be metabolized in the body to an active co-enzyme form. However, the cyanocobalamin form of B-12 does not occur in nature normally, but is a byproduct of the fact that other forms of B-12 are avid binders of cyanide (-CN) which they pick up in the process of activated charcoal purification of the vitamin after it is made by bacteria in the commercial process. Since the cyanocobalamin form of B-12 is deeply red coloured, easy to crystallize, and is not sensitive to air-oxidation, it is typically used as a form of B-12 for food additives and in many common multivitamins. However, this form is not perfectly synonymous with B-12, inasmuch as a number of substances (vitamers) have B-12 vitamin activity and can properly be labelled vitamin B-12, and cyanocobalamin is but one of them. (Thus, all cyanocobalamin is vitamin B-12, but not all vitamin B-12 is cyanocobalamin).

Vitamin B-12 is important for the normal functioning of the brain and nervous system and for the formation of blood. It is normally involved in the metabolism of every cell of the body, especially affecting the DNA synthesis and regulation but also fatty acid synthesis and energy production. However, many (though not all) of the effects of functions of B-12 can be replaced by sufficient quantities of folic acid (another B vitamin), since B-12 is used to regenerate folate in the body. Most "B-12 deficient symptoms" are actually folate deficient symptoms, since they include all the effects of pernicious anemia and megaloblastosis, which are due to poor synthesis of DNA when the body does not have a proper supply of folic acid for the production of thymine. When sufficient folic acid is available, all known B-12 related deficiency syndromes normalize, save those narrowly connected with the B-12 dependent enzymes Methylmalonyl Coenzyme A mutase (MUT), and 5-methyltetrahydrofolate-homocysteine methyltransferase (MTR), also known as methionine synthase; and the buildup of their respective substrates (methylmalonic acid, MMA) and homocysteine.

B-12 is the most chemically complex of all the vitamins. The structure of B-12 is based on a corrin ring, which is similar to the porphyrin ring found in heme, chlorophyll, and cytochrome. The central metal ion is **Co** (cobalt). Four of the six coordination sites are provided by the corrin ring, and a fifth by a dimethylbenzimidazole group. The sixth coordination site, the center of reactivity, is variable, being a cyano group (-CN), a hydroxyl group (-OH), a methyl group ($-CH_3$) or a 5'-deoxyadenosyl group (here the C5' atom of the deoxyribose forms the covalent bond with Co), respectively, to yield the four B-12 forms mentioned above. The covalent C-Co bond is one of first examples of carbon-metal bonds in biology. The hydrogenases and, by necessity, enzymes associated with cobalt utilization, involve metal-carbon bonds.

## Synthesis

Vitamin B-12 cannot be made by plants or animals. as the only type of organism that have the enzymes required for the synthesis of B-12 is bacteria. The total synthesis of B-12 was reported by Robert Burns Woodward and Albert Eschenmoser, and remains one of the classic feats of total synthesis.

Species from the following genera are known to synthesize B-12: *Aerobacter*, *Agrobacterium*, *Alcaligenes*, *Azotobacter*, *Bacillus*, *Clostridium*, *Corynebacterium*, *Flavobacterium*, *Micromonospora*, *Mycobacterium*, *Nocardia*, *Propionibacterium*, *Protaminobacter*, *Proteus*, *Pseudomonas*, *Rhizobium*, *Salmonella*, *Serratia*, *Streptomyces*, *Streptococcus* and *Xanthomonas*. Industrial production of B-12 is through fermentation of selected microorganisms.[10] The most used species are *Pseudomonas denitrificans* and *Propionibacterium shermanii*, often genetically engineered and grown under special conditions to enhance yield.

## Functions

Coenzyme B-12's reactive C-Co bond participates in two types of enzyme-catalyzed reactions.

1. Rearrangements in which a hydrogen atom is directly transferred between two adjacent atoms with concomitant exchange of the second substituent, X, which may be a carbon atom with substituents, an oxygen atom of an alcohol, or an amine.
2. Methyl ($-CH_3$) group transfers between two molecules.

In humans, only two corresponding coenzyme B-12-dependent enzymes are known:

1. Methylmalonyl Coenzyme A mutase (MUT) which uses the AdoB-12 form and reaction type 1 to catalyze a carbon skeleton rearrangement (the X group is -COSCoA). MUT's reaction converts MMl-CoA to Su-CoA, an important step in the extraction of energy from proteins and fats *(for more see MUT's reaction mechanism)*. This functionality is lost in vitamin B-12 deficiency, and can be measured clinically as an increased methylmalonic acid (MMA) level. Unfortunately, an elevated MMA, though sensitive to B-12 deficiency, is probably overly sensitive, and not all who have it actually have B-12 deficiency. For example, MMA is elevated in 90-98% of patients with B-12 deficiency; however 25-20% of patients over the age of 70 have elevated levels of MMA, yet 25-33% of them do not have B-12 deficiency. For this reason, MMA is not routinely recommended in the elderly. The "gold standard" test for B-12 deficiency continues to be low blood levels of the vitamin. The MUT function cannot be affected by folate supplementation, and which is necessary for myelin synthesis (see mechanism below) and certain other functions of the central nervous system. Other functions of B-12 related to DNA synthesis related to MTR dysfunction (see below) can often be corrected with supplementation with the vitamin folic acid, but not the elevated levels of homocysteine, which is normally converted to methionine by MTR.

2. 5-methyltetrahydrofolate-homocysteine methyltransferase (MTR), also known as methionine synthase. This is a methyl transfer enzyme, which uses the MeB-12 and reaction type 2 to catalyze the conversion of the amino acid Hcy back into Met *(for more see MTR's reaction mechanism)*.[13] This functionality is lost in vitamin B-12 deficiency, and can be measured clinically as an increased homocysteine level *in vitro*. Increased homocysteine can also be caused by a folic acid deficiency, since B-12 helps to regenerate the tetrahydrofolate (THF) active form of folic acid. Without B-12, folate is trapped as 5-methyl-folate, from which THF cannot be recovered unless a MTR process reacts the 5-methyl-folate with homocysteine to produce methionine and THF, thus decreasing the need for fresh sources of THF from the diet. THF may be produced in the conversion of homocysteine to methionine, or may be obtained in the diet. It is converted by a non-B12-dependent process to 5,10-methylene-THF, which is involved in the synthesis of thymine. Reduced availability of 5,10-methylene-THF results in problems with DNA synthesis, and ultimately in ineffective production cells with rapid turnover, in particular blood cells, and also intestinal wall cells which are responsible for absorption. The failure of blood cell production results in the once-dreaded and fatal disease, pernicious anemia. All of the DNA synthetic effects, including the megaloblastic anemia of pernicious anemia, resolve if sufficient folate is present (since levels of 5,10-methylene-THF still remain adequate with enough dietary folate). Thus the best known function of B-12 (that which is indirectly involved with DNA synthesis and restoration of cell-division and anemia) is actually a facultative function which is mediated by B-12 conservation of active folate which can be used for DNA production.

If folate is present in quantity, then of the two absolutely B-12 dependent reactions, the MUT reaction shows the most direct and characteristic secondary effects, focusing on the nervous system. Since the late 1990's folic acid has begun to be added to fortify flour in many countries, so that folate deficiency is now more rare. At the same time, since DNA synthetic-sensitive tests for anemia and erythrocyte size are routinely done in even simple medical test clinics (so that these folate mediated-biochemical effects are more often directly detected), the MTR dependent effects of B-12 deficiency are becoming apparent not as anemia (as they were classically), but now mainly as an elevation of homocysteine in the blood and urine (homocysteinuria). This condition may result in long-term damage to arteries and in clotting (stroke and heart attack), but is difficult to separate from other processes associated with atherosclerosis and aging.

The B-12 dependent MTR reactions may have neurological effects through an indirect mechanism. Adequate methionine (which must otherwise be obtained in the diet) is needed to make S-adenosyl-methionine, which is in turn necessary for methylation of myelin sheath phospholipids. In addition, SAMe is involved in the manufacture of certain neurotransmitters, catecholamines and in brain metabolism. These neurotransmitters are important for maintaining mood, possibly explaining why depression is associated with B-12 deficiency. Methylation of the myelin sheath phospholipids may also depend on adequate folate, which in turn is dependent on MTR recycling, unless ingested in relatively high amounts.

The specific myelin damage resulting from from B-12 deficiency has also been connected to B-12 reactions related to MUT, which is needed to convert methylmalonyl coenzyme A into succinyl coenzyme

A. Failure of this second reaction to occur results in elevated levels of methylmalonic acid (MMA), a myelin destabilizer. Excessive MAA will prevent normal fatty acid synthesis, or it will be incorporated into fatty acid itself rather than normal malonic acid. If this abnormal fatty acid subsequently is incorporated into myelin, the resulting myelin will be too fragile, and demyelination will occur. Although the precise methanism(s) are not known with certainty, the result is subacute combinded degeneration of central nervous system and spinal cord.[15] Whatever the cause, it is known that B-12 deficiency causes neuropathies, even if folic acid is present in good supply, and therefore anemia is not present.

## Human Absorption and Distribution

The human physiology of vitamin B-12 is complex, and therefore is prone to mishaps leading to vitamin B-12 deficiency. The vitamin as it occurs in foods enters the digestive tract bound to proteins, known as salivary R-binders. Stomach proteolysis of these proteins requires an acid pH, and also requires proper pancreatic release of proteolytic enzymes. (Even small amounts of B-12 taken in supplements bypasses these steps and thus any need for gastric acid, which may be blocked by antacid drugs).

The free B-12 then attaches to gastric intrinsic factor, which is generated by the gastric parietal cells. If this step fails due to gastric parietal cell atrophy (the problem in pernicious anemia), sufficient B-12 is not absorbed later on, unless administered orally in relatively massive doses (500 to 1000 mcg/day).

The conjugated vitamin B-12-intrinsic factor complex (IF/B-12) is then normally absorbed by the terminal ileum of the small bowel. Absorption of food vitamin B-12 therefore requires an intact and functioning stomach, exocrine pancreas, intrinsic factor, and small bowel. Problems with any one of these organs makes a vitamin B-12 deficiency possible.

Once the IF/B-12 complex is recognized by specialized ileal receptors, it is transported into the portal circulation. The vitamin is then transferred to transcobalamin II (TC-II/B12), which serves as the plasma transporter of the vitamin. Genetic deficiencies of this protein are known, also leading to functional B-12 deficiency.

For the vitamin to serve inside cells, the TC-II/B-12 complex must bind to a cell receptor, and be endocytosed. The transcobalamin-II is degraded within a lysozyme, and the B-12 is finally released into the cytoplasm, where it may be transformed into the proper coenzyme, by certain cellular enzymes (see above).

Hereditary defects in production of the transcobalamins and their receptors may produce functional deficiencies in B-12 and infantile megaloblastic anemia, and abnormal B-12 related biochemistry, even in some cases with normal blood B-12 levels

The total amount of vitamin B-12 stored in body is about 2,000-5,000 mcg in adults. Around 80% of this is stored in the liver[2]. 0.1 % of this is lost per day by secretions into the gut as not all these secretions are reabsorbed. How fast B-12 levels change depends on the balance between how much B12 is obtained from the diet, how much is secreted and how much is absorbed. B-12 deficiency may arise in a year if initial stores are low and genetic factors unfavourable or may not appear for decades. In infants, B-12 deficiency can appear much more quickly.

## History of B-12 as a Treatment for Pernicious Anemia

B-12 deficiency is the cause of pernicious anemia, a usually-fatal disease of unknown etiology when it was first described in medicine. The cure was discovered by accident. George Whipple had been inducing anemia in dogs by bleeding them, and then conducting experiments in which he fed them various foods to observe which diets allowed them fastest recovery from the anemia produced. In the process, he discovered that ingesting large amounts of liver seemed to most-rapidly cure the anemia of blood loss, and hypothesized that therefore liver ingestion be tried for pernicious anemia, an anemic disease of the time with no known cause or cure. He tried this and reported some signs of success in 1920. After a series of careful clinical studies George Minot and William Murphy set out to partly isolate the substance in liver which cured anemia in dogs, and found that it was iron. They found further that the partly isolated water-soluble liver-substance which cured pernicious anemia **in humans**, was something else entirely different — and which had no effect at all on canines under the conditions used. The specific factor treatment for pernicious anemia, found in liver juice, had been found by this coincidence. These experiments were reported by Minot and Murphy in 1926, marking the date of the first real progress with this disease, though for several years, patients were still required to eat large amounts of raw liver or to drink considerable amounts of liver juice.

In 1928, the chemist Edwin Cohn prepared a liver extract that was 50 to 100 times more potent than the natural liver products. The extract was the first workable treatment for the disease. For their initial work in pointing the way to a working treatment, Whipple, Minot, and Murphy shared the 1934 Nobel Prize in Physiology or Medicine.

The active ingredient in liver was not isolated until 1948 by the chemists Karl A. Folkers of the United States and Alexander R. Todd of Great Britain. The substance was a cobalamin called vitamin B-12. It could also be injected directly into muscle, making it possible to treat pernicious anemia more easily.[17]

The chemical structure of the molecule was determined by Dorothy Crowfoot Hodgkin and her team in 1956, based on crystallographic data. Eventually, methods of producing the vitamin in large quantities from bacteria cultures were developed in the 1950's, and these led to the modern form of treatment for the disease.

## Symptoms and Damage from Deficiency

Vitamin B-12 deficiency can potentially cause severe and irreversible damage, especially to the brain and nervous system. At levels only slightly lower than normal, a range of symptoms such as fatigue, depression, and poor memory may be experienced. However, these symptoms by themselves are too nonspecific to diagnose deficiency of the vitamin.

Vitamin B-12 deficiency has the following pathomorphology and symptoms

**Pathomorphology** includes: A spongiform state of neural tissue along with edema of fibers and deficiency of tissue. The myelin decays, along with axial fiber. In later phases, fibric sclerosis of nervous tissues occurs. Those changes apply to dorsal parts of the spinal cord, and to pyramidal tracts in lateral cords.

In the brain itself, changes are less severe: they occur as small sources of nervous fibers decay and accumulation of astrocytes, usually subcortically located, an also round hemorrhages with a torus of glial cells. Pathological changes can be noticed as well in the posterior roots of the cord and, to lesser extent, in peripheral nerves.

**Clinical Symptoms** : The main syndrome of vitamin B-12 deficiency is Addison's disease and Biermer's disease (pernicious anemia). It is characterized by a triad of symptoms:

(1) Anemia with bone marrow promegaloblastosis (Megaloblastic anemia)

(2) Gastrointestinal symptoms;

(3) Neurological symptoms.

Each of those symptoms can occur either alone or along with others. The neurological complex, defined as *myelosis funicularis*, consists of the following symptoms:

(1) Impaired perception of deep touch, pressure and vibration, abolishment of sense of touch, very annoying and persistent paresthesias;

(2) Ataxia of dorsal cord type;

(3) Decrease or abolishment of deep muscle-tendon reflexes;

(4) Pathological reflexes - Babinski, Rossolimo and others, also severe paresis.

During the course of disease, mental disorders can occur: irritablity, focus/concentration problems, depressive state with suicidal tendencies, paraphrenia complex. These symptoms may not reverse after correction of hematological abnormalities, and the chance of complete reversal decreases with the length of time the neurological symptoms have been present.

## Sources

Vitamin B-12 is naturally found in foods of animal origin including meat (especially liver and shellfish) and milk products. Animals, in turn, must obtain it directly or indirectly from bacteria, and these bacteria may inhabit a section of the gut which is posterior to the section where B-12 is absorbed. Thus, herbivorous animals must either obtain B-12 from bacteria in their rumens, or (if fermenting plant material in the hindgut) by reingestion of cecotrope feces. Eggs are often mentioned as a good B-12 source, but they also contain a factor that blocks absorption. Certain insects such as termites contain B-12 produced by their gut bacteria, in a manner analogous to ruminant animals. An NIH Fact Sheet lists a variety of food sources of vitamin B-12.

Plants only supply B-12 to humans when the soil containing B-12-producing microorganisms has not been washed from them. Vegan humans who eat only washed vegetables must take special care to supplement their diets accordingly. According to the U.K. Vegan Society, the only reliable vegans sources of B-12 are foods fortified with B-12 (including some plant milks, some soy products and some breakfast cereals), and B-12 supplements. Fortified breakfast cereals are a particularly valuable source of vitamin B-12 for vegetarians and vegans.

While lacto-ovo vegetarians usually get enough B-12 through consuming dairy products, vitamin B-12 may be found to be lacking in those practicing vegan diets who do not use multivitamin supplements or eat B-12 fortified foods, such as fortified breakfast cereals, fortified soy-based products, and fortified energy bars. Claimed sources of B-12 that have been shown through direct studies of vegans to be inadequate or unreliable include, laver (a seaweed), barley grass, and human gut bacteria. People on a vegan raw food diet are also susceptible to B-12 deficiency if no supplementation is used

The Vegan Society, the Vegetarian Resource Group, and the Physicians Committee for Responsible Medicine, among others, recommend that vegans either consistently eat foods fortified with B-12 or take a daily or weekly B-12 supplement.

Cyanocobalamin is converted to its active forms, first hydroxocobalamin and then methylcobalamin and adenosylcobalamin in the liver. The sublingual route, in which B-12 is presumably or supposedly absorbed more directly under the tongue, has not proven to be necessary or helpful. A 2003 study found no significant difference in absorption for serum levels from oral vs. sublingual delivery of 500 micrograms of cobalamin. However, if patient has inborn errors in the methyltransfer pathway (cobalamin C disease, combined methylmalonic aciduria and homocystinuria), treating with intravenous or intramuscular hydroxocobalamin is needed.

Vitamin B-12 can be supplemented in healthy subjects also by liquid, strip, nasal spray, or injection. B-12 is available singly or in combination with other supplements.

Injection is sometimes used in cases where digestive absorption is impaired, but there is some evidence that this course of action may not be necessary with modern high potency oral supplements (such as 500 to 1000 mcg or more). These supplements carry such large doses of the vitamin that the many different components of the B-12 absorption system are not required, and enough of the vitamin (only a few mcg a day) is obtained simply by mass-action transport across the gut. Even pernicious anemia can be treated entirely by the oral route.

For the much lower amounts of B-12 found in food sources, however, oral absorption is complex and requires stomach acid, and also specific intestinal transport proteins (intrinsic factor) produced in the stomach. Lack of function in these systems is the causes of much of the increased risk in many elderly persons who develop B-12 deficiency later in life. However, it can be treated with a simple high dose oral B-12 supplement. Cyanocobalamin is also sometimes added to beverages including Diet Coke Plus and many energy drinks, in some cases with over 80 times the recommended intake. However, 500 mcg would be needed to reverse biochemical signs of vitamin B-12 deficiency in older adults.

## Allergies

Vitamin B-12 supplements in theory should be avoided in people sensitive or allergic to cobalamin, cobalt, or any other product ingredients. However, direct allergy to a vitamin or nutrient is extremely rare, and if reported, other causes should be sought.

## Side Effects, Contraindications, and Warnings

- Dermatologic: Itching, rash, transitory exanthema, and urticaria have been reported. Vitamin B-12 (20 micrograms/day) and pyridoxine (80mg/day) has been associated with cases of rosacea fulminans, characterized by intense erythema with nodules, papules, and pustules. Symptoms may persist for up to 4 months after the supplement is stopped, and may require treatment with systemic corticosteroids and topical therapy.

- Gastrointestinal: Diarrhea has been reported.
- Hematologic: Peripheral vascular thrombosis has been reported. Treatment of vitamin B-12 deficiency can unmask polycythemia vera, which is characterized by an increase in blood volume and the number of red blood cells. The correction of megaloblastic anemia with vitamin B-12 can result in fatal hypokalemia and gout in susceptible individuals, and it can obscure folate deficiency in megaloblastic anemia. Caution is warranted.
- Leber's disease: Vitamin B-12 in the form of cyanocobalamin is contraindicated in early Leber's disease, which is hereditary optic nerve atrophy. Cyanocobalamin can cause severe and swift optic atrophy, but other forms of vitamin B-12 are available.[citation needed] However, the sources of this statement are not clear, while an opposing view concludes: "The clinical picture of optic neuropathy associated with vitamin B-12 deficiency shows similarity to that of Leber's disease optic neuropathy. Both involve the nerve fibres of the papillomacular bundle. The present case reports suggest that optic neuropathy in patients carrying a primary LHON mtDNA mutation may be precipitated by vitamin B-12 deficiency. Therefore, known carriers should take care to have an adequate dietary intake of vitamin B-12 and malabsorption syndromes like those occurring in familial pernicious anaemia or after gastric surgery should be excluded."

### Normal Requirement, and in Pregnancy and Breastfeeding

The Dietary Reference Intake for an adult ranges from 2 to 3 μg (micrograms). The recommended optimal daily intake (ODI) is 10 to 15 μg.

Vitamin B-12 is believed to be safe when used orally in amounts that do not exceed the recommended dietary allowance (RDA). The RDA for vitamin B-12 in pregnant women is 2.6 μg per day and 2.8 μg during lactation periods. There is insufficient reliable information available about the safety of consuming greater amounts of Vitamin B-12 during pregnancy.

### Other Medical Uses

Hydroxycobalamin, or hydoxocobalamin, also known as Vitamin B-12a, is used in Europe both for vitamin B-12 deficiency and as a treatment for cyanide poisoning, sometimes with a large amount (5-10 g) given intravenously, and sometimes in combination with sodium thiosulfate.[38] The mechanism of action is straightforward: the hydroxycobalamin hydroxide ligand is displaced by the toxic cyanide ion, and the resulting harmless B-12 complex is excreted in urine. In the United States, the FDA has approved in 2006 the use of hydroxocobalamin for acute treatment of cyanide poisoning.

## Interactions

### Interactions with Drugs

- Alcohol (ethanol): Excessive alcohol intake lasting longer than two weeks can decrease vitamin B-12 absorption from the gastrointestinal tract, leading to Korsakoff's Syndrome.
- Aminosalicylic acid (para-aminosalicylic acid, PAS, Paser): Aminosalicylic acid can reduce oral vitamin B-12 absorption, possibly by as much as 55%, as part of a general malabsorption

syndrome. Megaloblastic changes, and occasional cases of symptomatic anemia have occurred, usually after doses of 8 to 12 grams/day for several months. Vitamin B-12 levels should be monitored in people taking aminosalicylic acid for more than one month.

- Antibiotics: An increased bacterial load can bind significant amounts of vitamin B-12 in the gut, preventing its absorption. In people with bacterial overgrowth of the small bowel, antibiotics such as metronidazole (Flagyl®) can actually improve vitamin B-12 status. The effects of most antibiotics on gastrointestinal bacteria are unlikely to have clinically significant effects on vitamin B-12 levels.
- Hormonal contraception: The data regarding the effects of oral contraceptives on vitamin B-12 serum levels are conflicting. Some studies have found reduced serum levels in oral contraceptive users, but others have found no effect despite use of oral contraceptives for up to 6 months. When oral contraceptive use is stopped, normalization of vitamin B-12 levels usually occurs. Lower vitamin B-12 serum levels seen with oral contraceptives probably are not clinically significant.
- Chloramphenicol (Chloromycetin®): Limited case reports suggest that chloramphenicol can delay or interrupt the reticulocyte response to supplemental vitamin B-12 in some patients. Blood counts should be monitored closely if this combination cannot be avoided.
- Cobalt irradiation: Cobalt irradiation of the small bowel can decrease gastrointestinal (GI) absorption of vitamin B-12.
- Colchicine: Colchicine in doses of 1.9 to 3.9mg/day can disrupt normal intestinal mucosal function, leading to malabsorption of several nutrients, including vitamin B-12. Lower doses do not seem to have a significant effect on vitamin B-12 absorption after 3 years of colchicine therapy. The significance of this interaction is unclear. Vitamin B-12 levels should be monitored in people taking large doses of colchicine for prolonged periods.
- Colestipol (Colestid®), Cholestyramine (Questran®): These resins used for sequestering bile acids in order to decrease cholesterol, can decrease gastrointestinal (GI) absorption of vitamin B-12. It is unlikely that this interaction will deplete body stores of vitamin B-12 unless there are other factors contributing to deficiency. In a group of children treated with cholestyramine for up to 2.5 years there was not any change in serum vitamin B-12 levels. Routine supplements are not necessary.
- $H_2$-receptor antagonists: include cimetidine (Tagamet®), famotidine (Pepcid®), nizatidine (Axid®), and ranitidine (Zantac®). Reduced secretion of gastric acid and pepsin produced by $H_2$ blockers can reduce absorption of protein-bound (dietary) vitamin B-12, but not of supplemental vitamin B-12. Gastric acid is needed to release vitamin B-12 from protein for absorption. Clinically significant vitamin B-12 deficiency and megaloblastic anemia are unlikely, unless $H_2$ blocker therapy is prolonged (2 years or more), or the person's diet is poor. It is also more likely if the person is rendered achlorhydric (with complete absence of gastric acid secretion), which occurs more frequently with proton pump inhibitors than $H_2$ blockers. Vitamin B-12 levels should be monitored in people taking high doses of $H_2$ blockers for prolonged periods.

- Metformin (Glucophage®): Metformin may reduce serum folic acid and vitamin B-12 levels. These changes can lead to hyperhomocysteinemia, adding to the risk of cardiovascular disease in people with diabetes. There are also rare reports of megaloblastic anemia in people who have taken metformin for 5 years or more. Reduced serum levels of vitamin B-12 occur in up to 30% of people taking metformin chronically.[39][40] However, clinically significant deficiency is not likely to develop if dietary intake of vitamin B-12 is adequate. Deficiency can be corrected with vitamin B-12 supplements even if metformin is continued. The metformin-induced malabsorption of vitamin B-12 is reversible by oral calcium supplementation.[41] The general clinical significance of metformin upon B-12 levels is as yet unknown.[42]
- Neomycin: Absorption of vitamin B-12 can be reduced by neomycin, but prolonged use of large doses is needed to induce pernicious anemia. Supplements are not usually needed with normal doses.
- Nicotine: Nicotine can reduce serum vitamin B-12 levels. The need for vitamin B-12 supplementation has not been adequately studied.
- Nitrous oxide: Nitrous oxide inactivates the cobalamin form of vitamin B-12 by oxidation. Symptoms of vitamin B-12 deficiency, including sensory neuropathy, myelopathy, and encephalopathy, can occur within days or weeks of exposure to nitrous oxide anesthesia in people with subclinical vitamin B-12 deficiency. Symptoms are treated with high doses of vitamin B-12, but recovery can be slow and incomplete. People with normal vitamin B-12 levels have sufficient vitamin B-12 stores to make the effects of nitrous oxide insignificant, unless exposure is repeated and prolonged (such as recreational use). Vitamin B-12 levels should be checked in people with risk factors for vitamin B-12 deficiency prior to using nitrous oxide anesthesia. Chronic nitrous oxide B-12 poisoning (usually from use of nitrous oxide as a recreational drug), however, may result in B-12 functional deficiency even with normal measured blood levels of B-12 [43]
- Phenytoin (Dilantin®), phenobarbital, primidone (Mysoline®): These anticonvulsants have been associated with reduced vitamin B-12 absorption, and reduced serum and cerebrospinal fluid levels in some patients. This may contribute to the megaloblastic anemia, primarily caused by folate deficiency, associated with these drugs. It's also suggested that reduced vitamin B-12 levels may contribute to the neuropsychiatric side effects of these drugs. Patients should be encouraged to maintain adequate dietary vitamin B-12 intake. Folate and vitamin B-12 status should be checked if symptoms of anemia develop.
- Proton pump inhibitors (PPIs): The PPIs include omeprazole (Prilosec®, Losec®), lansoprazole (Prevacid®), rabeprazole (Aciphex®), pantoprazole (Protonix®, Pantoloc®), and esomeprazole (Nexium®). The reduced secretion of gastric acid and pepsin produced by PPIs can reduce absorption of protein-bound (dietary) vitamin B-12, but not supplemental vitamin B-12. Gastric acid is needed to release vitamin B-12 from protein for absorption. Reduced vitamin B-12 levels may be more common with PPIs than with H2-blockers, because they are more likely to produce achlorhydria (complete absence of gastric acid secretion). However, clinically significant vitamin B-12 deficiency is unlikely, unless PPI therapy is prolonged (2 years or more) or

dietary vitamin intake is low. Vitamin B-12 levels should be monitored in people taking high doses of PPIs for prolonged periods.

- Zidovudine (AZT, Combivir®, Retrovir®): Reduced serum vitamin B-12 levels may occur when zidovudine therapy is started. This adds to other factors that cause low vitamin B-12 levels in people with HIV, and might contribute to the hematological toxicity associated with zidovudine. However, data suggests vitamin B-12 supplements are not helpful for people taking zidovudine.

## Interactions with Herbs and Dietary Supplements

- Folic acid: Folic acid, particularly in large doses, can mask vitamin B-12 deficiency by completely correcting hematological abnormalities. In vitamin B-12 deficiency, folic acid can produce complete resolution of the characteristic megaloblastic anemia, while allowing potentially irreversible neurological damage (from continued inactivity of methylmalonyl mutase) to progress. Thus, vitamin B-12 status should be determined before folic acid is given as monotherapy.
- Potassium: Potassium supplements can reduce absorption of vitamin B-12 in some people. This effect has been reported with potassium chloride and, to a lesser extent, with potassium citrate. Potassium might contribute to vitamin B-12 deficiency in some people with other risk factors, but routine supplements are not necessary

# VITAMIN $B_5$

**Pantothenic acid**, also called **vitamin $B_5$** (a B vitamin), is a water-soluble vitamin required to sustain life (essential nutrient). Pantothenic acid is needed to form coenzyme-A (CoA), and is critical in the metabolism and synthesis of carbohydrates, proteins, and fats. In chemical structure, it is the amide between D-pantoate and beta-alanine. Its name is derived from the Greek *pantothen* (παντοθεν) meaning "from everywhere" and small quantities of pantothenic acid are found in nearly every food, with high amounts in whole-grain cereals, legumes, eggs, meat, and royal jelly. It is commonly found as its alcohol analog, the provitamin panthenol, and as calcium pantotherate.

## Biological Role

Only the dextrorotatory (D) isomer of pantothenic acid possesses biologic activity. The levorotatory (L) form may antagonize the effects of the dextrorotatory isomer.

Pantothenic acid is used in the synthesis of coenzyme A (abbreviated as CoA). Coenzyme A may act as an acyl group carrier to form acetyl-CoA and other related compounds; this is a way to transport carbon atoms within the cell. The transfer of carbon atoms by coenzyme A is important in cellular respiration, as well as the biosynthesis of many important compounds such as fatty acids, cholesterol, and acetylcholine.

Since pantothenic acid participates in a wide array of key biological roles, it is considered essential to all forms of life. As such, deficiencies in pantothenic acid may have numerous wide-ranging effects, as discussed below.

Pantothenic acid is vital for a healthy pregnancy.

## Sources

Small quantities of pantothenic acid are found in most foods, with high quantities found in whole grain and eggs. Pantothenic acid can also be found in many dietary supplements (as calcium-D-pantotherate), and some companies are now adding pantothenic acid to their beverages.

A recent study also suggests that gut bacteria in humans can generate pantothenic acid.

## Daily Requirement

Pantothenate in the form of pantethine is considered to be the more active form of the vitamin in the body, but is unstable at high temperatures or when stored for long periods, so calcium pantothenate is the more usual form of vitamin B5 when it is sold as a dietary supplement. Ten mg of calcium pantothenate is equivalent to 9.2 mg of pantothenic acid.

| Age group | Age | Requirements (in mg per day) |
|---|---|---|
| infants | 0-6 months | 1.7 |
| infants | 7-12 months | 2 |
| children | 4-8 years | 3 |
| children | 9-13.5 years | 4 |
| adolescents | 14-18 years | 5 |
| adults | 19 years and older | 5 |
| pregnant women | | 6 |
| breastfeeding women | | 7 |
| • United Kingdom RDA: 6 mg/day | | |

## Deficiency

Pantothenic acid deficiency is exceptionally rare and has not been thoroughly studied. In the few cases where deficiency has been seen (victims of starvation and limited volunteer trials), nearly all symptoms can be reversed with the return of pantothenic acid.

Symptoms of deficiency are similar to other vitamin B deficiencies. Most are minor, including fatigue, allergies, nausea, and abdominal pain. In a few rare circumstances more serious (but reversible) conditions have been seen, such as adrenal insufficiency and hepatic encephalopathy.

It has been noted that painful burning sensations of the feet were reported in tests conducted on volunteers. Deficiency of pantothenic acid may explain similar sensations reported in malnourished prisoners of war.

### Hair Care

Mouse models identified skin irritation and loss of hair colour as possible results of severe pantothenic acid deficiency. As a result, the cosmetic industry began adding pantothenic acid to various cosmetic products, including shampoo. These products, however, showed no benefits in human trials.[ Despite this, many cosmetic products still advertise pantothenic acid additives.

### Acne

Following from discoveries in mouse trials,[ in the late 1990s a small study was published promoting the use of pantothenic acid to treat acne vulgaris.

According to a study published in 1995 by Dr. Lit-Hung Leung,[ high doses of Vitamin $B_5$ resolved acne and decreased pore size. Dr. Leung also proposes a mechanism, stating that CoA regulates both hormones and fatty-acids, and without sufficient quantities of pantothenic acid, CoA will preferentially produce androgens. This causes fatty acids to build up and be excreted through sebaceous glands, causing acne. Leung's study gave 45 Asian males and 55 Asian females varying doses of 10-20g of pantothenic acid (100,000%-200,000% of the US Daily Value), 80% orally and 20% through topical cream. Leung noted improvement of acne within one week to one month of the start of the treatment

Critics are quick to point out the flaws in Dr. Leung's study, however. Dr. Leung's study was not a double-blind placebo controlled trial. To date, the only study looking at the effect of Vitamin $B_5$ on acne is Dr. Leung's, and few if any dermatologists prescribe high-dose pantothenic acid. Furthermore, there is no evidence documenting acetyl-CoA regulation of androgens instead of fatty acids in times of stress or limited availability, since fatty acids are also necessary for life.

### Diabetic Peripheral Polyneuropathy

28 out of 33 patients (84,8%) previously treated with alpha-lipoic acid for peripheral polyneuropathy reported further improvement after combination with pantothenic acid. The theoretical basis for this is that both substances intervene at different sites in pyruvate metabolism and are thus more effective than one substance alone. Additional clinical findings indicated that diabetic neuropathy may occur in association with a latent prediabetic metabolic disturbance, and that the symptoms of neuropathy can be favourably influenced by the described combination therapy, even in poorly controlled diabetes.[7]

## VITAMIN $B_6$

**Pyridoxine** is one of the compounds that can be called vitamin B6, along with Pyridoxal and Pyridoxamine. It differs from pyridoxamine by the substituent at the '4' position. It is often used as 'pyridoxine hydrochloride'.

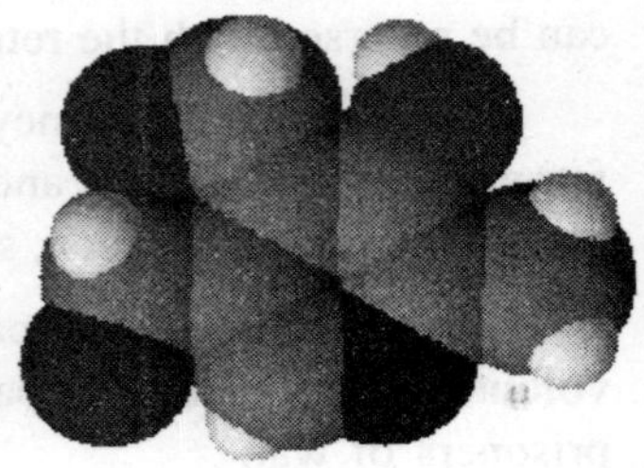

### Chemistry

It is based on a pyridine ring, with hydroxyl, methyl, and hydroxymethyl substituents. It is converted to the biologically active form pyridoxal 5-phosphate.

## Function in the Body

Pyridoxine assists in the balancing of sodium and potassium as well as promoting red blood cell production. It is linked to cardiovascular health by decreasing the formation of homocysteine. It has been suggested that Pyridoxine might help children with learning difficulties, and may also prevent dandruff, eczema, and psoriasis. In addition, pyridoxine can help balance hormonal changes in women and aid in immune system. Lack of pyridoxine may cause anemia, nerve damage, seizures, skin problems, and sores in the mouthIt is required for the production of the monoamine neurotransmitters serotonin, dopamine, noradrenaline and adrenaline, as it is the precursor to pyridoxal phosphate: cofactor for the enzyme aromatic amino acid decarboxylase. This enzyme is responsible for converting the precusors 5-hydroxytryptophan (5-HTP) into serotonin and levodopa (L-DOPA) into dopamine, noradrenaline and adrenaline. As such it has been implicated in the treatment of depression and anxiety.

A very good source of pyridoxine is dragon fruit from South East Asia

## Medicinal Uses

It is given to patients taking isoniazid to combat the toxic side effects of the drug. Pyridoxine is given 10-50 mg/day to patients on INH (Isoniazid) to prevent peripheral neuropathy and CNS effects that are associated with the use of isoniazid.

Vitamin $B_6$ can be compounded into a variety of different dosage forms. It can be used orally as a tablet, capsule, or solution. It can also be used as a nasal spray or for injection when in its solution form. The following is a procedure for producing a diluting solution of vitamin $B_6$, taken from the USP/NF.

"Diluting Solution: Dissolve 25g of edentate disodium in 1000mL of water and mix. Dissolve an accurately weighed quantity of USP Pyridoxine Hydrochloride RS in Diluting Solution, and dilute quantitatively, and stepwise if necessary, with Diluting Solution to obtain a Solution having a known concentration of about 0.024mg/mL."

# VITAMIN C OR ASCORBIC ACID

**Ascorbic acid** is a sugar acid with antioxidant properties. Its appearance is white to light-yellow crystals or powder. It is water-soluble. The L-enantiomer of ascorbic acid is commonly known as vitamin C. The name is derived from the alpha privative *a-* (meaning no) and *scorbuticus* (scurvy), the disease caused by a deficiency of vitamin C. In 1937 the Nobel Prize for chemistry was awarded to Walter Haworth for his work in determining the structure of ascorbic acid (shared with Paul Karrer, who received his award for work on vitamins), and the prize for Physiology or

Medicine that year went to Albert Szent-Györgyi for his studies of the biological functions of L-ascorbic acid. At the time of its discovery in the 1920s, it was called **hexuronic acid** by some researchers.

## Chemistry

### Acidity

Ascorbic acid, the formula of which is $C_6H_8O_6$, behaves as a vinylogous carboxylic acid, wherein the double bond ("vinyl") transmits electron pairs between the hydroxyl and the carbonyl. There are two resonance structures for the deprotonated form, differing in the position of the double bond.

Another way to look at ascorbic acid is to consider it as an enol. The deprotonated form is an enolate, which is usually strongly basic. However, the adjacent double bond stabilizes the deprotonated form.

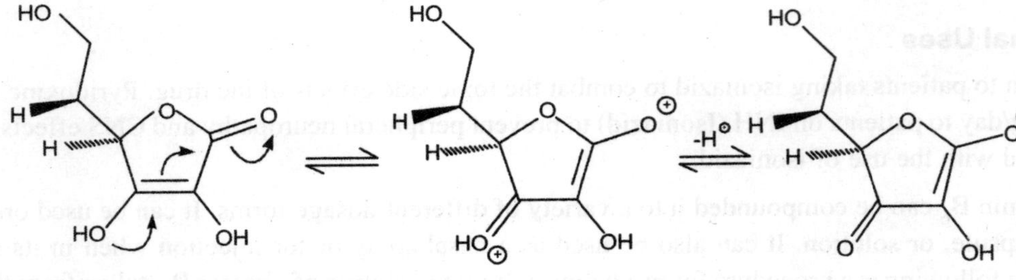

Movement of electron pairs in deprotonation

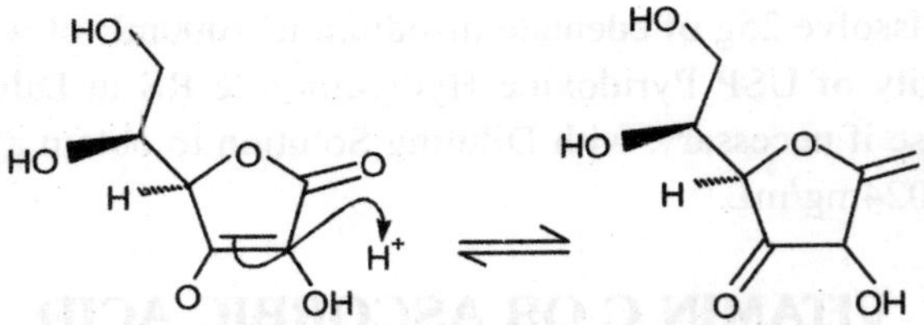

Nucleophilic attack of ascorbic enol on proton to give 1,3-diketone Ascorbic acid also rapidly interconverts into two unstable diketone tautomers by proton transfer, although it is the most stable in the enol form. The proton of the enol is lost, and reacquired by electrons from the double bond, to produce a diketone. This is an enol reaction. There are two possible forms: 1,2-diketone and1,3-diketone.

### Determination

The concentration of a solution of ascorbic acid can be determined in many ways, the most common ways involving titration with an oxidizing agent.

## DCPIP

A commonly-used oxidising agent is the dye 2,6-dichlorophenol-indophenol, or DCPIP for short. The blue dye is run into the ascorbic acid solution which turns colourless

## Iodine

Another method involves using iodine and a starch indicator, wherein iodine reacts with ascorbic acid, and, when all the ascorbic acid has reacted, the iodine is then in excess, forming a blue-black complex with the starch indicator. This indicates the end-point of the titration. As an alternative, ascorbic acid can be reacted with iodine in excess, followed by back titration with sodium thiosulfate while using starch as an indicator.

## Iodate and Iodine

The above method involving iodine requires making up and standardising the iodine solution. One way around this is to generate the iodine in the presence of the ascorbic acid by the reaction of iodate and iodide ion in acid solution.

## N-Bromosuccinimide

A much-less-common oxidising agent is *N*-bromosuccinimide, (NBS). In this titration, the NBS oxidises the ascorbic acid (in the presence of potassium iodide and starch). When the NBS is in excess (i.e., the reaction is complete), the NBS liberates the iodine from the potassium iodide, which then forms the blue/black complex with starch, indicating the end-point of the titration.

HO HO H O O HO OH

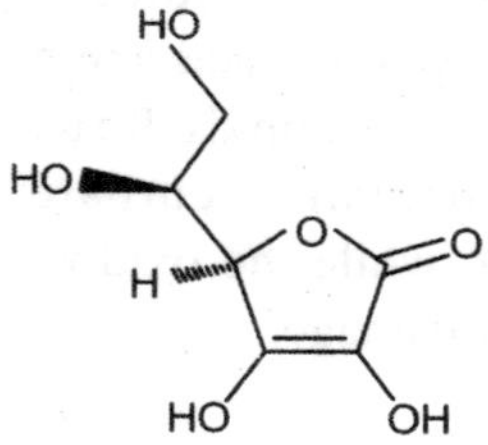

ascorbic acid (reduced form of Vitamin C)

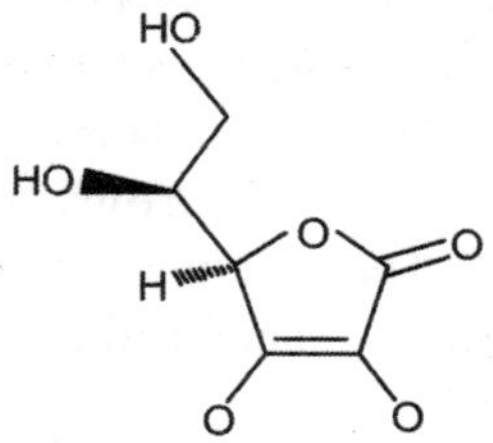

HO HO H O O O O

dehydroascorbic acid (oxidized form of Vitamin C)

Ascorbic acid is easily oxidized and so is used as a reductant in photographic developer solutions (among others) and as a preservative.

Exposure to oxygen, metals, light, and heat destroys ascorbic acid, so it must be stored in a dark, cold, and not metal container.

The oxidized form of ascorbic acid is known as dehydroascorbic acid.

The L-enantiomer of ascorbic acid is also known as vitamin C. The name "ascorbic" comes from its property of preventing and curing scurvy. Primates, including humans, and a few other species in all divisions of the animal kingdom, notably the guinea pig, have lost the ability to synthesize ascorbic acid, and must obtain it in their food.

Ascorbic acid and its sodium, potassium, and calcium salts are commonly used as antioxidant food additives. These compounds are water-soluble and thus cannot protect fats from oxidation: For this purpose, the fat-soluble esters of ascorbic acid with long-chain fatty acids (ascorbyl palmitate or ascorbyl stearate) can be used as food antioxidants. Eighty percent of the world's supply of ascorbic acid is produced in China

The relevant European food additive E numbers are:

1. E300 ascorbic acid,
2. E301 sodium ascorbate,
3. E302 calcium ascorbate,
4. E303 potassium ascorbate,
5. E304 fatty acid esters of ascorbic acid (i) ascorbyl palmitate (ii) ascorbyl stearate.

It can be added to water that has been treated with iodine to make it potable, neutralizing the unpleasant iodine taste, and increasing the health benefits of drinking water, although increasing the chance of tooth decay.

In plastic manufacturing, ascorbic acid can be used to assemble molecular chains more quickly and with less waste than traditional synthesis methods.

## Antioxidant Mechanism

Ascorbate acts as an antioxidant by being available for energetically favourable oxidation. Many oxidants (typically, reactive oxygen species) such as the hydroxyl radical (formed from hydrogen peroxide), contain an unpaired electron, and, thus, are highly reactive and damaging to humans and plants at the molecular level. This is due to their interaction with nucleic acid, proteins, and lipids. Reactive oxygen species oxidize (take electrons from) ascorbate first to monodehydroascorbate and then dehydroascorbate. The reactive oxygen species are reduced to water, while the oxidized forms of ascorbate are relatively stable and unreactive, and do not cause cellular damage.

## Ascorbic Acid Synthesis in Non-primates

Ascorbic acid is found in plants, animals, and single-cell organisms All living animals either make it, eat it, or die from scurvy due to lack of it. Reptiles and older orders of birds make ascorbic acid in their kidneys. Recent orders of birds and most mammals make ascorbic acid in their livers where the enzyme L-gulonolactone oxidase is required to convert glucose to ascorbic acid Humans, guinea pigs, and some other primates are not able to make L-gulonolactone oxidase because of a genetic defect and are therefore unable to make ascorbic acid in their livers. This genetic mutation occurred about 63 million years ago This would have had lethal consequences for the mutated primate were it not for the fact that it occurred to an arboreal animal living in a tropical environment where plenty of foodstuffs containing ascorbic acid were available throughout the year. Although ascorbic acid is a vital food nutrient for humans and is therefore termed a vitamin, it is a natural liver metabolite in most other animals.

# VITAMIN D

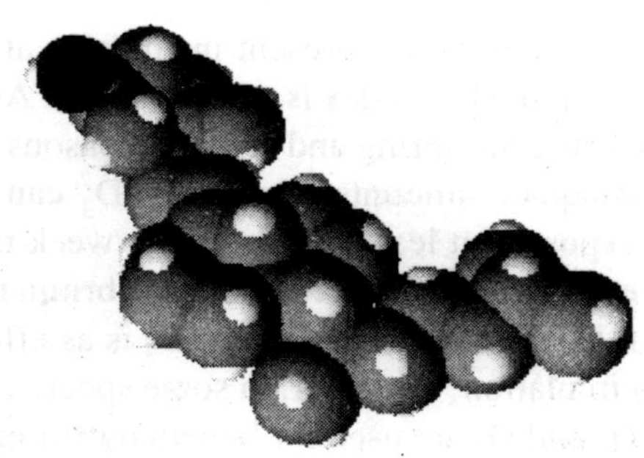

**Vitamin D** is a group of fat-soluble prohormones, the two major forms of which are vitamin $D_2$ (or ergocalciferol) and vitamin $D_3$ (or cholecalciferol). The term vitamin D also refers to metabolites and other analogues of these substances. Vitamin $D_3$ is produced in skin exposed to sunlight, specifically ultraviolet B radiation.

Vitamin D plays an important role in the maintenance of organ systems.

- Vitamin D regulates the calcium and phosphorus levels in the blood by promoting their absorption from food in the intestines, and by promoting re-absorption of calcium in the kidneys.
- It promotes bone formation and mineralization and is essential in the development of an intact and strong skeleton. However, at very high levels it will promote the resorption of bone.
- It inhibits parathyroid hormone secretion from the parathyroid gland.
- Vitamin D affects the immune system by promoting immunosuppression, phagocytosis, and anti-tumor activity.

Vitamin D deficiency can result from inadequate intake coupled with inadequate sunlight exposure, disorders that limit its absorption, conditions that impair conversion of vitamin D into active metabolites, such as liver or kidney disorders, or, rarely, by a number of hereditary disorders.[2] Deficiency results in impaired bone mineralization, and leads to bone softening diseases, rickets in children and osteomalacia in adults, and possibly contributes to osteoporosis. Research has indicated that vitamin D deficiency is linked to colon cancer; conflicting evidence links vitamin D deficiency to other forms of cancer.

## Forms

Several forms (vitamers) of vitamin D have been discovered. The two major forms are vitamin $D_2$ or ergocalciferol, and vitamin $D_3$ or cholecalciferol.

- **Vitamin $D_1$**: molecular compound of ergocalciferol with lumisterol, 1:1
- **Vitamin $D_2$**: ergocalciferol or calciferol (made from ergosterol)
- **Vitamin $D_3$**: cholecalciferol (made from 7-dehydrocholesterol in the skin).
- **Vitamin $D_4$**: 22-dihydroergocalciferol
- **Vitamin $D_5$**: sitocalciferol (made from 7-dehydrositosterol)

Chemically, the various forms of vitamin D are secosteroids; i.e., broken-open steroids.[3] The structural difference between vitamin $D_2$ and vitamin $D_3$ is in their side chains. The side chain of $D_2$ contains a double bond between carbons 22 and 23, and a methyl group on carbon 24. Vitamin $D_2$ is derived from fungal and plant sources, and is not produced by the human body. Vitamin $D_3$ is derived from animal sources and is made in the skin when 7-dehydrocholesterol reacts with UVB ultraviolet light at wavelengths between 270–300 nm, with peak synthesis occurring between 295-297 nm. These wavelengths are present in sunlight at sea level when the sun is more than 45° above the horizon, or when the UV index is greater than 3. At this solar elevation, which occurs daily within the tropics, daily during the spring and summer seasons in temperate regions, and almost never within the arctic circles, adequate amounts of vitamin $D_3$ can be made in the skin only after ten to fifteen minutes of sun exposure at least two times per week to the face, arms, hands, or back without sunscreen. With longer exposure to UVB rays, an equilibrium is achieved in the skin, and the vitamin simply degrades as fast as it is generated. In humans, $D_3$ is as effective than $D_2$ at increasing the levels of vitamin D hormone in circulation; However, in some species, such as rats, vitamin $D_2$ is more effective than $D_3$. Both vitamin $D_2$ and $D_3$ are used for human nutritional supplementation, and pharmaceutical forms include calcitriol (1alpha, 25-dihydroxycholecalciferol), doxercalciferol and calcipotriene.

## Biochemistry

Vitamin D is a prohormone, meaning that it has no hormone activity itself, but is converted to the active hormone 1,25-D through a tightly regulated synthesis mechanism. Production of vitamin D in nature always appears to require the presence of some UV light; even vitamin D in foodstuffs is ultimately derived from organisms, from mushrooms to animals, which are not able to synthesize it except through the action of sunlight at some point in the synthetic chain. For example, fish contain vitamin D only because they ultimately exist on calories from ocean algae which synthesize vitamin D in shallow waters from the action of solar UV.

## Production in the Skin

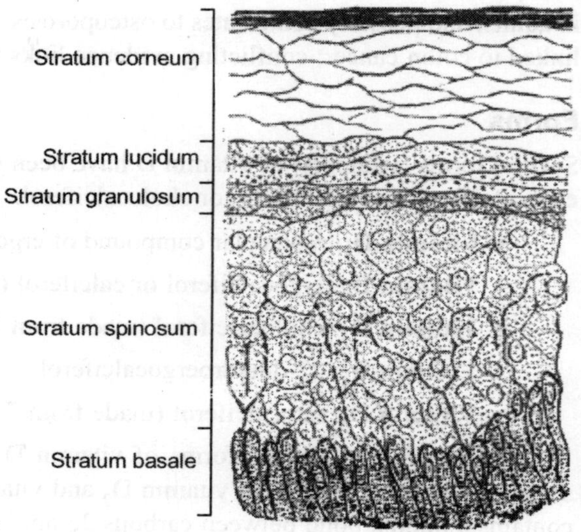


The epidermal strata of the skin. Production is greatest in the stratum basale (colored red in the illustration) and stratum spinosum (colored orange).The skin consists of two primary layers: the inner layer called the dermis, composed largely of connective tissue, and the outer thinner epidermis. The epidermis consists of five *strata*; from outer to inner they are: the stratum corneum, stratum lucidum, stratum granulosum, stratum spinosum, and stratum basale. Vitamin $D_3$ is produced photochemically in the skin from 7-dehydrocholesterol. The highest concentrations of 7-dehydrocholesterol are found in the epidermal layer of skin, specifically in the stratum basale and stratum spinosum.[4] The production of pre-vitamin $D_3$ is therefore greatest in these two layers, whereas production in the other layers is reduced. Synthesis in the skin involves UVB radiation which effectively penetrates only the epidermal layers of skin. While 7-Dehydrocholesterol absorbs UV light at wavelengths between 270–300 nm, optimal synthesis occurs in a narrow band of

UVB spectra between 295-300 nm. Peak isomerization is found at 297 nm. This narrow segment is sometimes referred to as D-UV. The two most important factors that govern the generation of pre-vitamin $D_3$ are the quantity (intensity) and quality (appropriate wavelength) of the UVB irradiation reaching the 7-dehydrocholesterol deep in the stratum basale and stratum spinosum. A critical determinant of vitamin $D_3$ production in the skin is the presence and concentration of melanin. Melanin functions as a light filter in the skin, and therefore the concentration of melanin in the skin is related to the ability of UVB light to penetrate the epidermal strata and reach the 7-dehydrocholesterol-containing stratum basale and stratum spinosum. Under normal circumstances, ample quantities of 7-dehydrocholesterol (about 25-50 mg/cm² of skin) are available in the stratum spinosum and stratum basale of human skin to meet the body's vitamin D requirements,[4] and melanin content does not alter the amount of vitamin D that can be produced. Thus, individuals with higher skin melanin content will simply require more time in sunlight to produce the same amount of vitamin D as individuals with lower melanin content. As noted below, the amount of time an individual requires to produce a given amount of Vitamin D may also depend upon the person's distance from the equator and on the season of the year.

## Synthesis Mechanism

1. Vitamin $D_3$ is synthesized from 7-dehydrocholesterol, a derivative of cholesterol, which is then photolyzed by ultraviolet light in 6-electron conrotatory electrocyclic reaction. The product is *pre-vitamin* $D_3$.

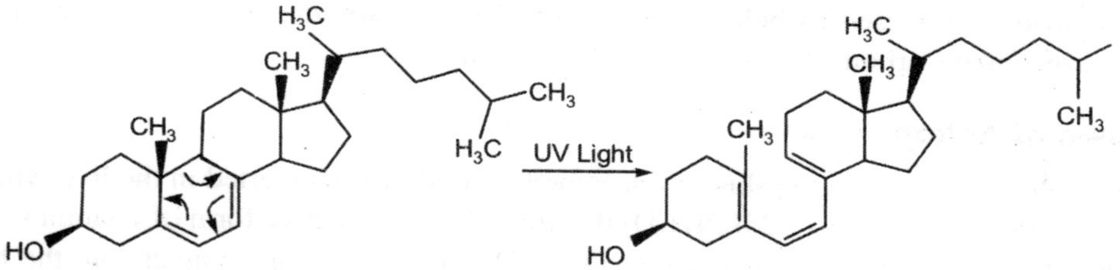

H₃C CH₃ CH₃ CH₃ H₃C HO UV Light H₃C CH₃ CH₃ CH₃ HO

2. Pre-vitamin $D_3$ then spontaneously isomerizes to Vitamin $D_3$ in a antarafacial hydride [1,7]Sigmatropic shift.

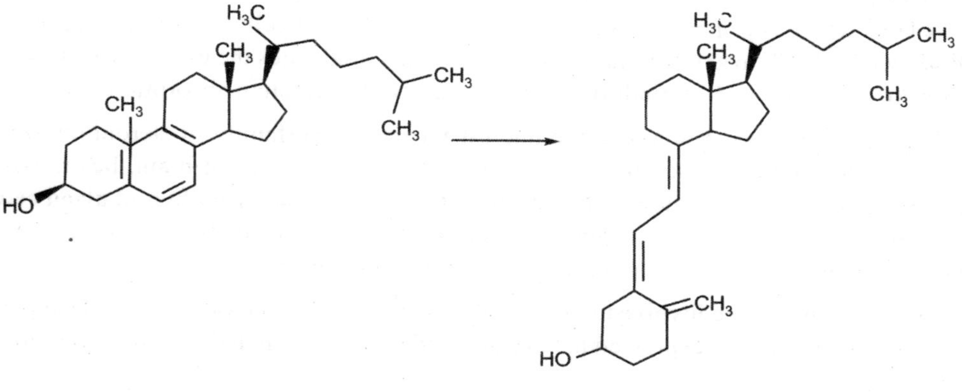

H₃C CH₃ CH₃ CH₃ CH₃ HO H₃C CH₃ CH₃ CH₃ CH₃ HO

**3. Whether it is made in the skin or ingested, vitamin $D_3$ (cholecalciferol) is then hydroxylated in the liver to 25-hydroxycholecalciferol (25(OH)$D_3$ or calcidiol) by the enzyme 25-hydroxylase produced by hepatocytes, and stored until it is needed.**

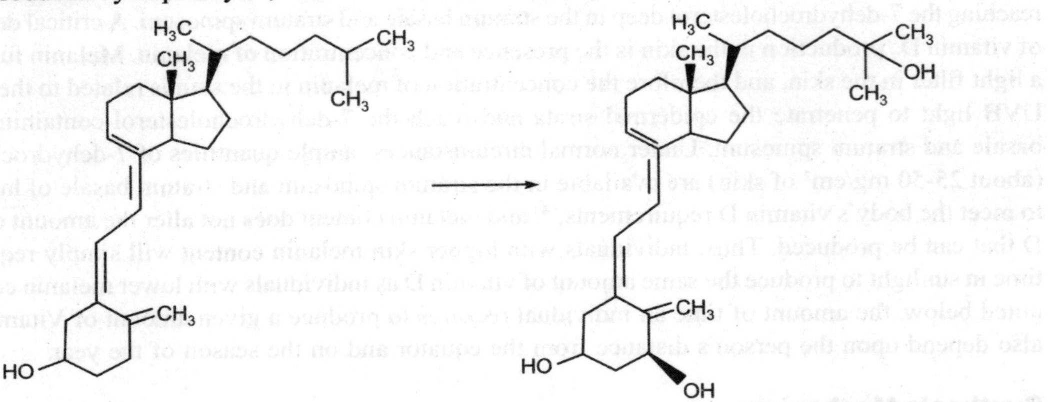

**25-hydroxycholecalciferol is further hydroxylated in the kidneys by the enzyme 1α-hydroxylase, into two dihydroxylated metabolites, the main biologically active hormone 1,25-dihydroxycholecalciferol (1,25(OH)2$D_3$ or calcitriol) and 24R,25(OH)2$D_3$. This conversion occurs in a tightly regulated fashion.**

**Calcitriol is represented below right (hydroxylated Carbon 1 is on the lower ring at right, hydroxylated Carbon 25 is at the upper right end).**

## Mechanism of Action

**Once vitamin D is produced in the skin or consumed in food, it is converted in the liver and kidney to form 1,25 dihydroxyvitamin D, (1,25$(OH)_2$D) the physiologically active form of vitamin D (when "D" is used without a subscript it refers to either $D_2$ or $D_3$). Following this conversion, the hormonally active form of vitamin D is released into the circulation, and by binding to a carrier protein in the plasma, vitamin D binding protein (VDBP), it is transported to various target organs.[3]**

**The hormonally active form of vitamin D mediates its biological effects by binding to the vitamin D receptor (VDR), which is principally located in the nuclei of target cells.[3] The binding of calcitriol to the VDR allows the VDR to act as a transcription factor that modulates the gene expression of transport proteins (such as TRPV6 and calbindin), which are involved in calcium absorption in the intestine.**

**The Vitamin D receptor belongs to the nuclear receptor superfamily of steroid/thyroid hormone receptors, and VDR are expressed by cells in most organs, including the brain, heart, skin, gonads, prostate, and breast. VDR activation in the intestine, bone, kidney, and parathyroid gland cells leads to the maintenance of calcium and phosphorus levels in the blood (with the assistance of parathyroid hormone and calcitonin) and to the maintenance of bone content.[12]**

**The VDR is known to be involved in cell proliferation, differentiation. Vitamin D also affects the immune system, and VDR are expressed in several white blood cells including monocytes and activated T and B cells.**

## Nutrition

Few foods are naturally rich in vitamin D, and most vitamin D intake is in the form of fortified products including milk, soy milk and cereal grains.

A blood calcidiol (25-hydroxy-vitamin D) level is the accepted way to determine vitamin D nutritional status. The optimal level of serum 25-hydroxyvitamin D remains a point for debate among medical scientists. The U.S. Dietary Reference Intake for adequate intake (AI) of vitamin D for infants, children and men and women aged 19–50 is 5 micrograms/day (200 IU/day). Adequate intake increases to 10 micrograms/day (400 IU/day) for men and women aged 51–70 and up to 15 micrograms/day (600 IU/day) past the age of 70. These dose rates will be too low during winter months above 30° latitude. In the absence of sun exposure, 1000 IU of cholecalciferol is required daily for children. 2000-4000 IU of vitamin D may be required for adults absent summer UVB. Milk and cereal grains are often fortified with vitamin D.In light of its apparent health benefits, The Canadian Cancer Society recommends that non-white adults take 1,000 IU daily year-round and whites take that amount in fall and winter. The Canadian Pediatric Society recommends 2,000 IU daily for pregnant and breastfeeding women.

## In Food

Season, geographic latitude, time of day, cloud cover, smog, and sunscreen affect UV ray exposure and vitamin D synthesis in the skin, and it is important for individuals with limited sun exposure to include good sources of vitamin D in their diet.

In some countries, foods such as milk, yogurt, margarine, oil spreads, breakfast cereal, pastries, and bread are fortified with vitamin $D_2$ and/or vitamin $D_3$, to minimize the risk of vitamin D deficiency.[15] In the United States and Canada, for example, fortified milk typically provides 100 IU per glass, or one quarter of the estimated adequate intake for adults over the age of 50.

Fatty fish, such as salmon, are natural sources of vitamin D.

Fortified foods represent the major dietary sources of vitamin D, as very few foods naturally contain significant amounts of vitamin D.

Natural sources of vitamin D include:

- Fish liver oils, such as cod liver oil, 1 Tbs. (15 mL) provides 1,360 IU
- Fatty fish species, such as:
  - Herring, 3 oz provides 1383 IU
  - Catfish, 3 oz provides 425 IU
  - Salmon, cooked, 3.5 oz provides 360 IU
  - Mackerel, cooked, 3.5 oz, 345 IU
  - Sardines, canned in oil, drained, 1.75 oz, 250 IU
  - Tuna, canned in oil, 3 oz, 200 IU
  - Eel, cooked, 3.5 oz, 200 IU
- Mushrooms provide over 2700 IU per serving (approx. 3 oz or 1/2 cup) of vitamin $D_2$, if exposed to just 5 minutes of UV light after being harvested; this is one of a few natural sources of vitamin D for vegans.
- One whole egg, 20 IU

## Deficiency

Vitamin D deficiency can result from: inadequate intake coupled with inadequate sunlight exposure, disorders that limit its absorption, conditions that impair conversion of vitamin D into active metabolites, such as liver or kidney disorders, or, rarely, by a number of hereditary disorders. Deficiency results in impaired bone mineralization, and leads to bone softening diseases, rickets in children and osteomalacia in adults, and possibly contributes to osteoporosis.

## Diseases Caused by Deficiency

Calcitriol (1,25-dihydroxycholecalciferol). Active form. Note extra OH groups at upper right and lower left.

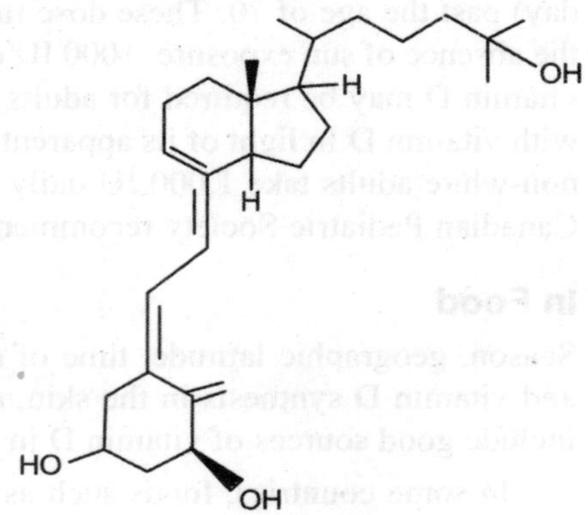

The role of diet in the development of rickets was determined by Edward Mellanby between 1918–1920. In 1921 Elmer McCollum identified an anti-rachitic substance found in certain fats could prevent rickets. Because the newly discovered substance was the fourth vitamin identified, it was called vitamin D.[17] The 1928 Nobel Prize in Chemistry was awarded to Adolf Windaus, who discovered the steroid 7-dehydrocholesterol, the precursor of vitamin D.

Vitamin D deficiency is known to cause several bone diseases including:

- Rickets, a childhood disease characterized by impeded growth, and deformity, of the long bones.
- Osteomalacia, a bone-thinning disorder that occurs exclusively in adults and is characterized by proximal muscle weakness and bone fragility.
- Osteoporosis, a condition characterized by reduced bone mineral density and increased bone fragility.

Prior to the fortification of milk products with vitamin D, rickets was a major public health problem. In the United States, milk has been fortified with 10 micrograms (400 IU) of vitamin D per quart since the 1930s, leading to a dramatic decline in the number of rickets cases

Vitamin D malnutrition may also be linked to an increased susceptibility to several chronic diseases such as high blood pressure, tuberculosis, cancer, periodontal disease, multiple sclerosis, chronic pain, depression, schizophrenia, seasonal affective disorder, and several autoimmune diseases including type 1 diabetes (see role in immunomodulation).

## Groups at Greater Risk of Deficiency

Vitamin D requirements increase with age, while the ability of skin to convert 7-dehydrocholesterol to pre-vitamin $D_3$ decreases. In addition the ability of the kidneys to convert calcidiol to its active form also decreases with age, prompting the need for increased vitamin D supplementation in elderly individuals. One consensus concluded that for optimal prevention of osteoporotic fracture the blood calcidiol concentration should be higher than 30 ng/mL, which is equal to 75 nmol/L.[21] One billion people in the world are currently Vitamin D deficient, if 75 nmol/L is used as cutoff value for insufficiency.

The American Pediatric Associations advises vitamin D supplementation of 200 IU/day (5ìg/d) from birth onwards. (1 IU Vitamin D is the biological equivalent of 0.025 μg cholecalciferol/ergocalciferol). The Canadian Paediatric Society recommends that pregnant or breastfeeding women consider taking 2000 IU/day, that all babies who are exclusively breastfed receive a supplement of 400 IU/day, and that babies living above 55 degrees latitude get 800 IU/day from October to April.[24] Health Canada recommends 400IU/day (10ìg/d). While infant formula is generally fortified with vitamin D, breast milk does not contain significant levels of vitamin D, and parents are usually advised to avoid exposing babies to prolonged sunlight. Therefore, infants who are exclusively breastfed are likely to require vitamin D supplementation beyond early infancy, especially at northern latitudes. Liquid "drops" of vitamin D, as a single nutrient or combined with other vitamins, are available in water based or oil-based preparations ("Baby Drops" in North America, or "Vigantol oil" in Europe). However, babies may be safely exposed to sunlight for short periods; as little as 10 minutes a day without a hat can suffice, depending on location and season. The vitamin D found in supplements and infant formula is less easily absorbed than that produced by the body naturally and carries a risk of overdose that is not present with natural exposure to sunlight.

Obese individuals may have lower levels of the circulating form of vitamin D, probably because of reduced bioavailability, and are at higher risk of deficiency. To maintain blood levels of calcium, therapeutic vitamin D doses are sometimes administered (up to 100,000 IU or 2.5 mg daily) to patients who have had their parathyroid glands removed (most commonly renal dialysis patients who have had tertiary hyperparathyroidism, but also to patients with primary hyperparathyroidism) or with hypoparathyroidism.[26] Patients with chronic liver disease or intestinal malabsorption disorders may also require larger doses of vitamin D (up to 40,000 IU or 1 mg (1000 micrograms) daily).

The use of sunscreen with a sun protection factor (SPF) of 8 inhibits more than 95% of vitamin D production in the skin.[12][27] Recent studies showed that, following the successful "Slip-Slop-Slap" health campaign encouraging Australians to cover up when exposed to sunlight to prevent skin cancer, an increased number of Australians and New Zealanders became vitamin D deficient. Ironically, there are indications that vitamin D deficiency may lead to skin cancer. To avoid vitamin D deficiency dermatologists recommend supplementation along with sunscreen use.

The reduced pigmentation of light-skinned individuals tends to allow more sunlight to be absorbed even at higher latitudes, thereby reducing the risk of vitamin D deficiency. However, at higher latitudes (above 30°) during the winter months, the decreased angle of the sun's rays, reduced daylight hours, protective clothing during cold weather, and fewer hours of outside activity, diminish absorption of sunlight and the production of vitamin D. Because melanin acts like a sun-block, prolonging the time required to generate vitamin D, dark-skinned individuals, in particular, may require extra vitamin D to avoid deficiency at higher latitudes. In June 2007, The Canadian Cancer Society began recommending that all adult Canadians consider taking 1000 IU of vitamin D during the fall and winter months (when typically the country's northern latitude prevents sufficient sun-stimulated production of vitamin D). At latitudes below 30° where sunlight and day-length are more consistent, vitamin D supplementation may not be required. Individuals clad in full body coverings during all their outdoor activity, most notably women wearing burquas in daylight, are at risk of vitamin D deficiency. This poses a lifestyle-related health risk mostly for female residents of conservative Muslim nations in the Middle East, but also for strict adherents in other parts of the world.

## Overdose

Vitamin D stored in the human body as calcidiol (25-hydroxy-vitamin D) has a large volume of distribution and a long half-life (about 20 to 29 days). However, the synthesis of bioactive vitamin D hormone is tightly regulated and vitamin D toxicity usually occurs only if excessive doses (prescription forms or rodenticide analogs) are taken. Although normal food and pill vitamin D concentration levels are too low to be toxic in adults, because of the high vitamin A content in codliver oil, it is possible to reach toxic levels of **vitamin A** (but not **vitamin D**) via this route, if taken in multiples of the normal dose in an attempt to increase the intake of vitamin D. Most historical cases of vitamin D overdose have occurred due to manufacturing and industrial accidents.

Exposure to sunlight for extended periods of time does not cause vitamin D toxicity. This is because within about 20 minutes of ultraviolet exposure in light skinned individuals (3–6 times longer for pigmented skin) the concentration of vitamin D precursors produced in the skin reach an equilibrium, and any further vitamin D that is produced is degraded. Maximum endogenous production with full body exposure to sunlight is 250 μg (10,000 IU) per day.

The exact long-term safe dose of vitamin D is not entirely known, but dosages up to 60 micrograms (2,400 IU) /day in healthy adults are believed to be safe., and all known cases of vitamin D toxicity with hypercalcemia have involved intake of or over 1,000 micrograms (40,000 IU)/day. The U.S. Dietary Reference Intake Tolerable Upper Intake Level (UL) of vitamin D for children and adults is 50 micrograms/day (2,000 IU/day). In adults, sustained intake of 2500 micrograms/day (100,000 IU) can produce toxicity within a few months. For infants (birth to 12 months) the tolerable UL is set at 25 micrograms/day (1000 IU/day), and vitamin D concentrations of 1000 micrograms/day (40,000 IU) in infants has been shown to produce toxicity within 1 to 4 months. In the United States, overdose exposure of vitamin D was reported by 284 individuals in 2004, leading to 1 death. The *Nutrition Desk Reference* states "The threshold for toxicity is 500 to 600 micrograms [vitamin D] per kilogram body weight per day." The US EPA published an oral LD50 of 619 mg/kg for female rats.

Serum levels of calcidiol (25-hydroxy-vitamin D) are typically used to diagnose vitamin D overdose. In healthy individuals, calcidiol levels are normally between 25 to 40 ng/mL (60 to 100 nmol/L), but these levels may be as much as 15-fold greater in cases of vitamin D toxicity. Serum levels of bioactive vitamin D hormone (1,25(OH2)D) are usually normal in cases of vitamin D overdose.

Some symptoms of vitamin D toxicity are a result of hypercalcemia (an elevated level of calcium in the blood) caused by increased intestinal calcium absorption. Vitamin D toxicity is known to be a cause of high blood pressure. Gastrointestinal symptoms of vitamin D toxicity can include anorexia, nausea, and vomiting. These symptoms are often followed by polyuria (excessive production of urine), polydipsia (increased thirst), weakness, nervousness, pruritus (itch), and eventually renal failure. Other signals of kidney disease including elevated protein levels in the urine, urinary casts, and a build up of wastes in the blood stream can also develop. In one study, hypercalciuria and bone loss occurred in four patients with documented vitamin D toxicity. Another study showed elevated risk of ischaemic heart disease when 25D was above 89 ng/mL.

Vitamin D toxicity is treated by discontinuing vitamin D supplementation, and restricting calcium intake. If the toxicity is severe blood calcium levels can be further reduced with corticosteroids or bisphosphonates. In some cases kidney damage may be irreversible.

Seniors who consume high levels of calcium and vitamin D are much more likely to have larger brain lesions that can lead to cognitive impairment, depression or stroke.

## Role in Immunomodulation

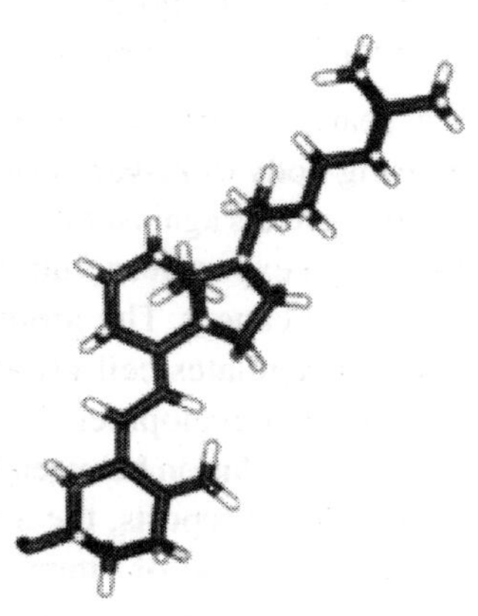

Cholecalciferol ($D_3$)

The hormonally active form of vitamin D mediates immunological effects by binding to nuclear vitamin D receptors (VDR) which are present in most immune cell types including both innate and adaptive immune cells. The VDR is expressed constitutively in monocytes and in activated macrophages, dendritic cells, NK cells, T and B cells. In line with this observation, activation of the VDR has potent anti-proliferative, pro-differentiative, and immunomodulatory functions including both immune-enhancing and immunosuppressive effects.

Effects of VDR-ligands, such as vitamin D hormone, on T-cells include suppression of T cell activation and induction of regulatory T cells, as well as effects on cytokine secretion patterns. VDR-ligands have also been shown to affect maturation, differentiation, and migration of dendritic cells, and inhibits DC-dependent T cell activation, resulting in an overall state of immunosuppression.

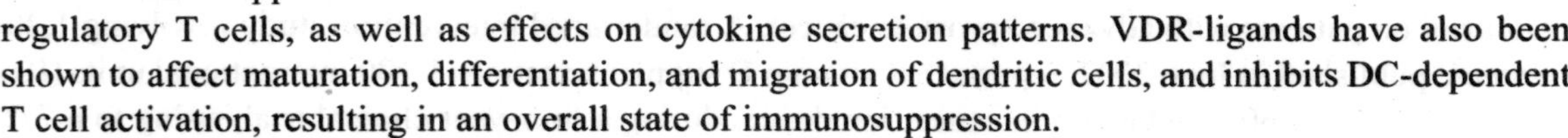

VDR ligands have also been shown to increase the activity of natural killer cells, and enhance the phagocytic activity of macrophages. Active vitamin D hormone also increases the production of cathelicidin, an antimicrobial peptide that is produced in macrophages triggered by bacteria, viruses, and fungi. Vitamin D deficiency tends to increase the risk of infections, such as influenza and tuberculosis. In a 1997 study, Ethiopian children with rickets were 13 times more likely to get pneumonia than children without rickets.

These immunoregulatory properties indicate that ligands with the potential to activate the VDR, including supplementation with calcitriol (as well as a number of synthetic modulators), may have therapeutic clinical applications in the treatment of; inflammatory diseases (rheumatoid arthritis, psoriatic arthritis), dermatological conditions (psoriasis, actinic keratosis), osteoporosis, cancers (prostate, colon, breast, myelodysplasia, leukemia, head and neck squamous cell carcinoma, and basal cell carcinoma), and autoimmune diseases (systemic lupus erythematosus, type I diabetes, multiple sclerosis) and in preventing organ transplant rejection. However the effects of supplementation with vitamin D, as yet, remain unclear, and supplementation may be inadvisable for individuals with sarcoidosis and other diseases involving vitamin D hypersensitivity.

A 2006 study published in the Journal of the American Medical Association, reported evidence of a link between Vitamin D deficiency and the onset of Multiple Sclerosis; the authors posit that this is due to the immune-response suppression properties of Vitamin D.

## Role in Cancer Prevention and Recovery

The vitamin D hormone, calcitriol, has been found to induce death of cancer cells *in vitro* and *in vivo*. Although the anti-cancer activity of vitamin D is not fully understood, it is thought that these effects are mediated through vitamin D receptors expressed in cancer cells, and may be related to its immunomodulatory abilities. The anti-cancer activity of vitamin D observed in the laboratory has prompted some to propose that vitamin D supplementation might be beneficial in the treatment or prevention of some types of cancer.

A search of primary and review medical literature published between 1970 and 2007 found an increasing body of research supporting the hypothesis that the active form of vitamin D has significant, protective effects against the development of cancer. Epidemiological studies show an inverse association between sun exposure, serum levels of 25(OH)D, and intakes of vitamin D and risk of developing and/or surviving cancer. The protective effects of vitamin D result from its role as a nuclear transcription factor that regulates cell growth, differentiation, apoptosis and a wide range of cellular mechanisms central to the development of cancer. In 2005, scientists released a metastudy which demonstrated a beneficial correlation between vitamin D intake and prevention of cancer. Drawing from a meta-analysis of 63 published reports, the authors showed that intake of an additional 1,000 international units (IU) (or 25 micrograms) of vitamin D daily reduced an individual's colon cancer risk by 50%, and breast and ovarian cancer risks by 30%.] Research has also shown a beneficial effect of high levels of calcitriol on patients with advanced prostate cancer. A randomized intervention study involving 1,200 women, published in June 2007, reports that vitamin D supplementation (1,100 international units (IU)/day) resulted in a 60% reduction in cancer incidence, during a four-year clinical trial, rising to a 77% reduction for cancers diagnosed *after* the first year (and therefore excluding those cancers more likely to have originated prior to the vitamin D intervention). In 2006, a study at Northwestern University found that taking the U.S. RDA of vitamin D (400 IU per day) cut the risk of pancreatic cancer by 43% in a sample of more than 120,000 people from two long-term health surveys.

A 2006 study using data on over 4 million cancer patients from 13 different countries showed a marked difference in cancer risk between countries classified as sunny and countries classified as less–sunny for a number of different cancers. Research has also suggested that cancer patients who have surgery or treatment in the summer — and therefore make more endogenous vitamin D — have a better chance of surviving their cancer than those who undergo treatment in the winter when they are exposed to less sunlight

However, a large scientific review undertaken by the National Cancer Institute found no link between baseline vitamin D status and overall cancer mortality. They did find that vitamin D was beneficial in preventing colorectal cancer, which showed an inverse relationship with blood levels "80 nmol/L or higher associated with a 72% risk reduction".

### Role in Coronary Disease Prevention

Research indicates that vitamin D may play a role in preventing or reversing coronary disease. As with cancer incidence, a qualitative inverse correlations was found between coronary disease incidence and serum vitamin D levels of 32.0 versus 35.5 ng/mL. Cholesterol levels were found reduced in gardeners in the UK in the summer months. Heart attacks peak in winter and decline in summer in temperate but not tropical latitudes.

The issue of vitamin D in heart health may not yet be finally settled. Exercise may account for some of the benefit attributed to vitamin D, since Vitamin D levels are higher in physically active persons.[70] Moreover, there may be an upper limit after which cardiac benefits decline. One study found an elevated risk of ischaemic heart disease in Southern India in individuals whose vitamin D levels were above 89 ng/mL. Vitamin D intakes were 50% higher in men who qualified for a disability pension due to heart problems than in those not on disability, with the increase seen to occur at an intake of 1,200 i.u. daily. Earlier studies also found reason for concern. Infants with high blood calcium syndrome, related to high levels of vitamin D often have high cholesterol levels. Studies in pigs suggest that too much vitamin D might weaken arteries

The early studies may be flawed. Vitamin D is reported to cause falsely high blood cholesterol readings taken using the Zlatkis-Zak reaction, a little-used method. If this method was used, vitamin D would have raised the readings but not the underlying levels. Additionally, vitamin D levels soar in Caucasians working outdoors, as farmers in the Netherlands might be expected to. Similarly, South Indian patients may routinely experience exposure to abundant sun. If so, the dietary contributions to vitamin D status may have been far less than the amount contributed by sun exposure, and the supplemented and unsupplemented groups' vitamin D status would not have been as different as assumed. Almost all the farmers and South Indians would have had abundant vitamin D levels at the end of the summer, and this contrasts with urban dwellers today, many of whom are vitamin D deficient.PMID 9504937These sun-living groups results would not generalize to sun-deprived urban dwellers. Among a group with heavy sun exposure, taking supplemental vitamin D may have resulted in blood levels over the ideal range, while urban dwellers not taking supplemental vitamin D may fall under the levels recognized as ideal, and being above or below the preferable levels may cause adverse affects on the health of each group.

## VITAMIN E

### History

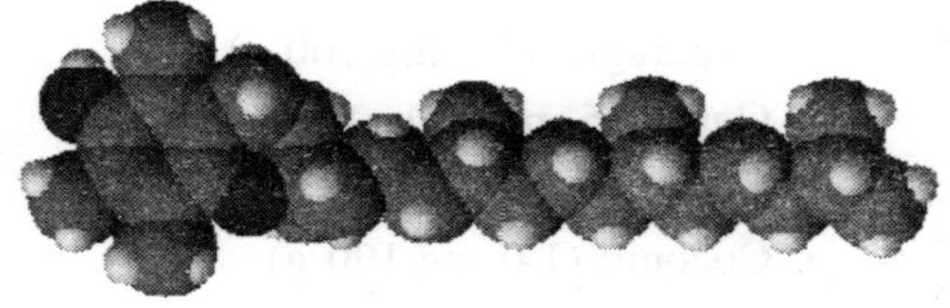

During feeding experiments with rats Herbert McLean Evans concluded in 1922 that besides vitamins B and C, an unknown vitamin existed. Although every other nutrition was present, the rats were not fertile. This condition could be changed by additional feeding with wheat germ. It took several years until 1936 when the substance was isolated from wheat germ and the formula $C_{29}H_{50}O_2$ was determined. Evans also found that the compound reacted like an alcohol and concluded that one of the oxygen atoms was part of an OH (hydroxyl) group. As noted in the introduction, the vitamin was given its name by Evans from Greek words meaning "to bear young" with the addition of the -ol as an alcohol. The structure was determined shortly thereafter in 1938. **Tocopherol**, a class of chemical compounds of which many have vitamin E activity, describes a series of organic compounds consisting of various methylated phenols. Because the vitamin activity was first identified in 1936 from a dietary fertility factor in rats, it was given the name "tocopherol" from the Greek words "τοκηετοσ" [child], and "φερονaλ", [to bear or carry] meaning in sum "to carry a pregnancy," with the ending "-ol" signifying its status as a chemical alcohol.

Tocotrienols, which are related compounds, may also have vitamin E activity. All of these various derivatives with vitamin activity, may correctly be referred to as "vitamin E." Tocopherols and tocotrienols are fat-soluble antioxidants.

The compound **á-tocopherol**, a common form of tocopherol added to food products, is denoted by the E number **E307**.

## Sources

In foods, the most abundant sources of vitamin E are vegetable oils such as palm oil, sunflower, corn, soybean, and olive oil. Nuts, sunflower seeds, seabuckthorn berries, kiwi fruit, and wheat germ are also good sources. Other sources of vitamin E are whole grains, fish, peanut butter, goats milk, and green leafy vegetables. Fortified breakfast cereals are also an important source of vitamin E in the United States. Although originally extracted from wheat germ oil, most natural vitamin E supplements are now derived from vegetable oils, usually soybean oil.

The content of Vitamin E for rich sources follows:[8]

- Wheat germ oil (215.4 mg/100 g)
- Sunflower oil (55.8 mg/100 g)
- Hazelnut (26.0 mg/100 g)
- Walnut oil (20.0 mg/100 g)
- Peanut oil (17.2 mg/100 g)
- Olive oil (12.0 mg/100 g)
- Peanut (9.0 mg/100 g)
- Pollard (2.4 mg/100 g)
- Corn (2.0 mg/100 g)
- Asparagus (1.5 mg/100 g)
- Oats (1.5 mg/100 g)
- Chestnut (1.2 mg/100 g)
- Coconut (1.0 mg/100 g)
- Tomatoes (0.9 mg/100 g)
- Carrots (0.6 mg/100 g)
- Goat's milk (0.1 mg/100ml)

## Forms

Natural vitamin E exists in eight different forms, four tocopherols and four tocotrienols. All feature a chromanol ring, with a hydroxyl group that can donate a hydrogen atom to reduce free radicals and a hydrophobic side chain which allows for penetration into biological membranes. Both the tocopherols and tocotrienols occur in alpha, beta, gamma and delta forms, determined by the number of methyl groups on the chromanol ring. Each form has slightly different biological activity.[2]

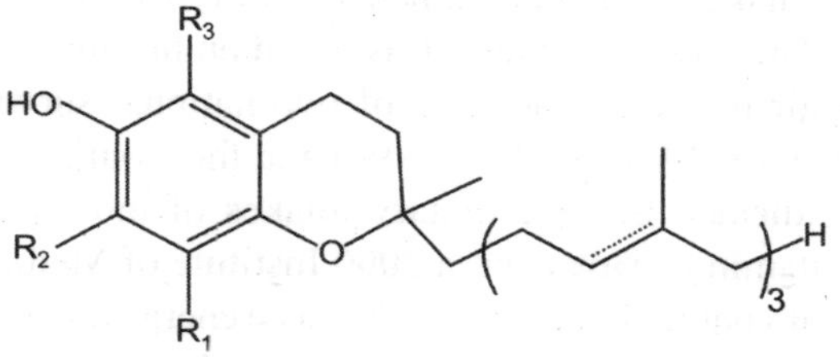

α-tocopherol, $R_1 = R_2 = R_3 = CH_3$
α-tocopherol, $R_1 = R_2 = R_3 = CH_3$

β-tocopherol, $R_1 = R_3 = CH_3$; $R_2 = H$
β-tocopherol, $R_1 = R_3 = CH_3$; $R_2 = H$

γ-tocopherol, $R_1 = R_2 = CH_3$ $R_3 = H$
γ-tocopherol, $R_1 = R_2 = CH_3$ $R_3 = H$

δ-tocopherol, $R_1 = R_2 = R_3 = H$
δ-tocopherol, $R_1 = R_2 = R_3 = H$

As a food additive, tocopherol is labelled with these E numbers: **E307** (α-tocopherol), **E308** (α-tocopherol), and **E309** (δ-tocopherol). Vitamin E capsules are sometimes used as visible markers in magnetic resonance imaging

## Alpha-Tocopherol

Alpha-tocopherol is traditionally considered the most active biological antioxidant in humans. The measurement of "vitamin E" activity in international units (IU) was based on fertility enhancement by the prevention of spontaneous abortions in pregnant rats relative to alpha tocopherol. It increases naturally to about 150% of normal in the maternal circulation during human pregnancies.

1 IU of vitamin E is defined as the biological equivalent of 0.667 milligrams of *RRR*-alpha-tocopherol (formerly named d-alpha-tocopherol or sometimes ddd-alpha-tocopherol). 1 IU is also defined as 1 milligram of the 8-isomer mix *all-rac*-alpha-tocopheryl acetate (commercially now "called" dl-alpha-tocopheryl acetate, even though it is more precisely dl,dl,dl-alpha-tocopherol acetate).

The original dl,l,l semi-synthetic molecular mix, originally called dl-alpha-tocopherol acetate also, but actually more precisely known as 2-ambo-alpha-tocopherol, is no longer manufactured.

## Other R, R, R Tocopherol

The other R, R, R tocopherol vitamins are slowly being recognized as research begins to elucidate their additional roles in the human body. Many naturopathic and orthomolecular medicine advocates suggest that vitamin E supplements contain at least 20% by weight of the other natural vitamin E isomers.

### Tocotrienols

Tocotrienols, with four d- isomers, also belong to the vitamin E family. The four tocotrienols have structures corresponding to the four tocopherols, except with an unsaturated bond in each of the three isoprene units that form the hydrocarbon tail. Tocopherols have a saturated phytyl tail.

### Recommended Amounts

The U.S. Dietary Reference Intake (DRI) Recommended Daily Amount (RDA) for a 25-year old male for Vitamin E is 15 mg/day. The DRI for vitamin E is based on the alpha-tocopherol form because it is the most active form as originally tested. Results of two national surveys, the National Health and Nutrition Examination Survey (NHANES III 1988-91) and the Continuing Survey of Food Intakes of Individuals (1994 CSFII) indicated that the dietary intakes of most Americans do not provide the recommended amounts of vitamin E. However, a 2000 Institute of Medicine (IOM) report on vitamin E states that intake estimates of vitamin E may be low because energy and fat intake is often underreported in national surveys and because the kind and amount of fat added during cooking is often not known. The IOM states that most North American adults get enough vitamin E from their normal diets to meet current recommendations. However, they do caution individuals who consume low fat diets because vegetable oils are such a good dietary source of vitamin E. "Low-fat diets can substantially decrease vitamin E intakes if food choices are not carefully made to enhance alpha-tocopherol intakes". Vitamin E supplements are absorbed best when taken with meals.

Because vitamin E can act as an anticoagulant and may increase the risk of bleeding problems, many agencies have set an upper tolerable intake level (UL) for vitamin E at 1,000 mg (1,500 IU) per day.

### Deficiency

There are three specific situations when a vitamin E deficiency is likely to occur. It is seen in persons who cannot absorb dietary fat, has been found in premature, very low birth weight infants (birth weights less than 1500 grams, or 3.5 pounds), and is seen in individuals with rare disorders of fat metabolism. A vitamin E deficiency is usually characterized by neurological problems due to poor nerve conduction.

Individuals who cannot absorb fat may require a vitamin E supplement because some dietary fat is needed for the absorption of vitamin E from the gastrointestinal tract. Anyone diagnosed with cystic fibrosis, individuals who have had part or all of their stomach removed, and individuals with malabsorptive problems such as Crohn's disease, liver disease or pancreatic insufficiency may not absorb fat and should discuss the need for supplemental vitamin E with their physician (3). People who cannot absorb fat often pass greasy stools or have chronic diarrhea and bloating.

Very low birth weight infants may be deficient in vitamin E. A neonatologist, a pediatrician specializing in the care of newborns, typically evaluates the nutritional needs of premature infants.

Abetalipoproteinemia is a rare inherited disorder of fat metabolism that results in poor absorption of dietary fat and vitamin E. The vitamin E deficiency associated with this disease causes problems such as poor transmission of nerve impulses, muscle weakness, and degeneration of the retina that can cause blindness. Individuals with abetalipoproteinemia may be prescribed special vitamin E supplements by a physician to treat this disorder. In addition, there is a rare genetic condition termed isolated vitamin E deficiency or ataxia with isolated with vitamin E deficiency, caused by mutations in the tocopherol transfer protein gene. These individuals have an extremely poor capacity to absorb vitamin E and develop neurological complications that are reversible by supplementation with high doses of vitamin E.

Also, in adults, erythrocyte membrane fragility results as the erythrocytes are oxidized.

## Supplements

Commercial vitamin E supplements can be classified into several distinct categories:

- Fully synthetic vitamin E, "dl-alpha-tocopherol", the most inexpensive, most commonly sold supplement form usually as the acetate ester;
- Semi-synthetic "natural source" vitamin E esters, the "natural source" forms used in tablets and multiple vitamins. These are highly fractionated d-alpha tocopherol or its esters, often made by synthetic methylation of gamma and beta d,d,d tocopherol vitamers extracted from plant oils.
- Less fractionated "natural mixed tocopherols" and high d-gamma-tocopherol fraction supplements

## Synthetic all-racemic

Synthetic vitamin E derived from petroleum products is manufactured as all-racemic alpha tocopheryl acetate with three chiral centers. This mixture has only one alpha tocopherol molecule (moiety) in 8 molecules as actual R, R,R-alpha tocopherol.

The 8-isomer *all-rac* vitamin E is always marked on labels simply as **dl-tocopherol** or **dl-tocopheryl acetate**, even though it is (if fully written out) actually dl,dl,dl-tocopherol. The present largest manufacturers of this type are DSM and BASF.

(An earlier semisynthetic vitamin E actually contained 50% d,d,d-alpha tocopherol moiety and 50% l,d,d-alpha-tocopherol moiety, as synthesized by an earlier process which started with a plant sterol intermediate with the correct chirality in the tail, and thus resulted in a racemic mixture at only one chiral center. This form, known as 2-ambo tocopherol, is no longer made.)

Natural alpha-tocopherol is the RRR-alpha (or ddd-alpha) form. The synthetic dl,dl,dl-alpha ("dl-alpha") form is not as active as the natural ddd-alpha ("d-alpha") tocopherol form. This is mainly due to nearly complete vitamin inactivity of the 4 possible chiral alpha epimers which are represented by the *l* or *S* enantiomer at the first chiral center (an S or l configuration between the chromanol ring and the tail, i.e., the SRR, SRS, SSR, and SSS epimers). Unnatural 2R chiral forms with natural R configuration at this key center, but S at the other centers in the tail (RSR, RRS), appear to retain substantial RRR vitamin activity because they are recognized by the alpha tocopherol transport protein, and thus maintained in the plasma, where the other four epimers are not. Thus, the synthetic all-rac-á-tocopherol probably has only about half the vitamin activity of RRR-á-tocopherol in humans, even though the ratio of activities of the 8 isomer sythetic mixture to the natural vitamin is 1 to 1.36 in the rat pregnancy model. [9]

Although it is clear that synthetic form is not as active as the natural alpha tocopherol form, in the ratios discussed above, specific information on any side effects of the seven synthetic vitamin E epimers is not readily available. Naturopathic and orthomolecular medicine advocates have held that none of the synthetic vitamin E forms have merit for cancer, circulatory and heart diseases, but hold this opinion without being able to point to definitive studies of the matter.

## Natural and Semi-synthetic

Semisynthetic "natural source" vitamin E, manufacturers convert the common natural beta, gamma and delta tocopherol isomers into the alpha form by adding methyl groups to yield d-alpha tocopherol. The largest manufacturers of this form are presently Cognis,Cargill and Archer Daniels Midland. Manufacturers also commonly convert the alcohol form of the vitamins to esters using acetic or succinic acid. These tocopheryl esters are more stable and are easy to use in tablets and multiple vitamin pills.

Because only alpha tocopherols were officially counted as "vitamin E" in supplements, refiners and manufacturers faced enormous economic pressure to esterify and methylate the other natural tocopherol isomers, d-beta-, d-gamma- and d-delta-tocopherol into d-alpha tocopheryl acetate or succinate. However, these alpha tocopheryl esters have been shown to be variably and less efficiently absorbed in humans than in the original normative tests using rats.[10] In the healthy human body, the semisynthetic forms are easily de-esterified over several days, primarily in the liver, but not for common problems in premature babies, aged or ill patients.

Tocopheryl nicotinate and tocopheryl linolate esters are used in cosmetics and some pharmaceuticals.

## Mixed Tocopherols

"Mixed tocopherols" in the US contain at least 20% w/w other natural R, R,R- tocopherols, i.e. R, R,R-alpha-tocopherol content plus at least 25% R, R,R-beta-, R, R,R-gamma-, R, R,R-delta-tocopherols.

Some premium brands may contain 200% w/w or more of the other tocopherols and measurable tocotrienols. Some mixed tocopherols with higher gamma-tocopherol content are marketed as "High Gamma-Tocopherol". The label should report each component in milligrams, except R, R,R-alpha-tocopherol may still be reported in IU. Mixed tocopherols can also be found in various nutritional supplements manufactured by high end supplement companies.

## Other Uses

Conventional medical studies on vitamin E, as of 2006 and as below, use either a synthetic all-racemic ("d, l-") alpha tocopheryl ester (acetate or succinate) or a semi-synthetic d-alpha tocopheryl ester (acetate or succinate). Proponents of megavitamin, orthomolecular and naturally based therapies have advocated, for the last two thirds of a century, and have used the *natural tocopherols*, often mixed tocopher*ols* with an additional 25% - 200% w/w d-beta-, d-gamma-,[11][12] and d-delta-tocopherol. Based on various clinical, experimental, patent, and individual data, natural health proponents have long held that the other poorly studied tocopherols, especially the abundant d-gamma-tocopherol, in combination with other antioxidants such as selenium, coQ10, vitamin C, vitamin $K_2$, mixed carotenoids, and lipoic acid, provide unique biochemical benefits. The methodology, interpretation and reporting of conventional vitamin E studies have even become contentious within conventional medicine circles.

## Controversy

"Megadoses" of Vitamin E are not recommended by many government agencies, due to a possible increased risk of bleeding. Two meta-analyses have concluded that synthetic and semisynthetic vitamin E supplements (alpha tocopheryl esters) increase mortality, although these meta-analyses have been repeatedly challenged in the nutrition literature for cofounders and selection bias.

A 2005 meta-analysis by Miller found that high-dosage vitamin E supplements may increase all-cause mortality. "High dose" vitamin E esters (>400 units/day) were also associated with an increased risk in all-cause mortality of 39 per 10,000 persons, and a statistically significant relation existed between dose and mortality, with increased risk at doses exceeding 150 units per day. These trials included synthetic beta-carotene and other cofounders.

The Miller study was rebutted by Houston in the Journal of the American Nutraceutical Association.[16] Furthermore, Rosenberg concluded that "toxicity symptoms have not been reported even at intakes of 800 IU per kilogram of body weight daily for 5 months" according to the Food and Nutrition Board (Rosenberg, et all), an amount that corresponds to 60,000 IU per day for a 75 kg adult.

A review of a number of randomized controlled trials in the scientific literature by the Cochrane Collaboration published in JAMA in 2007 also found an increase in mortality, of 4% (Relative Risk 1.04, 95% confidence interval 1.01-1.07), or 400 per 10,000 persons.

## As a Preservative

Vitamin E is widely used in industry as an inexpensive preservative (namely for cosmetics and foods). Vitamin E containing products are commonly used in the belief that vitamin E is good for the skin; many cosmetics include it, often labelled as tocopherol acetate, tocopheryl linoleate or tocopheryl nicotinate. Individuals can still experience allergic reactions to some tocopheryl esters or develop a rash and hives that may spread over the entire body from the use of topical products with alpha tocopheryl esters.

## Reduce Scarring

Topical use of Vitamin E is often claimed by manufacturers of skin creams and lotions to play a role in encouraging skin healing and reducing scarring after injuries such as burns on the basis of limited research but the weak evidence of a benefit of silicon gel sheeting with or without added Vitamin E is limited by the poor quality of the research.[22] Indeed one study found that it did not improve or worsened the cosmetic appearance in 90% of patients, with a third developing contact dermatitis.

## During Pregnancy

Recent studies into the use of both vitamin C and the single isomer vitamin E esters as possible help in preventing oxidative stress leading to pre-eclampsia has failed to show significant benefits,[24] but did increase the rate of babies born with a low birthweight in one study. However, earlier work that suggested vitamin K (similar structures to natural E isomers) and C together have 91% benefit in nausea and vomiting remains unaddressed.

## Heart Disease

Preliminary research has led to a widely held belief that vitamin E may help prevent or delay coronary heart disease, but larger controlled studies have not shown any benefit. Many researchers advance the belief that oxidative modification of LDL-cholesterol (sometimes called "bad" cholesterol) promotes blockages in coronary arteries that may lead to atherosclerosis and heart attacks. Vitamin E may help prevent or delay coronary heart disease by limiting the oxidation of LDL-cholesterol. Vitamin E also may help prevent the formation of blood clots, which could lead to a heart attack. Observational studies have associated lower rates of heart disease with higher vitamin E intake. A study of approximately 90,000 nurses suggested that the incidence of heart disease was 30% to 40% lower among nurses with the highest intake of vitamin E from diet and supplements. The range of intakes from both diet and supplements in this group was 21.6 to 1,000 IU (32 to 1,500 mg), with the median intake being 208 IU (139 mg). A 1994 review of 5,133 Finnish men and women aged 30 - 69 years suggested that increased dietary intake of vitamin E was associated with decreased mortality (death) from heart disease.

But even though these observations are promising, randomized clinical trials have consistently shown lack of benefit to the role of vitamin E supplements in heart disease. The Heart Outcomes Prevention Evaluation (HOPE) Study followed almost 10,000 patients for 4.5 years who were at high risk for heart attack or stroke. In this intervention study the subjects who received 265 mg (400) IU of vitamin E daily did not experience significantly fewer cardiovascular events or hospitalizations for heart failure or chest pain when compared to those who received a sugar pill. The researchers suggested that it is unlikely that the vitamin E supplement provided any protection against cardiovascular disease in the HOPE study. This study is continuing, to determine whether a longer duration of intervention with vitamin E supplements will provide any protection against cardiovascular disease.

Furthermore, meta analysis of several trials of antioxidants, including vitamin E, have not shown any benefit to vitamin E supplementation for preventing coronary heart disease. One study suggested that Vitamin E (as alpha-tocopherol only) supplementation may increase the risk for heart failure. Supplementing alpha-tocopherol without gamma-tocopherol is known to lead to reduced serum gamma- and delta-tocopherol concentrations.

Orthomolecular and naturopathic medicine use much different types of vitamin E, the natural mixed tocopherols, and other supportive cofactors such as, selenium, vitamin C, carnitine, lysine, and co-Q10 for various cardiovascular diseases..

On September 10, 2007, the American Heart Association (in its journal Circulation) stated that women taking regular doses of vitamin E or Tocopherol were 21% less likely to suffer a blood clot. Dr. Robert Glynn of Harvard Medical School said (it was an interesting finding but not yet proven and) further research must confirm the link in the prevention of venous thromboembolism, and patients must not stop taking prescribed blood thinners.

## Cancer

Antioxidants such as vitamin E help protect against the damaging effects of free radicals, which may contribute to the development of chronic diseases such as cancer. Vitamin E also may block the formation of nitrosamines, which are carcinogens formed in the stomach from nitrites consumed in the diet. It also may protect against the development of cancers by enhancing immune function. To date, human trials and surveys that have tried to associate vitamin E with incidence of cancer remain generally inconclusive.

Some evidence associates higher intake of vitamin E with a decreased incidence of prostate cancer (see ATBC study) and breast cancer. Some studies correlate additional cofactors, such as specific vitamin E isomers, e.g. gamma-tocopherol, and other nutrients, e.g. selenium, with dramatic risk reductions in prostate cancer. However, an examination of the effect of dietary factors, including vitamin E, on incidence of postmenopausal breast cancer in over 18,000 women from New York State did not associate a greater vitamin E intake with a reduced risk of developing breast cancer. A study of the effect on lung cancer in smokers also showed no benefit.

A study of women in Iowa provided evidence that an increased dietary intake of vitamin E may decrease the risk of colon cancer, especially in women under 65 years of age. On the other hand, vitamin E intake was not statistically associated with risk of colon cancer in almost 2,000 adults with cancer who were compared to controls without cancer. At this time there is limited evidence to recommend vitamin E supplements for the prevention of cancer.

Recent studies have found that increased intake of vitamin E, especially among smokers may be responsible for an increase in the incidence of lung cancer, with one study finding an increase in the incidence of lung cancer by 7% for each 100IU of vitamin E taken daily.

## Cataracts

A cataract is a condition of clouding of the tissue of the lens of the eye. They increase the risk of disability and blindness in aging adults. Antioxidants are being studied to determine whether they can help prevent or delay cataract growth. Observational studies have found that lens clarity, which is used to diagnose cataracts, was better in regular users of vitamin E supplements and in persons with higher blood levels of vitamin E. A study of middle aged male smokers, however, did not demonstrate any effect from vitamin E supplements on the incidence of cataract formation. The effects of smoking, a major risk factor for developing cataracts, may have overridden any potential benefit from the vitamin E, but the conflicting results also indicate a need for further studies before researchers can confidently recommend extra vitamin E for the prevention of cataracts.

It is important to note that the term “cataract” may be used in common parlance for an opacity involving any tissue of the eye, for example a corneal scar. Thus a character in theater or on television who is blind from cataracts might have white instead of clear corneas, covering over the iris and pupil. Since the lens is behind the pupil, real cataracts are difficult to see without special instrumentation, so people with cataracts have rather normally appearing eyes.

### Age-related Macular Degeneration (AMD)

Age-related macular degeneration (AMD) is the leading cause of visual impairment and blindness in the United States and the developed world among people 65 years and older. It has been shown that vitamin E alone does not attenuate the development or progression of AMD.

However, studies focusing on efficacy of Vitamin E combined with other antioxidants, like zinc and vitamin C, indicate a protective effect against the onset and progression of AMD

### Glaucoma

A 2007 study published in the *European Journal of Ophthalmology* found that, along with other treatments for glaucoma, adding alpha-tocopherol appeared to help protect the retina from glaucomatous damage. Groups receiving 300 mg and 600 mg per day of alpha-tocopherol, delivered orally, showed statistically significant decreases in the resistivity index in the posterior ciliary arteries and in the pulsatility index in the ophthalmic arteries, after six and twelve months of therapy. Alpha-tocopherol-treated patients also had significantly lower differences in mean visual field deviations.

### Alzheimer's Disease

Alzheimer's disease is a wasting disease of the brain. An observational trial conducted by The Johns Hopkins University Bloomberg School of Public Health found that when vitamin E is taken daily in large doses (400-1000IU) in combination with vitamin C (500-1000mg) the onset of Alzheimer's was reduced between 64 and 78%.

### Parkinson's Disease

In May 2005, *The Lancet Neurology* published a study suggesting that vitamin E may help protect against Parkinson's disease. Individuals with moderate to high intakes of dietary vitamin E were found to have a lower risk of Parkinson's. No conclusion was drawn about whether supplemental vitamin E has the same effect, however.

## VITAMIN K

### Discovery

In 1929, Danish scientist Henrik Dam investigated the role of cholesterol by feeding chickens a cholesterol-depleted diet After several weeks, the animals developed hemorrhages and started bleeding. These defects could not be restored by adding purified cholesterol to the diet. It appeared that - together with the cholesterol - a second compound had been extracted from the food, and this compound was called the coagulation vitamin. The new vitamin received the letter K because the initial discoveries were reported in a German journal, in which it was designated as *Koagulationsvitamin*. Edward Adelbert Doisy of Saint Louis University did much of the research that led to the discovery of

the structure and chemical nature of Vitamin K. Dam and Doisy shared the 1943 Nobel Prize for medicine for their work on Vitamin K. Several laboratories synthesized the compound in 1939.

For several decades the vitamin K-deficient chick model was the only method of quantitating vitamin K in various foods: the chicks were made vitamin K-deficient and subsequently fed with known amounts of vitamin K-containing food. The extent to which blood coagulation was restored by the diet was taken as a measure for its vitamin K content.

The first published report of successful treatment with vitamin K of life-threatening hemorrhage in a jaundiced patient with prothrombin deficiency was made in 1938 at the University of Iowa Department of Pathology by Drs. Harry Pratt Smith, Emory Warner, Kenneth Brinkhous, and Walter Seegers

**Vitamin K** (K from "Koagulations-Vitamin" in German and Danish) denotes a group of lipophilic, hydrophobic vitamins that are needed for the posttranslational modification of certain proteins, mostly required for blood coagulation. Chemically they are 2-methyl-1,4-naphthoquinone derivatives.

Vitamin $K_2$ (menaquinone, menatetrenone) is normally produced by bacteria in the intestines, and dietary deficiency is extremely rare unless the intestines are heavily damaged or are unable to absorb the molecule

## Chemical Structure

All members of the vitamin K group of vitamins share a methylated naphthoquinone ring structure, and vary in the aliphatic side chain attached at the 3-position (see figure 1). Phylloquinone (also known as vitamin $K_1$) invariably contains in its side chain four isoprenoid residues, one of which is unsaturated.

Menaquinones have side chains composed of a variable number of unsaturated isoprenoid residues; generally they are designated as MK-n, where n specifies the number of isoprenoids.

It is generally accepted that the naphthoquinone is the functional group, so that the mechanism of action is similar for all K-vitamins. Substantial differences may be expected, however, with respect to intestinal absorption, transport, tissue distribution, and bio-availability. These differences are caused by the different lipophilicity of the various side chains, and by the different food matrices in which they occur.

## Physiology

Vitamin K is involved in the carboxylation of certain glutamate residues in proteins to form gamma-carboxyglutamate residues (abbreviated Gla-residues). The modified residues are situated within specific protein domains called Gla domains. Gla-residues are usually involved in binding calcium. The Gla-residues are essential for the biological activity of all known Gla-proteins.[2]

At this time 14 human proteins with Gla domains have been discovered, and they play key roles in the regulation of three physiological processes:

- Blood coagulation: (prothrombin (factor II), factors VII, IX, X, protein C, protein S and protein Z).
- Bone metabolism: osteocalcin, also called bone Gla-protein (BGP), and matrix gla protein MGP).
- Vascular biology.

### Recommended Amounts

The U.S. Dietary Reference Intake (DRI) for an Adequate Intake (AI) of Vitamin K for a 25-year old male is 120 micrograms/day. No Tolerable Upper Intake Level (UL) has been set. The human body stores Vitamin K, so it is not necessary to take Vitamin K daily According to the above mentioned page the human body does not store Vitamin K in large quantities and the body does require daily intake of Vitamin K.

### Sources of Vitamin K

Vitamin K is found chiefly in leafy green vegetables, particularly the dark green ones such as spinach and kale; *Brassica* (*e.g.* cabbage, cauliflower, broccoli, and brussels sprouts) are also high in Vitamin K as are some fruits such as avocado and kiwifruit. By way of reference, two tablespoons of parsley contain 153% of the recommended daily amount of vitamin K.. Some vegetable oils, notably soybean, contain vitamin K, but at levels that would require relatively large caloric consumption to meet the USDA recommended levels.

Phylloquinone (vitamin K1) is the major dietary form of vitamin K.

### Role in Disease

Vitamin K-deficiency may occur by disturbed intestinal uptake (such as would occur in a bile duct obstruction), by therapeutic or accidental intake of vitamin K-antagonists or, very rarely, by nutritional vitamin K-deficiency. As a result, Gla-residues are inadequately formed and the Gla-proteins are insufficiently active. Lack of control of the three processes mentioned above may lead to the following: risk of massive, uncontrolled bleeding, cartilage calcification and severe malformation of developing bone, or deposition of insoluble calcium salts in the walls of arteries. The deposition of calcium in soft tissues, including arterial walls, is quite common, especially in those suffering from atherosclerosis, suggesting that Vitamin K deficiency is more common than previously thought. Menaquinone, but not phylloquinone, intake is associated with reduced risk of CHD mortality, all-cause mortality and severe aortic calcification.

Postmenopausal and elderly women in Thailand have high risk of Vitamin K(2) deficiency, comparing to the normal value of young, reproductive females. Current dosage recommendations for Vitamin K may be too low.

### Use on Newborn Babies

In some countries, injections of Vitamin K are routinely given to newborn babies. Vitamin K is used as prophylactic measure to prevent late-onset haemorrhagic disease (HDN). However, HDN is a relatively rare problem, and many parents now choose for their babies not to have such an injection.

## Biochemistry

### Function in the Cell

The precise function of vitamin K was not discovered until 1974, when three laboratories isolated the vitamin K-dependent coagulation factor prothrombin (Factor II) from cows that received a high dose of a vitamin K antagonist, warfarin. It was shown that while warfarin-treated cows had a form of prothrombin that contained 10 glutamate amino acid residues near the amino terminus of this protein, the normal (untreated) cows contained 10 unusual residues which were chemically identified as gamma-carboxyglutamate, or Gla. The extra carboxyl group in Gla made clear that vitamin K plays a role in a carboxylation reaction during which Glu is converted into Gla.

The biochemistry of how Vitamin K is used to convert Glu to Gla has been elucidated over the past thirty years in academic laboratories throughout the world. Within the cell, Vitamin K undergoes electron reduction to a reduced form of Vitamin K (called Vitamin K hydroquinone) by the enzyme Vitamin K epoxide reductase (or VKOR). Another enzyme then oxidizes Vitamin K hydroquinone to allow carboxylation of Glu to Gla; this enzyme is called the gamma-glutamyl carboxylaseor the Vitamin K-dependent carboxylase. The carboxylation reaction will only proceed if the carboxylase enzyme is able to oxidize Vitamin K hydroquinone to vitamin K epoxide at the same time; the carboxylation and epoxidation reactions are said to be coupled reactions. Vitamin K epoxide is then re-converted to Vitamin K by the Vitamin K epoxide reductase. These two enzymes comprise the so-called Vitamin K cycle. One of the reasons why Vitamin K is rarely deficient in a human diet is because Vitamin K is continually recycled in our cells.

Warfarin and other coumadin drugs block the action of the Vitamin K epoxide reductase. This results in decreased concentrations of Vitamin K and Vitamin K hydroquinone in the tissues, such that the carboxylation reaction catalyzed by the glutamyl carboxylase is inefficient. This results in the production of clotting factors with inadequate Gla. Without Gla on the amino termini of these factors, they no longer bind stably to the blood vessel endothelium and cannot activate clotting to allow formation of a clot during tissue injury. As it is impossible to predict what dose of Warfarin will give the desired degree of suppression of the clotting, Warfarin treatment must be carefully monitored to avoid over-dosing. See Warfarin.

## Gla-proteins

At present, the following human Gla-containing proteins have been characterized to the level of primary structure: the blood coagulation factors II (prothrombin), VII, IX, and X, the anticoagulant proteins C and S, and the Factor X-targeting protein Z. The bone Gla-protein osteocalcin, the calcification inhibiting matrix gla protein (MGP), the cell growth regulating growth arrest specific gene 6 protein (Gas6), and the four transmembrane Gla proteins (TMGPs) the function of which is at present unknown. Gas6 can function as a growth factor that activates the Axl receptor tyrosine kinase and stimulates cell proliferation or prevents apoptosis in some cells. In all cases in which their function was known, the presence of the Gla-residues in these proteins turned out to be essential for functional activity.

Gla-proteins are known to occur in a wide variety of vertebrates: mammals, birds, reptiles, and fish. The venom of a number of Australian snakes acts by activating the human blood clotting system. Remarkably, in some cases activation is accomplished by snake Gla-containing enzymes that bind to the endothelium of human blood vessels and catalyze the conversion of procoagulant clotting factors into activated ones, leading to unwanted and potentially deadly clotting.

Another interesting class of invertebrate Gla-containing proteins is synthesized by the fish-hunting snail *Conus geographus*. These snails produce a venom containing hundreds of neuro-active peptides,

**or conotoxins, which is sufficiently toxic to kill an adult human. Several of the conotoxins contain 2-5 Gla residues.**

## Function in Bacteria

**Many bacteria, such as *Escherichia coli* found in the large intestine, can synthesize Vitamin $K_2$ (menaquinone) but not Vitamin $K_1$ (phylloquinone). In these bacteria, menaquinone will transfer two electrons between two different small molecules, in a process called anaerobic respiration. For example, a small molecule with an excess of electrons (also called an electron donor) such as lactate, formate, or NADH, with the help of an enzyme, will pass two electrons to a menaquinone. The menaquinone, with the help of another enzyme, will in turn transfer these 2 electrons to a suitable oxidant, such fumarate or nitrate (also called an electron acceptor). Adding two electrons to fumarate or nitrate will convert the molecule to succinate or nitrite + water, respectively. Some of these reactions generate a cellular energy source, ATP, in a manner similar to eukaryotic cell aerobic respiration, except that the final electron acceptor is not molecular oxygen, but say fumarate or nitrate (In aerobic respiration, the final oxidant is molecular oxygen ($O_2$) , which accepts four electrons from an electron donor such as NADH to be converted to water.) *Escherichia coli* can carry out aerobic respiration and menaquninone-mediated anaerobic respiration.**

## Carcinogenicity

**Vitamin K substances are IARC Group 3 carcinogens. One study conducted in the United Kingdom in 1970 found a nearly two-fold increase of leukaemia in children administered synthetic Vitamin $K_1$ phytomenadione intramuscularly but later studies have failed to find whether Vitamin K is carcinogenic or not.**

# VITAMIN H OR $B_7$

**Biotin**, **also known as vitamin H or $B_7$, has the chemical formula $C_{10}H_{16}N_2O_3S$ (Biotin; Coenzyme R, Biopeiderm), is a water-soluble B-complex vitamin which is composed of an ureido (tetrahydroimidizalone) ring fused with a tetrahydrothiophene ring. A valeric acid substituent is attached to one of the carbon atoms of the tetrahydrothiophene ring. Biotin is a cofactor in the metabolism of fatty acids and leucine, and in gluconeogenesis.**

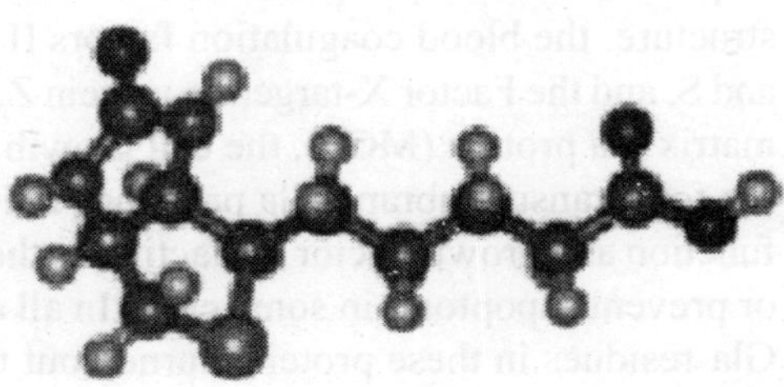

## General Overview

**Biotin is necessary for cell growth, the production of fatty acids, and the metabolism of fats and amino acids. It plays a role in the Citric acid cycle, which is the process by which biochemical energy is generated during aerobic respiration. Biotin not only assists in various metabolic**

reactions, but also helps to transfer carbon dioxide. Biotin is also helpful in maintaining a steady blood sugar level. Biotin is often recommended for strengthening hair and nails. Consequently, it is found in many cosmetic and health products for the hair and skin.

Deficiency is extremely rare, as intestinal bacteria generally produce an excess of the body's daily requirement. For that reason, statutory agencies in many countries (e.g., the Australian Department of Health and Aging) do not prescribe a recommended daily intake.

## Uses

### Hair Problems

Biotin supplements are often recommended as a natural product to counteract the problem of hair loss in both children and adults. There are, however, no studies that show any benefit in any case where the subject is not actually biotin deficient. The signs and symptoms of biotin deficiency include hair loss which progresses in severity to include loss of eye lashes and eye brows in severely deficient subjects. Some shampoos are available that contain biotin, but it is doubtful whether they would have any useful effect, as biotin is not absorbed well through the skin.

### Cradle Cap (Seborrheic Dermatitis)

Children with a rare inherited metabolic disorder called phenylketonuria (PKU; in which one is unable to break down the amino acid phenylalanine) often develop skin conditions such as eczema and seborrheic dermatitis in areas of the body other than the scalp. The scaly skin changes that occur in people with PKU may be related to poor ability to use biotin. Increasing dietary biotin has been known to improve seborrheic dermatitis in these cases.

### Diabetes

People with type 2 diabetes often have low levels of biotin. Biotin may be involved in the synthesis and release of insulin. Preliminary studies in both animals and people suggest that biotin may help improve blood sugar control in those with diabetes, particularly type 2 diabetes.

### Deficiency

Biotin deficiency is relatively rare and mild, and can be addressed with supplementation. Such deficiency can be caused by the excessive consumption of raw egg whites, which contain high levels of the protein avidin, which binds biotin strongly. Avidin is inactivated by cooking, while the biotin remains intact.

Biotinidase deficiency is not due to inadequate biotin, but rather to a deficiency in the enzymes which process it.

### Biochemistry

Biotin is a cofactor responsible for carbon dioxide transfer in several carboxylase enzymes:

- Acetyl-CoA carboxylase alpha
- Acetyl-CoA carboxylase beta
- Methylcrotonyl-CoA carboxylase
- Propionyl-CoA carboxylase
- Pyruvate carboxylase

The attachment of biotin to various chemical sites, called biotinylation, can be used as an important laboratory technique to study various processes including protein localization, protein interactions, DNA transcription and replication. Biotin itself is known to biotinylate histones, but is not found naturally in chromatin. Holocarboxylase synthetase is involved in the binding of biotin.

Biotin binds very tightly to the tetrameric protein avidin (also streptavidin and neutravidin), with a dissociation constant $K_d$ in the order of $10^{-15}$ mol/L (Bonjour, 1977; Green 1975; and Roth, 1985). This is often used in different biotechnological applications. Until 2005, very harsh conditions were required to break the biotin-streptavidin bond.

## Laboratory Uses

In the biochemistry laboratory, biotin is often chemically linked, or tagged, to a molecule or protein for biochemical assays. This process is called biotinylation. Since avidins bind preferentially to biotin, biotin-tagged molecules can be extracted from a sample by mixing them with beads with covalently-attached avidin, and washing away anything unbound to the beads.

For example, biotin can be attached to a molecule of interest (e.g. a protein), and this modified molecule will be mixed with a complex mixture of proteins. Avidin or streptavidin beads are added to the mixture, and the biotinylated molecule will bind to the beads. Any other proteins binding to the biotinylated molecule will also stay with the beads. All other unbound proteins can be washed away, and the scientist can use a variety of methods to determine which proteins have bound to the biotinylated molecule.

Biotinylated antibodies are used to capture avidin or streptavidin in both the ELISPOT and ELISA techniques.

# NIACIN

## History

Niacin was first described by Weidel in 1873 in his studies of nicotine. The original preparation remains useful: the oxidation of nicotine using nitric acid. Niacin was extracted from livers by Conrad Elvehjem who later identified the active ingredient, then referred to as the "pellagra-preventing factor" and the "anti-blacktongue factor."[5] When the biological significance of nicotinic acid was realized, it was thought appropriate to choose a name to dissociate it from nicotine, in order to

avoid the perception that vitamins or niacin-rich food contains nicotine. The resulting name 'niacin' was derived from "ni"cotinic "ac"id + vitam"in."

Niacin is referred to as Vitamin $B_3$ because it was the third of the B vitamins to be discovered. It has historically been referred to as "vitamin PP."

**Niacin**, also known as nicotinic acid and vitamin $B_3$, is the organic compound with the formula $HO_2CC_5H_4N$. This water-soluble, colourless solid is a derivative of pyridine, featuring a carboxylic acid functional group at the 3-position. The designation *vitamin $B_3$* also includes the corresponding amide nicotinamide ("niacinamide"), wherein the $CO_2H$ group has been replaced by a $CONH_2$ group. Niacin is a precursor to NADH, NAD, $NAD^+$, and NADP, which play essential metabolic roles in living cells. Among its many functions, niacin derivatives are involved with detoxification of xenochemicals, DNA repair, and the production of steroid hormones in the adrenal gland.

## Food Sources

**Animal products:**
- liver, heart and kidney
- chicken
- beef
- fish: tuna, salmon
- milk
- eggs

**Fruits and vegetables:**
- leaf vegetables
- broccoli
- tomatoes
- carrots
- dates
- sweet potatoes
- asparagus
- avocados

**Seeds:**
- nuts
- whole grain products
- legumes
- saltbush seeds

**Fungi:**
- mushrooms
- brewer's yeast

## Biosynthesis

The liver can synthesize niacin from the essential amino acid tryptophan (see below), but the synthesis is extremely inefficient; 60 mg of tryptophan are required to make one milligram of niacin.

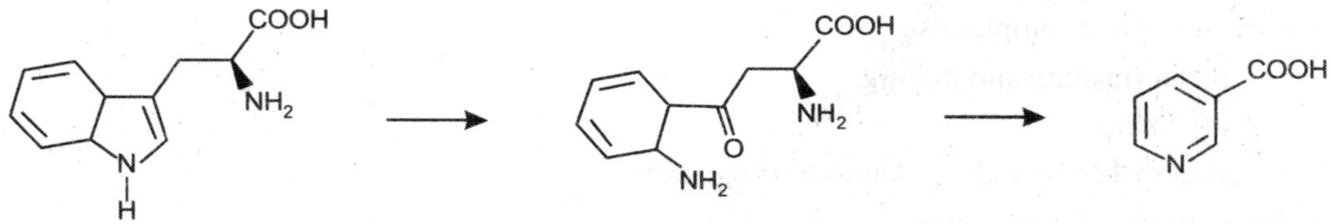

Biosynthesis: Tryptophan → kynurenine → niacin

The 5-membered aromatic heterocycle of the essential amino acid, tryptophan, is cleaved and rearranged with the alpha amino group of tryptophan into the 6-membered aromatic heterocycle of niacin by the following reaction:

## Dietary Needs

Severe deficiency of niacin in the diet causes the disease pellagra, whereas mild deficiency slows the metabolism, causing decreased tolerance to cold. Dietary niacin deficiency tends to occur only in areas where people eat corn (maize), the only grain low in niacin, as a staple food, *and* that do not use lime during meal/flour production. Alkali lime releases the tryptophan from the corn in a process called nixtamalization so that it can be absorbed in the intestine, and converted to niacin.

The recommended daily allowance of niacin is 2-12 mg/day for children, 14 mg/day for women, 16 mg/day for men, and 18 mg/day for pregnant or breast-feeding women.

## Pharmacological Uses

Niacin, when taken in large doses, blocks the breakdown of fats in adipose tissue, thus altering blood lipid levels. Niacin is used in the treatment of hyperlipidemia because it reduces very-low-density lipoprotein (VLDL), a precursor of low-density lipoprotein (LDL) or "bad" cholesterol. Because niacin blocks breakdown of fats, it causes a decrease in free fatty acids in the blood and, as a consequence, decreased secretion of VLDL and cholesterol by the liver.

By lowering VLDL levels, niacin also *increases* the level of high-density lipoprotein (HDL) or "good" cholesterol in blood, and therefore it is sometimes prescribed for patients with low HDL, who are also at high risk of a heart attack. An extended release formulation of niacin for this indication is marketed by Abbott Laboratories under the trade name **Niaspan** by prescription in the United States; Slo-Niacin, made by Upsher-Smith Laboratories, is available over the counter.

Niacin is sometimes consumed in large quantities by people who wish to fool drug screening tests, particularly for lipid soluble drugs such as marijuana. It is believed to "promote metabolism" of the drug and cause it to be "flushed out." Scientific studies have shown it does not affect drug screenings, but can pose a risk of overdose, causing arrhythmias, metabolic acidosis, hyperglycemia, and other serious problems (see below).

## Toxicity

People taking pharmacological doses of niacin (1.5 - 6 g per day) often experience a syndrome of side-effects that can include one or more of the following:

- dermatological complaints
  - facial flushing and itching
  - dry skin
  - skin rashes including acanthosis nigricans
- gastrointestinal complaints
  - dyspepsia (indigestion)

- liver toxicity
  - fulminant hepatic failure
- hyperglycemia
- cardiac arrhythmias
- birth defects

Facial flushing is the most commonly-reported side-effect. It lasts for about 15 to 30 minutes, and is sometimes accompanied by a prickly or itching sensation, particularily in areas covered by clothing. This effect is mediated by prostaglandins and can be blocked by taking 300 mg of aspirin half an hour before taking niacin, or by taking one tablet of ibuprofen per day. Taking the niacin with meals also helps reduce this side-effect. After 1 to 2 weeks of a stable dose, most patients no longer flush. Slow- or "sustained"-release forms of niacin have been developed to lessen these side-effects. One study showed the incidence of flushing was 4.5x lower (1.9 vs. 8.6 episodes in the first month) with a sustained-release formulation.

Doses above 2 g per day have been associated with liver damage, particularly with slow-release formulations.

High-dose niacin may also elevate blood sugar, thereby worsening diabetes mellitus. Hyperuricemia is another side-effect of taking high-dose niacin; thus niacin may worsen gout

Niacin at doses used in lowering cholesterol has been associated with birth defects in laboratory animals and should not be taken by pregnant women.

Niacin at extremely high doses can have life-threatening acute toxic reactions. One patient suffered vomiting after taking eleven 500-milligram niacin tablets over 36 hours, and another was unresponsive for several minutes after taking five 500-milligram tablets over two days. Extremely high doses of niacin can also cause niacin maculopathy, a thickening of the macula and retina which leads to blurred vision and blindness.

### Inositol Hexanicotinate

One popular form of dietary supplement is inositol hexanicotinate, usually sold as "flush-free" or "no-flush" niacin (although those terms are also used for regular sustained-release.) While this form of niacin does not cause the flushing associated with the nicotinic acid form, it is not clear whether it is pharmacologically equivalent in its positive effect.

## FOLIC ACID

### History

A key observation by researcher Lucy Wills in 1931 led to the identification of folate as the nutrient needed to prevent anemia during pregnancy. Dr. Wills demonstrated that anemia could be reversed with brewer's yeast. Folate was identified

as the corrective substance in brewer's yeast in the late 1930s and was extracted from spinach leaves in 1941. It was first synthesized in 1946 by Yellapragada Subbarao. **Folic acid** and **folate** (the anion form) are forms of the water-soluble Vitamin $B_9$. These occur naturally in food and can also be taken as supplements. Folate gets its name from the Latin word *folium* ("leaf").

## Folate in Foods

Leafy vegetables such as spinach, turnip greens, lettuces, dried beans and peas, fortified cereal products, sunflower seeds and certain other fruits and vegetables are rich sources of folate. Some breakfast cereals (ready-to-eat and others) are fortified with 25% to 100% of the recommended dietary allowance (RDA) for folic acid. A table of selected food sources of folate and folic acid can be found at the USDA National Nutrient Database for Standard Reference. Folate is also found in Vegemite, with an average serving (5gm) containing 100µg.

## Biological Roles

Folate is necessary for the production and maintenance of new cells. This is especially important during periods of rapid cell division and growth such as infancy and pregnancy. Folate is needed to replicate DNA. Thus folate deficiency hinders DNA synthesis and cell division, affecting most clinically the bone marrow, a site of rapid cell turnover. Because RNA and protein synthesis are not hindered by folate deficiency, large red blood cells called megaloblasts are produced, resulting in megaloblastic anemia.[2] Both adults and children need folate to make normal red blood cells and prevent anemia.[3]

## Biochemistry

In the form of a series of tetrahydrofolate compounds, folate derivatives are substrates in a number of single-carbon-transfer reactions, and also are involved in the synthesis of dTMP (2'-deoxythymidine-5'-phosphate) from dUMP (2'-deoxyuridine-5'-phosphate). It helps convert vitamin B12 to one of its coenzyme forms and helps synthesize the DNA required for all rapidly growing cells.The pathway leading to the formation of tetrahydrofolate ($FH_4$) begins when folate (F) is reduced to dihydrofolate ($FH_2$), which is then reduced to tetrahydrofolate ($FH_4$). Dihydrofolate reductase catalyses both steps.[4]Methylene tetrahydrofolate ($CH_2FH_4$) is formed from tetrahydrofolate by the addition of methylene groups from one of three carbon donors: formaldehyde, serine, or glycine. Methyl tetrahydrofolate ($CH_3$–$FH_4$) can be made from methylene tetrahydrofolate by reduction of the methylene group; formyl tetrahydrofolate (CHO-$FH_4$, folinic acid) results from oxidation of methylene tetrahydrofolate.

In other words:

$$F \rightarrow FH_2 \rightarrow FH_4 \rightarrow CH_2{=}FH_4 \rightarrow \text{1-carbon chemistry}$$

A number of drugs interfere with the biosynthesis of folic acid and tetrahydrofolate. Among them are the dihydrofolate reductase inhibitors (such as trimethoprim and pyrimethamine), the sulfonamides (competitive inhibitors of para-aminobenzoic acid in the reactions of dihydropteroate synthetase), and the anticancer drug methotrexate (inhibits both folate reductase and dihydrofolate reductase).

**1998 RDAs for Folate**

| Men (19+) | Women | | |
|---|---|---|---|
| | (19+) | Pregnancy | Breast feeding |
| 400 μg | 400 μg | 600 μg | 500 μg |

1 μg of food folate = 0.6 μg folic acid from supplements and fortified foods.

The National Health and Nutrition Examination Survey (NHANES III 1988-91) and the Continuing Survey of Food Intakes by Individuals (1994-96 CSFII) indicated that most adults did not consume adequate folate. However, the folic acid fortification program in the United States has increased folic acid content of commonly eaten foods such as cereals and grains, and as a result diets of most adults now provide recommended amounts of folate equivalents.

## Folate Deficiency

### Human Reproduction

Folic acid is very important for all women who may become pregnant. Adequate folate intake during the periconceptional period, the time just before and just after a woman becomes pregnant, helps protect against a number of congenital malformations including neural tube defects.[8] Neural tube defects result in malformations of the spine (spina bifida), skull, and brain (anencephaly). The risk of neural tube defects is significantly reduced when supplemental folic acid is consumed in addition to a healthy diet prior to and during the first month following conception. Women who could become pregnant are advised to eat foods fortified with folic acid or take supplements in addition to eating folate-rich foods to reduce the risk of some serious birth defects. Taking 400 micrograms of synthetic folic acid daily from fortified foods and/or supplements has been suggested. The Recommended Dietary Allowance (RDA) for folate equivalents for pregnant women is 600-800 micrograms, twice the normal RDA of 400 micrograms for women who are not pregnant.

Recent research has shown that it is also very important for men who are planning on fathering children, reducing birth defect risks.

### Folic Acid Supplements and Masking of $B_{12}$ Deficiency

There has been concern about the interaction between vitamin $B_{12}$ and folic acid.[13]Folic acid supplements can correct the anemia associated with vitamin $B_{12}$ deficiency. Unfortunately, folic acid will not correct changes in the nervous system that result from vitamin $B_{12}$ deficiency. Permanent nerve damage could theoretically occur if vitamin $B_{12}$ deficiency is not treated. Therefore, intake of supplemental folic acid should not exceed 1000 micrograms (1000 mcg or 1 mg) per day to prevent folic acid from masking symptoms of vitamin $B_{12}$ deficiency. In fact, to date the evidence that such masking actually occurs is scarce, and there is no evidence that folic acid fortification in Canada or the US has increased the prevalence of vitamin $B_{12}$ deficiency or its consequences. However one recent study has demonstrated that high folic or folate levels when combined with low $B_{12}$ levels are associated with significant cognitive impairment among the elderly. If the observed relationship for seniors between folic acid intake, $B_{12}$ levels, and cognitive impairment is replicated and confirmed, this is likely to re-open the debate on folic acid fortification in food, even though public health policies tend generally to support the developmental needs of infants and children over slight risks to other population groups. In any case, it is important for older adults to be aware of the relationship between folic acid and vitamin $B_{12}$ because they are at greater risk of having a vitamin $B_{12}$ deficiency. If you are 50 years of age or older, ask your physician to check your $B_{12}$ status before you take a supplement that contains folic acid.

### Health Risk of too much Folic Acid

The risk of toxicity from folic acid is low. The Institute of Medicine has established a tolerable upper intake level (UL) for folate of 1 mg for adult men and women, and a UL of 800 µg for pregnant and lactating (breast-feeding) women less than 18 years of age. Supplemental folic acid should not exceed the UL to prevent folic acid from masking symptoms of vitamin $B_{12}$ deficiency. Research suggests high levels of folic acid can interfere with some antimalarial treatments. A 10000-patient study at Tufts University in 2007 concluded that excess folic acid worsens the effects of $B_{12}$ deficiency and in fact may affect the absorption of $B_{12}$.

## Some Current Issues and Controversies about Folate

### Dietary Fortification of Folic Acid

Since the discovery of the link between insufficient folic acid and neural tube defects (NTDs), governments and health organisations worldwide have made recommendations concerning folic acid *supplementation* for women intending to become pregnant. For example, the United States Public Health Service (see External links) recommends an extra 0.4 mg/day, which can be taken as a pill. However, many researchers believe that supplementation in this way can never work effectively enough since about half of all pregnancies in the U.S. are unplanned and not all women will comply with the recommendation. This has led to the introduction in many countries of *fortification,* where folic acid is added to flour with the intention of everyone benefiting from the associated rise in blood folate levels. This is controversial, with issues having been raised concerning individual liberty, and the masking effect of folate fortification on pernicious anaemia (vitamin $B_{12}$ deficiency). However, most North and South American countries now fortify their flour, along with a number of Middle Eastern countries and Indonesia. Mongolia and a number of ex-Soviet republics are amongst those having widespread voluntary fortification; about five more countries (including Morocco, the first African country) have agreed but not yet implemented fortification. In the UK the Food Standards Agency has recommended fortification. To date, no EU country has yet mandated fortification Australia is considering fortification, but a period for comments ending 2006-07-31 attracted strong opposition from industry as well as academia. Recent debate has emerged in the United Kingdomand Australia regarding the inclusion of folic acid in products such as bread and flour.

# LIPIDS

**Lipids** are broadly defined as any fat-soluble (lipophilic), naturally-occurring molecule, such as fats, oils, waxes, cholesterol, sterols, fat-soluble vitamins (such as vitamins A, D, E and K), monoglycerides, diglycerides, phospholipids, and others. The main biological functions of lipids include energy storage, acting as structural components of cell membranes, and participating as important signaling molecules.Although the term *lipid* is sometimes used as a synonym for fats, fats are a subgroup of lipids called triglycerides and should not be confused with the term fatty acid. Lipids also encompass molecules such as fatty acids and their derivatives (including tri-, di-, and monoglycerides and phospholipids), as well as other sterol-containing metabolites such as cholesterol. Lipids are a diverse group of compounds that have many key biological functions, such as acting as structural components of cell membranes, serving as energy storage sources and participating in signaling pathways. Lipids may be broadly defined as hydrophobic or amphiphilic small molecules that originate entirely or in part from two distinct types of biochemical subunits or "building blocks": ketoacyl and isoprene groups. Using this approach, lipids may be divided into eight categories : fatty acyls, glycerolipids, glycerophospholipids, sphingolipids, saccharolipids and polyketides (derived from condensation of ketoacyl subunits); and sterol lipids and prenol lipids (derived from condensation of isoprene subunits).

- **Fatty acyls** (including fatty acids) are a diverse group of molecules synthesized by chain-elongation of an acetyl-CoA primer with malonyl-CoA or methylmalonyl-CoA groups.[2][3] The fatty acyl structure represents the major lipid building block of complex lipids and therefore is one of the most fundamental categories of biological lipids. The carbon chain may be saturated or unsaturated, and may be attached to functional groups containing oxygen, halogens, nitrogen and sulfur. Examples of biologically interesting fatty acyls are the eicosanoids which are in turn derived from arachidonic acid which include prostaglandins, leukotrienes, and thromboxanes. Other major lipid classes in the fatty acyl category are the fatty esters and fatty amides. Fatty esters include important biochemical intermediates such as wax esters, fatty acyl thioester coenzyme A derivatives, fatty acyl thioester ACP derivatives and fatty acyl carnitines. The fatty amides include N-acyl ethanolamines such as anandamide.

- **Glycerolipids** are composed mainly of mono-, di- and tri-substituted glycerols,[4] the most well-known being the fatty acid esters of glycerol (triacylglycerols), also known as triglycerides. these comprise the bulk of storage fat in animal tissues. Additional subclasses are represented by glycosylglycerols, which are characterized by the presence of one or more sugar residues attached to glycerol via a glycosidic linkage. Examples of structures in this category are the digalactosyldiacylglycerols found in plant membranes and seminolipid from mammalian spermatazoa.

- **Glycerophospholipids**, also referred to as phospholipids, are ubiquitous in nature and are key components of the lipid bilayer of cells, as well as being involved in metabolism and signaling. Glycerophospholipids[5] may be subdivided into distinct classes, based on the nature of the polar headgroup at the *sn*-3 position of the glycerol backbone in eukaryotes and eubacteria or the *sn*-1 position in the case of archaebacteria. Examples of glycerophospholipids found in biological membranes are phosphatidylcholine (also known as PC or GPCho, and lecithin), phosphatidylethanolamine (PE or GPEtn) and phosphatidylserine (PS or GPSer). In addition to serving as a primary component of cellular membranes and binding sites for intra- and intercellular proteins, some glycerophospholipids in eukaryotic cells, such as phosphatidylinositols and phosphatidic acids are either precursors of, or are themselves, membrane-derived second messengers. Typically one or both of these hydroxyl groups are acylated with long-chain fatty acids, but there are also alkyl-linked and 1Z-alkenyl-linked (plasmalogen) glycerophospholipids, as well as dialkylether variants in prokaryotes.

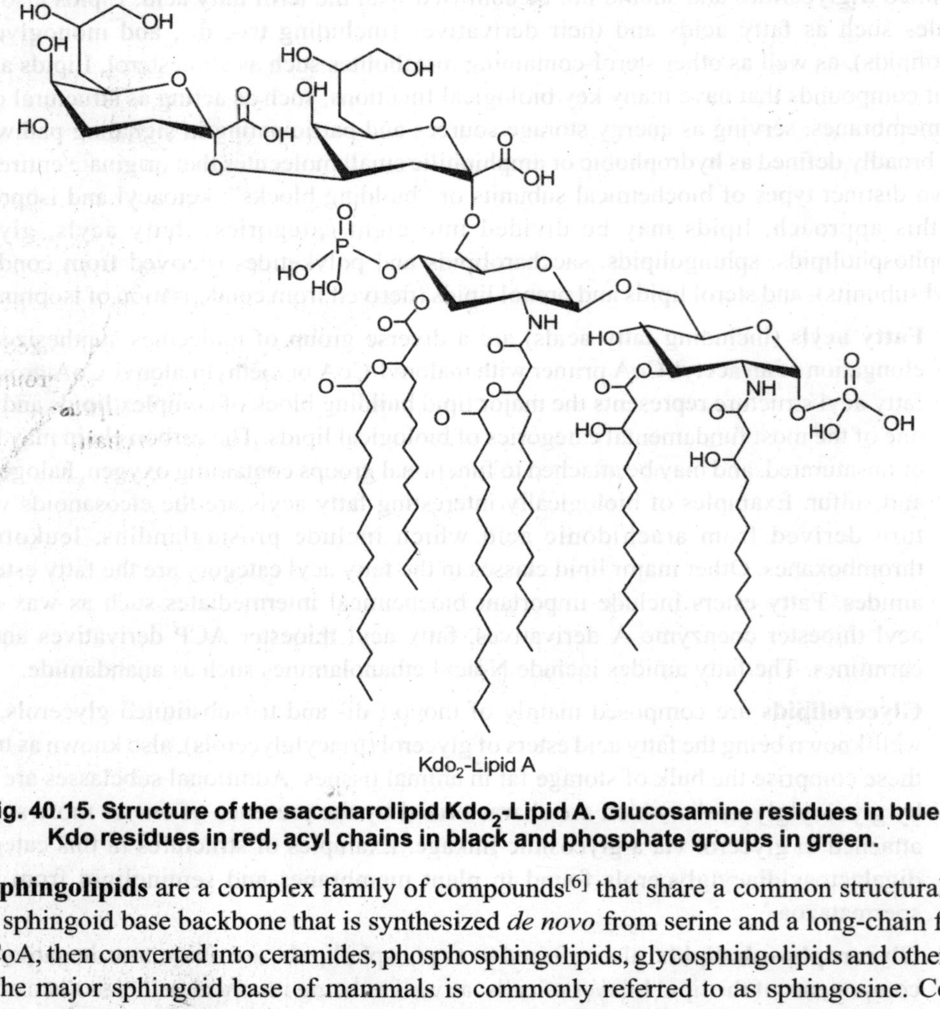

$Kdo_2$-Lipid A

**Fig. 40.15. Structure of the saccharolipid $Kdo_2$-Lipid A. Glucosamine residues in blue, Kdo residues in red, acyl chains in black and phosphate groups in green.**

- **Sphingolipids** are a complex family of compounds[6] that share a common structural feature, a sphingoid base backbone that is synthesized *de novo* from serine and a long-chain fatty acyl CoA, then converted into ceramides, phosphosphingolipids, glycosphingolipids and other species. The major sphingoid base of mammals is commonly referred to as sphingosine. Ceramides (N-acyl-sphingoid bases) are a major subclass of sphingoid base derivatives with an amide-linked fatty acid. The fatty acids are typically saturated or mono-unsaturated with chain lengths from 14 to 26 carbon atoms. The major phosphosphingolipids of mammals are sphingomyelins (ceramide phosphocholines), whereas insects contain mainly ceramide phosphoethanolamines and fungi have phytoceramidephosphoinositols and mannose containing headgroups. The Glycosphingolipids are a diverse family of molecules composed of one or more sugar residues linked via a glycosidic bond to the sphingoid base. Examples of these are the simple and complex glycosphingolipids such as cerebrosides and gangliosides.

- **Sterol lipids**, such as cholesterol and its derivatives are an important component of membrane lipids, along with the glycerophospholipids and sphingomyelins. The steroids, which also contain the same fused four-ring core structure, have different biological roles as hormones and signaling molecules. The C18 steroids include the estrogen family whereas the C19 steroids comprise the androgens such as testosterone and androsterone. The C21 subclass includes the progestogens as well as the glucocorticoids and mineralocorticoids. The secosteroids, comprising various forms of vitamin D, are characterized by cleavage of the B ring of the core structure. Other examples of sterols are the bile acids and their conjugates, which in mammals are oxidized derivatives of cholesterol and are synthesized in the liver.
- **Prenol lipids** are synthesized from the 5-carbon precursors isopentenyl diphosphate and dimethylallyl diphosphate that are produced mainly via the mevalonic acid (MVA) pathway.[9] The simple isoprenoids (linear alcohols, diphosphates, etc.) are formed by the successive addition of C5 units, and are classified according to number of these terpene units. Structures containing greater than 40 carbons are known as polyterpenes. Carotenoids are important simple isoprenoids that function as anti-oxidants and as precursors of vitamin A. Another biologically important class of molecules is exemplified by the quinones and hydroquinones, which contain an isoprenoid tail attached to a quinonoid core of non-isoprenoid origin. Vitamin E and vitamin K, as well as the ubiquinones, are examples of this class. Bacteria synthesize polyprenols (called bactoprenols) in which the terminal isoprenoid unit attached to oxygen remains unsaturated, whereas in animal polyprenols (dolichols) the terminal isoprenoid is reduced.
- **Saccharolipids** describe compounds in which fatty acids are linked directly to a sugar backbone, forming structures that are compatible with membrane bilayers. In the saccharolipids, a sugar substitutes for the glycerol backbone that is present in glycerolipids and glycerophospholipids. The most familiar saccharolipids are the acylated glucosamine precursors of the Lipid A component of the lipopolysaccharides in Gram-negative bacteria. Typical lipid A molecules are disaccharides of glucosamine, which are derivatized with as many as seven fatty-acyl chains. The minimal lipopolysaccharide required for growth in *E. coli* is $Kdo_2$-Lipid A, a hexa-acylated disaccharide of glucosamine that is glycosylated with two 3-deoxy-D-manno-octulosonic acid (Kdo) residues.
- **Polyketides** are synthesized by polymerization of acetyl and propionyl subunits by classic enzymes as well as iterative and multimodular enzymes that share mechanistic features with the fatty acid synthases. They comprise a very large number of secondary metabolites and natural products from animal, plant, bacterial, fungal and marine sources, and have great structural diversity.[11] Many polyketides are cyclic molecules whose backbones are often further modified by glycosylation, methylation, hydroxylation, oxidation, and/or other processes. Many commonly used anti-microbial, anti-parasitic, and anti-cancer agents are polyketides or polyketide derivatives, such as erythromycins, tetracylines, avermectins, and antitumor epothilones.

# Biological Functions

## Membranes

The glycerophospholipids are the main structural component of biological membranes, such as the cellular plasma membrane and the intracellular membranes of organelles. In animal cells the plasma membrane physically separates the intracellular components from the extracellular environment. All eukaryotic cells are compartmentalized into membrane-bound organelles which carry out different functions. These glycerophospholipids are amphipathic molecules that contain a glycerol core linked to two fatty acid-derived "tails" by ester or, more rarely, ether linkages and to one "head" group by a phosphate ester linkage. While glycerophospholipids are the major component of biological membranes, other non-glyceride lipid components such as sphingomyelin and sterols (mainly cholesterol in animal cell membranes) are also found in biological membranes. In plants and algae, the galactosyldiacylglycerols, and sulfoquinovosyldiacylglycerol, which lack a phosphate group, are important components of membranes of chloroplasts and related organelles and are the most abundant lipids in photosynthetic tissues, including those of higher plants, algae and certain bacteria.

A biological membrane is a form of lipid bilayer, as is a liposome. The formation of lipid bilayers is an energetically-preferred process when the glycerophospholipids described above are in an aqueous environment. In an aqueous system, the polar heads of lipids orientate towards the polar, aqueous environment, while the hydrophobic tails minimise their contact with water. The lipophilic tails of lipids (U) tend to cluster together, forming a lipid bilayer (1) or a micelle (2). Other aggregations are also observed and form part of the polymorphism of amphiphile (lipid) behaviour. The polar heads (P) face the aqueous environment, curving away from the water. Phase behaviour is a complicated area within biophysics and is the subject of current academic research. Micelles and bilayers form in the polar medium by a process known as the hydrophobic effect. When dissolving a lipophilic or amphiphilic substance in a polar environment, the polar molecules (i.e. water in an aqueous solution) become more ordered around the dissolved lipophilic substance, since the polar molecules cannot form hydrogen bonds to the lipophilic areas of the amphiphile. So in an aqueous environment the water molecules form an ordered "clathrate" cage around the dissolved lipophilic molecule.

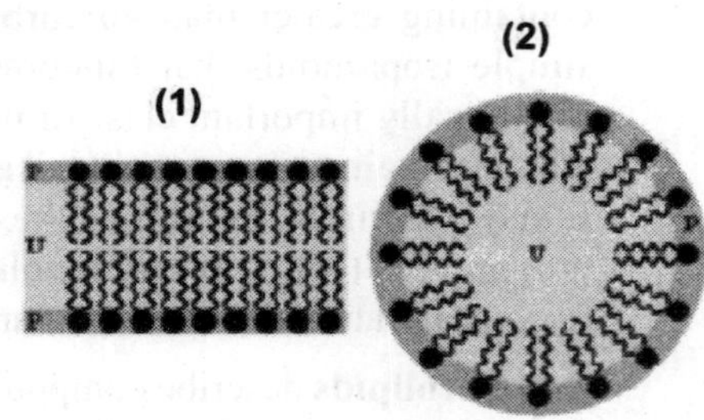


**Fig. 40.16. Self-organization of phospholipids. A lipid bilayer is shown on the left and a micelle on the right**

## Energy Storage and Metabolism

Triacylglycerols, stored in adipose tissue, are a major form of energy storage in animals. Animals use triglycerides for energy storage because of its high caloric content (9 KCal/g), whereas plants, which do not require energy for movement, can afford to store food for energy in a less compact but more easily accessible form, such as starch (carbohydrate). Triglycerides and phospholipids are broken down into free fatty acids by the action of lipases. Beta oxidation is the process by which fatty acids, in the form of acyl-CoA molecules, are broken down in the mitochondria and/or in peroxisomes to generate acetyl-CoA. The acetyl CoA is then ultimately converted into ATP, $CO_2$, and $H_2O$ using the citric acid cycle and the electron transport chain. Conversely, fatty acid biosynthesis (Lipogenesis) takes place in the cytoplasm, using acetyl-CoA (derived from carbohydrates, amino acids or fatty acids) as the precursor. The fatty acids may be subsequently converted to triacylglycerols that are packaged in lipoproteins (VLDL's) and secreted from the liver.

## Signaling

In recent years, evidence has emerged showing that lipid signaling is a vital part of the cell signaling Lipid signaling may occur via activation of GPCR's or nuclear receptors, and members of several different lipid categories have been identified as signaling molecules and cellular messengers. These include sphingosine-1-phosphate, a sphingolipid derived from ceramide that is a potent messenger molecule involved in regulating calcium mobilization, cell growth, apoptosis; diacylglycerol(DAG) and the phosphatidylinositol phosphates (PIPs), involved in calcium-mediated activation of protein kinase C; the prostaglandins, arachidonic acid -derived fatty acids involved in inflammation and immunity; the steroid hormones such as estrogen, testosterone and cortisol, which modulate a host of functions such as reproduction, metabolism and blood pressure; and the oxysterols such as 25-hydroxy-cholesterol that are Liver X receptor (LXR) agonists.

## Other Functions

The "fat-soluble" vitamins (A, D, E and K) which are isoprene-based lipids are essential nutrients stored in the liver and fatty tissues. These have a diverse range of functions discussed elsewhere. Acyl-carnitines are involved in the transport and metabolism of fatty acids in and out of mitochondria, where they undergo beta oxidation. Polyprenols and their phosphorylated derivatives also play important transport roles, in this case the transport of oligosaccharides across membranes. Polyprenol phosphate sugars and polyprenol diphosphate sugars function in extra-cytoplasmic glycosylation reactions, in extra-cellular polysaccharide biosynthesis (for instance peptidoglycan polymerization in bacteria), and in eukaryotic protein N-glycosylation.[19] Cardiolipins are a subclass of glycerophospholipids containing four acyl chains and three glycerol groups that are particularly abundant in the inner mitochondrial membrane. They are believed to activate enzymes involved with oxidative phosphorylation.

## Nutrition and Health

Lipids play diverse and important roles in nutrition and health. Many lipids are absolutely essential for life. However, there is also considerable awareness that abnormal levels of certain lipids, particularly cholesterol (in hypercholesterolemia) and trans fatty acids, are risk factors for heart disease amongst others.Humans have a requirement for certain essential fatty acids, such as linoleic acid (an omega-6 fatty acid) and alpha-linolenic acid (an omega-3 fatty acid) in the diet because they cannot be synthesized from simple precursors in the diet. Both of these fatty acids are 18-carbon polyunsaturated fatty acids differing in the number and position of the double bonds. Most vegetable oils are rich in linoleic acid (safflower, sunflower, and corn oils). Alpha-linolenic acid is found in the green leaves of plants, and in selected seeds, nuts and legumes (flax, canola, walnuts and soy). Fish oils are particularly rich in the longer-chain omega-6 fatty acids eicosapentaenoic acid (EPA) and docosahexaenoic acid (DHA). Most of the lipid found in food is in the form of triacylglycerols, cholesterol and phospholipids.Most of the saturated fatty acids (as triacylglycerols) in the diet are incorporated into adipose tissue stores, because the absence of double bonds allows a higher energy yield per carbon than is obtained from oxidation of unsaturated fatty acids. The longer chain fatty acids are incorporated into cell membranes as phospholipids regardless of degree of saturation. Since dietary fatty acids are exchanged with membrane fatty acids, dietary fat composition is reflected in membrane lipid composition. Thus dietary fatty acids can influence cell function through effects on membrane properties. Dietary fat provides an average energy intake which is approximately twice that of carbohydrate or protein. A minimum amount of dietary fat is necessary to facilitate absorption of fat-soluble vitamins (A, D, E and K) and carotenoids. A minimal amount of body fat is also necessary to provide insulation that prevents heat loss and protects vital organs from shock due to ordinary activities.High fat intake contributes to increased risk of obesity, diabetes and atherosclerosis. Atherosclerosis is the primary cause of coronary and cardiovascular diseases and is primary due to the buildup of plaque on the inside walls of arteries. Plaque is made up of cholesterol-rich low density lipoproteins (LDL), macrophages, smooth muscle cells, platelets, and other substances. In North America and most other western countries, atherosclerosis is the leading cause of illness and death, almost doubling the number of deaths from cancers. Despite significant medical advances, coronary artery disease and atherosclerotic stroke are responsible for more deaths than all other causes combined.A substantial amount of scientific evidence supports the impact of dietary fatty acids on cardiovascular health. Saturated fats have a profound hypercholesterolemic (increase blood cholesterol levels) effect and tend to increase plasma LDL. They are found predominantly in animal products (butter, cheese and meat) but coconut oil and palm oil are common vegetable sources. Intake of monounsaturated fats in oils such as olive oil is thought to be preferable to consumption of polyunsaturated fats in oils such as corn oil because the monounsaturated fats apparently do not lower high-density-lipoprotein (HDL) cholesterol levels. Keeping cholesterol in the normal range not only helps prevent heart attacks and strokes but may also prevent the progression of atherosclerosis. "Statins" are a class of drugs that lowers the level of cholesterol in the blood by inhibiting the enzyme HMG-CoA reductase. This is a key enzyme involved in the biosynthesis of cholesterol in the liver.

## HORMONES

An **hormone** (from Greek *ορμη-* "impetus") is a chemical messenger that carries a signal from one cell (or group of cells) to another via the blood. All multicellular organisms produce hormones (including plants and algae - *see phytohormone*). In general, hormones regulate the function of their target cells, i.e., cells that express a receptor for the hormone. The action, or net effect of hormones is determined by a number of factors including its pattern of secretion and the response of the receiving tissue - the signal transduction response. Endocrine hormone molecules are secreted (released) directly into the bloodstream, while exocrine hormones (or ectohormones) are secreted directly into a duct, and from the duct they either flow into the bloodstream or they flow from cell to cell by diffusion in a process known as paracrine signalling.

### Hierarchical Nature of Hormonal Control

Hormonal regulation of some physiological activities involves a hierarchy of cell types acting on each other either to stimulate or to modulate the release and action of a particular hormone. The secretion of hormones from successive levels of endocrine cells is stimulated by chemical signals originating from cells higher up the hierarchical system. The master coordinator of hormonal activity in mammals is the hypothalamus, which acts on input that it receives from the central nervous system.[3]

Other hormone secretion occurs in response to local conditions, such as the rate of secretion of parathyroid hormone by the parathyroid cells in response to fluctuations of ionized calcium levels in extracellular fluid.

### Hormone Signaling

Hormonal signalling across this hierarchy involves the following:

1. **Biosynthesis** of a particular hormone in a particular tissue
2. **Storage and secretion** of the hormone
3. **Transport** of the hormone to the target cell(s)
4. **Recognition** of the hormone by an associated cell membrane or intracellular receptor protein.
5. **Relay and amplification** of the received hormonal signal via a signal transduction process: This then leads to a cellular response. The reaction of the target cells may then be recognized by the original hormone-producing cells, leading to a down-regulation in hormone production. This is an example of a homeostatic negative feedback loop.
6. **Degradation** of the hormone.

As can be inferred from the hierarchical diagram, hormone biosynthetic cells are typically of a specialized cell type, residing within a particular endocrine gland (e.g., the thyroid gland, the ovaries, or the testes). Hormones may exit their cell of origin via exocytosis or another means of membrane transport. However, the hierarchical model is an oversimplification of the hormonal signaling process. Cellular recipients of a particular hormonal signal may be one of several cell types that reside within a number of different tissues, as is the case for insulin, which triggers a diverse range of systemic physiological effects. Different tissue types may also respond differently to the same hormonal signal. Because of this, hormonal signaling is elaborate and hard to dissect.

### Interactions with Receptors

Most hormones initiate a cellular response by initially combining with either a specific intracellular or cell membrane associated receptor protein. A cell may have several different receptors that recognize the same hormone and activate different signal transduction pathways, or alternatively different hormones and their receptors may invoke the same biochemical pathway.

For many hormones, including most protein hormones, the receptor is membrane associated and embedded in the plasma membrane at the surface of the cell. The interaction of hormone and receptor typically triggers a cascade of secondary effects within the cytoplasm of the cell, often involving phosphorylation or dephosphorylation of various other cytoplasmic proteins, changes in ion channel permeability, or increased concentrations of intracellular molecules that may act as secondary messengers (e.g. cyclic AMP). Some protein hormones also interact with intracellular receptors located in the cytoplasm or nucleus by an intracrine mechanism.

For hormones such as steroid or thyroid hormones, their receptors are located intracellularly within the cytoplasm of their target cell. In order to bind their receptors these hormones must cross the cell membrane. The combined hormone-receptor complex then moves across the nuclear membrane into the nucleus of the cell, where it binds to specific DNA sequences, effectively amplifying or suppressing the action of certain genes, and affecting protein synthesis.[4] However, it has been shown that not all steroid receptors are located intracellularly, some are plasma membrane associated.[5]

An important consideration, dictating the level at which cellular signal transduction pathways are activated in response to a hormonal signal is the effective concentration of hormone-receptor complexes that are formed. Hormone-receptor complex concentrations are effectively determined by three factors:

1. The number of hormone molecules available for complex formation
2. The number of receptor molecules available for complex formation and
3. The binding affinity between hormone and receptor.

The number of hormone molecules available for complex formation is usually the key factor in determining the level at which signal transduction pathways are activated. The number of hormone molecules available being determined by the concentration of circulating hormone, which is in turn influenced by the level and rate at which they are secreted by biosynthetic cells. The number of receptors at the cell surface of the receiving cell can also be varied as can the affinity between the hormone and its receptor.

## Physiology of Hormones

Most cells are capable of producing one or more molecules, which act as signalling molecules to other cells, altering their growth, function, or metabolism. The classical hormones produced by cells in the endocrine glands mentioned so far in this article are cellular products, specialized to serve as regulators at the overall organism level. However they may also exert their effects solely within the tissue in which they are produced and originally released.

The rate of hormone biosynthesis and secretion is often regulated by a homeostatic negative feedback control mechanism. Such a mechanism depends on factors which influence the metabolism and excretion of hormones. Thus, higher hormome concentration alone can not trigger the negative feedback mechanism. Negative feedback must be triggered by overproduction of an "effect" of the hormone.

Hormone secretion can be stimulated and inhibited by:

- Other hormones (*stimulating*- or *releasing*-hormones)
- Plasma concentrations of ions or nutrients, as well as binding globulins
- Neurons and mental activity
- Environmental changes, e.g., of light or temperature

One special group of hormones is the tropic hormones that stimulate the hormone production of other endocrine glands. For example, thyroid-stimulating hormone (TSH) causes growth and increased activity of another endocrine gland, the thyroid, which increases output of thyroid hormones.

A recently-identified class of hormones is that of the "hunger hormones" - ghrelin, orexin and PYY 3-36 - and "satiety hormones" - e.g., leptin, obestatin, nesfatin-1.

In order to release active hormones quickly into the circulation, hormone biosynthetic cells may produce and store biologically inactive hormones in the form of pre- or prohormones. These can then be quickly converted into their active hormone form in response to a particular stimulus.

## Hormone Effects

Hormone effects vary widely, but can include:

- stimulation or inhibition of growth,
- In puberty hormones can affect mood and mind
- induction or suppression of apoptosis (programmed cell death)
- activation or inhibition of the immune system
- regulating metabolism
- preparation for a new activity (e.g., fighting, fleeing, mating)
- preparation for a new phase of life (e.g., puberty, caring for offspring, menopause)
- controlling the reproductive cycle

In many cases, one hormone may regulate the production and release of other hormones

Many of the responses to hormone signals can be described as serving to regulate metabolic activity of an organ or tissue.

## Chemical Classes of Hormones

Vertebrate hormones fall into three chemical classes:

- Amine-derived hormones are derivatives of the amino acids tyrosine and tryptophan. Examples are catecholamines and thyroxine.
- Peptide hormones consist of chains of amino acids. Examples of small peptide hormones are TRH and vasopressin. Peptides composed of scores or hundreds of amino acids are referred to as proteins. Examples of protein hormones include insulin and growth hormone. More complex protein hormones bear carbohydrate side chains and are called glycoprotein hormones. Luteinizing hormone, follicle-stimulating hormone and thyroid-stimulating hormone are glycoprotein hormones.
- Lipid and phospholipid-derived hormones derive from lipids such as linoleic acid and arachidonic acid and phospholipids. The main classes are the steroid hormones that derive from cholesterol and the eicosanoids. Examples of steroid hormones are testosterone and cortisol. Sterol hormones such as calcitriol are a homologous system. The adrenal cortex and the gonads are primary sources of steroid hormones. Examples of eicosanoids are the widely studied prostaglandins.

## Pharmacology

Many hormones and their analogues are used as medication. The most commonly-prescribed hormones are estrogens and progestagens (as methods of hormonal contraception and as HRT), thyroxine (as levothyroxine, for hypothyroidism) and steroids (for autoimmune diseases and several respiratory disorders). Insulin is used by many diabetics. Local preparations for use in otolaryngology often contain pharmacologic equivalents of adrenaline, while steroid and vitamin D creams are used extensively in dermatological practice.

A "pharmacologic dose" of a hormone is a medical usage referring to an amount of a hormone far greater than naturally occurs in a healthy body. The effects of pharmacologic doses of hormones may be different from responses to naturally-occurring amounts and may be therapeutically useful. An example is the ability of pharmacologic doses of glucocorticoid to suppress inflammation.

## Important Human Hormones

Spelling is not uniform for many hormones. Current North American and international usage is estrogen, gonadotropin, while British usage retains the Greek diphthong in oestrogen and the unvoiced aspirant h in gonadotrophin.

| Structure | Name | Abbre-viation | Tissue | Cells | Mechanism | Target Tissue | Effect |
|---|---|---|---|---|---|---|---|
| amine - tryptophan | Melatonin (N-acetyl-5-methoxy tryptamine) | | pineal gland | pinealocyte | | | antioxidant and causes drowsiness |
| amine - tryptophan | Serotonin | 5-HT | CNS, GI tract | enterochro-maffin cell | | | Controls mood, appetite, and sleep less active form of thyroid hormone: increase the basal metabolic rate & sensitivity to catecholamines, affect protein synthesis |
| amine - tyrosine | Thyroxine (or tetraiodo-thyronine) (a thyroid - hormone) | T4 | thyroid gland | thyroid epithelial cell | direct | | potent form of thyroid hormone: increase the basal metabolic rate & sensitivity to catecholamines, |
| amine - tyrosine | Triiodothy-ronine (a thyroid hormone) | T3 | thyroid gland | thyroid epithelial cell | direct | | affect protein synthesis Fight-or-flight response: |
| amine - tyrosine (cat) | Epinephrine (or adrenaline) | EPI | adrenal medulla | chromaffin cell | | | Boosts the supply of oxygen and glucose to the brain and muscles (by |

| Structure | Name | Abbre-viation | Tissue | Cells | Mechanism | Target Tissue | Effect |
|---|---|---|---|---|---|---|---|
| amine - tyrosine (cat) | Norepinephrine (or noradrenaline) | NRE | adrenal medulla | chromaffin cell | | | increasing heart rate and stroke volume, vasodilation, increasing catalysis of glycogen in liver, breakdown of lipids in fat cells. dilate the pupils Suppress non-emergency bodily processes (e.g. digestion) Suppress immune system Fight-or-flight response: |
| amine - tyrosine (cat) | Dopamine (or prolactin inhibiting hormone | DPM, PIH or DA | kidney, hypothalamus | Chromaffin cells in kidney Dopamine neurons of the arcuate nucleus in hypothalamus | | | Boosts the supply of oxygen and glucose to the brain and muscles (by increasing heart rate and stroke volume, vasoconstriction and increased blood pressure, breakdown of lipids in fat cells. Increase skeletal muscle readiness. |
| peptide | Antimullerian hormone (or mullerian inhibiting factor or hormone) | AMH | testes | Sertoli cell | | | Increase heart rate and blood pressure Inhibit release of prolactin and TRH from anterior pituitary |
| peptide | Adiponectin | Acrp30 | adipose tissue | | | | Inhibit release of prolactin and TRH from anterior pituitary |
| peptide | Adrenocor-ticotropic hormone (or corticotropin) | ACTH | anterior pituitary | corticotrope | cAMP | | synthesis of corticosteroids (glucocorticoids and androgens) in adrenocortical cells |
| peptide | Angiotensinogen and angiotensin | AGT | liver | | IP3 | | vasoconstriction release of aldosterone from adrenal cortex dipsogen. |
| peptide | Antidiuretic hormone (or vasopressin, arginine vasopressin) | ADH | posterior pituitary | Parvocellular neurosecretory neurons in hypothalamus Magnocellular neurosecretory cells in posterior pituitary | varies | | retention of water in kidneys moderate vasoconstriction Release ACTH in anterior pituitary |

| Structure | Name | Abbre-viation | Tissue | Cells | Mechanism | Target Tissue | Effect |
|---|---|---|---|---|---|---|---|
| peptide | Atrial-natriuretic peptide (or atriopeptin) | ANP | heart | | cGMP | | Construct bone, reduce blood Ca2+ |
| peptide | Calcitonin | CT | thyroid gland | parafollicular cell | cAMP | | Release of digestive enzymes from pancreas |
| peptide | Cholecystokinin | CCK | duodenum | | | | Release of bile from gallbladder hunger suppressant |
| peptide | Corticotropin-releasing hormone | CRH | hypothalamus | | cAMP | | Release ACTH from anterior pituitary |
| peptide | Erythropoietin | EPO | kidney | Extraglomerular mesangial cells | | | Stimulate erythrocyte production<br>In female: stimulates maturation of Graafian follicles in ovary. |
| peptide | Follicle-stimulating hormone | FSH | anterior pituitary | gonadotrope | cAMP | | In male: spermatogenesis, enhances production of androgen-binding protein by the Sertoli cells of the testes |
| peptide | Gastrin | GRP | stomach, duodenum | G cell | | | Secretion of gastric acid by parietal cells |
| peptide | Ghrelin | | stomach | P/D1 cell | | | Stimulate appetite, secretion of growth hormone from anterior pituitary gland |
| peptide | Glucagon | GCG | pancreas | alpha cells | cAMP | | glycogenolysis and gluconeogenesis in liver |
| peptide | Gonadotropin-releasing hormone | GnRH | hypothalamus | | IP3 | | increases blood glucose level<br>Release of FSH and LH from anterior pituitary. |
| peptide | Growth hormone-releasing hormone | GHRH | hypothalamus | | IP3 | | Release GH from anterior pituitary |
| peptide | Human chorionic gonadotropin | hCG | placenta | syncytio trophoblast cells | cAMP | | promote maintenance of corpus luteum during beginning of pregnancy |

| Structure | Name | Abbre-viation | Tissue | Cells | Mechanism | Target Tissue | Effect |
|---|---|---|---|---|---|---|---|
| peptide | Human placental lactogen | HPL | placenta | | | | Inhibit immune response, towards the human embryo. |
| peptide | Growth hormone | GH or hGH | anterior pituitary | somatotropes | | | increase production of insulin and IGF-1 increase insulin resistance and carbohydrate intolerance stimulates growth and cell reproduction |
| peptide | Inhibin | | testes, ovary, fetus | Sertoli cells of testes granulosa cells of ovary trophoblasts in fetus | anterior pituitary | Inhibit production of FSH | Release Insulin-like growth factor 1 from liver |
| peptide | Insulin | INS | pancreas | beta cells | tyrosine kinase | | Intake of glucose, glycogenesis and glycolysis in liver and muscle from blood |
| peptide | Insulin-like growth factor (or somatomedin) | IGF | liver | Hepatocytes | tyrosine kinase | | intake of lipids and synthesis of triglycerides in adipocytes Other anabolic effects insulin-like effects |
| peptide | Leptin | LEP | adipose tissue | | | | regulate cell growth and development decrease of appetite and increase of metabolism. |
| peptide | Luteinizing hormone | LH | anterior pituitary | gonadotropes | cAMP | | In female: ovulation In male: stimulates Leydig cell production of testosterone |
| peptide | Melanocyte stimulating hormone | MSH or α-MSH | anterior pituitary/ pars intermedia | Melanotroph | cAMP | | melanogenesis by melanocytes in skin and hair release breast milk |
| peptide | Oxytocin | OXT | posterior pituitary | Magnocellular neurosecretory cells | IP3 | | Contraction of cervix and vagina Involved in orgasm, trust between people.[6] and circadian homeostasis (body temperature, activity level, wakefulness) [7]. |

| Structure | Name | Abbre-viation | Tissue | Cells | Mechanism | Target Tissue | Effect |
|---|---|---|---|---|---|---|---|
| | | | | | | | increase blood $Ca^{2+}$:<br>• indirectly stimulate osteoclasts<br>• $Ca^{2+}$ reabsorption in kidney<br>• activate vitamin D |
| peptide | Parathyroid hormone | PTH | parathyroid gland | parathyroid chief cell | cAMP | | (Slightly) decrease blood phosphate:<br>• (decreased reuptake in kidney but increased uptake from bones<br>• activate vitamin D) |
| peptide | Prolactin | PRL | anterior pituitary, uterus | lactotrophs of anterior pituitary<br>Decidual cells of uterus | | | milk production in mammary glands<br>sexual gratification after sexual acts |
| peptide | Relaxin | RLN | uterus | Decidual cells | | | Unclear in humans<br>Secretion of bicarbonate from liver, pancreas and duodenal Brunner's glands |
| peptide | Secretin | SCT | duodenum | S cell | | | Enhances effects of cholecystokinin Stops production of gastric juice |
| peptide | Somatostatin | SRIF | hypothalamus, islets of Langerhans, gastrointestinal system | delta cells in islets<br>Neuroendocrince cells of the Periventricular nucleus in hypothalamus | | | Inhibit release of GH and TRH from anterior pituitary<br>Suppress release of gastrin, cholecystokinin (CCK), secretin, motilin, vasoactive intestinal peptide (VIP), gastric inhibitory polypeptide (GIP), enteroglucagon in gastrointestinal system<br>Lowers rate of gastric emptying<br>Reduces smooth muscle contractions and blood flow within the intestine [8]<br>Inhibit release of insulin from beta cells [9]<br>Inhibit release of glucagon from beta |

| Structure | Name | Abbreviation | Tissue | Cells | Mechanism | Target Tissue | Effect |
|---|---|---|---|---|---|---|---|
| | | | | | | | cells [9] Suppress the exocrine secretory action of pancreas. |
| peptide | Thrombopoietin | TPO | liver, kidney, striated muscle | Myocytes | | megakaryocytes | produce platelets[10] |
| peptide | Thyroid-stimulating hormone (or thyrotropin) | TSH | anterior pituitary | thyrotropes | cAMP | thyroid gland | secrete thyroxine (T4) and triiodothyronine (T3) |
| peptide | Thyrotropin-releasing hormone | TRH | hypothalamus | Parvocellular neurosecretory neurons | IP3 | anterior pituitary | Release thyroid-stimulating hormone (primarily) Stimulate prolactin release Stimulation of gluconeogenesis |
| steroid - glu. | Cortisol | | adrenal cortex (zona fasciculata and zona reticularis cells) | | direct | | Inhibition of glucose uptake in muscle and adipose tissue Mobilization of amino acids from extrahepatic tissues Stimulation of fat breakdown in adipose tissue anti-inflammatory and immunosuppressive |
| steroid - min. | Aldosterone | | adrenal cortex (zona glomerulosa) | | direct | | Increase blood volume by reabsorption of sodium in kidneys (primarily) Potassium and $H^+$ secretion in kidney. Anabolic: growth of muscle mass and strength, increased bone density, growth and strength, |
| steroid - sex (and) | Testosterone | | testes | Leydig cells | direct | | Virilizing: maturation of sex organs, formation of scrotum, deepening of voice, growth of beard and axillary hair. |
| steroid - sex (and) | Dehydroepiandrosterone | DHEA | testes, ovary, kidney | Zona fasciculata and Zona reticularis cells of kidney theca cells of ovary Leydig cellss of testes | direct | | Virilization, anabolic |

| Structure | Name | Abbre-viation | Tissue | Cells | Mechanism | Target Tissue | Effect |
|---|---|---|---|---|---|---|---|
| steroid - sex (and) | Androstenedione | | adrenal glands, gonads | | direct | | Substrate for estrogen |
| steroid - sex (and) | Dihydrotesto sterone | DHT | multiple | | direct | | |
| steroid - sex (est) | Estradiol | E2 | females: ovary, males testes | females: granulosa cells, males: Sertoli cell | direct | | Females: Structural: • promote formation of female secondary sex characteristics • accelerate height growth • accelerate metabolism (burn fat) • reduce muscle mass • stimulate endometrial growth • increase uterine growth • maintenance of blood vessels and skin • reduce bone resorption, increase bone formation Protein synthesis: • increase hepatic production of binding proteins Coagulation: • increase circulating level of factors 2, 7, 9, 10, antithrombin III, plasminogen • increase platelet adhesiveness Increase HDL, triglyceride, height growth Decrease LDL, fat depositition Fluid balance: • salt (sodium) and water retention • increase growth hormone • increase cortisol, SHBG Gastrointestinal tract: • reduce bowel motility • increase cholesterol in bile Melanin: • increase pheomelanin, |

| Structure | Name | Abbre-viation | Tissue | Cells | Mechanism | Target Tissue | Effect |
|---|---|---|---|---|---|---|---|
| | | | | | | | reduce eumelanin Cancer: support hormone-sensitive breast cancers [11] Suppression of production in the body of estrogen is a treatment for these cancers. Lung function: • promote lung function by supporting alveoli[12]. |
| steroid - sex (est) | Estrone | | ovary | granulosa cells, Adipocytes | direct | | Males: Prevent apoptosis of germ cells[13] |
| steroid - sex (est) | Estriol | | placenta | syncytiotro phoblast | direct | | Support pregnancy[14]: |
| steroid - sex (pro) | Progesterone | | ovary, adrenal glands, placenta (when pregnant) | Granulosa cells theca cells of ovary | direct | | Convert endometrium to secretory stage Make cervical mucus permeable to sperm. Inhibit immune response, e.g. towards the human embryo. Decrease uterine smooth muscle contractility[14] Inhibit lactation Inhibit onset of labor. Support fetal production of adrenal mineralo- and glucosteroids. Other: Raise epidermal growth factor-1 levels Increase core temperature during ovulation[15] Reduce spasm and relax smooth muscle (widen bronchi and regulate mucus) Antiinflammatory Reduce gall-bladder activity[16] Normalize blood clotting and vascular tone, zinc and copper levels, cell oxygen levels, and use of fat stores |

| Structure | Name | Abbre-viation | Tissue | Cells | Mechanism | Target Tissue | Effect |
|---|---|---|---|---|---|---|---|
| | | | | | | | for energy. Assist in thyroid function and bone growth by osteoblasts<br>Relsilience in bone, teeth, gums, joint, tendon, ligament and skin Healing by regulating collagen<br>Nerve function and healing by regulating myelin Prevent endometrial cancer by regulating effects of estrogen.<br>Active form of vitamin D3 |
| sterol | Calcitriol (1,25-dihydroxyvitamin D3) | | skin/proximal tubule of kidneys | | direct | | Increase absorption of calcium and phosphate from gastrointestinal tract and kidneys inhibit release of PTH |
| sterol | Calcidiol (25-hydroxyvitamin D3) | | skin/proximal tubule of kidneys | | direct | | Inactive form of Vitamin D3 |
| eicosanoid | Prostaglandins | PG | seminal vesicle | | | | Release prolactin from anterior pituitary |
| eicosanoid | Leukotrienes | LT | | white blood cells | | | lipolysis and steroidogenesis, stimulates melanocytes to produce melanin |
| eicosanoid | Prostacyclin | PGI2 | endothelium | | | | |
| eicosanoid | Thromboxane | TXA2 | | platelets | | | (To a minor degree than ANP) reduce blood pressure by: reducing systemic vascular resistance, reducing blood water, sodium and fats |
| | Prolactin releasing hormone | PRH | hypothalamus | | | | |
| | Lipotropin | PRH | anterior pituitary | Corticotropes | | | |
| | Brain natriuretic peptide | BNP | heart | Cardiac myocytes | | | increased food intake and decreased physical activity |
| | Neuropeptide Y | NPY | Stomach | | | | stimulate gastric acid secretion |
| | Histamine | | Stomach | ECL cells | | | Smooth muscle contraction of stomach [17] |
| | Endothelin | | Stomach | X cells | | | |
| | Pancreatic polypeptide | | Pancreas | PP cells | | | Unknown |
| | Renin | | Kidney | Juxtaglomerular cells | | | Activates the renin-angiotensin system by producing angiotensin I of angiotensinogen |
| | Enkephalin | | Kidney | Chromaffin cells | | | Regulate pain |

# STEROIDS

A **steroid** is a terpenoid lipid characterized by a carbon skeleton with four fused rings, generally arranged in a 6-6-6-5 fashion.

Steroids vary by the functional groups attached to these rings and the oxidation state of the rings. Hundreds of distinct steroids are found in plants, animals, and fungi. All steroids are made in cells either from the sterol lanosterol (animals and fungi) or the sterol cycloartenol (plants). Both sterols are derived from the cyclization of the triterpene squalene.

## Origin

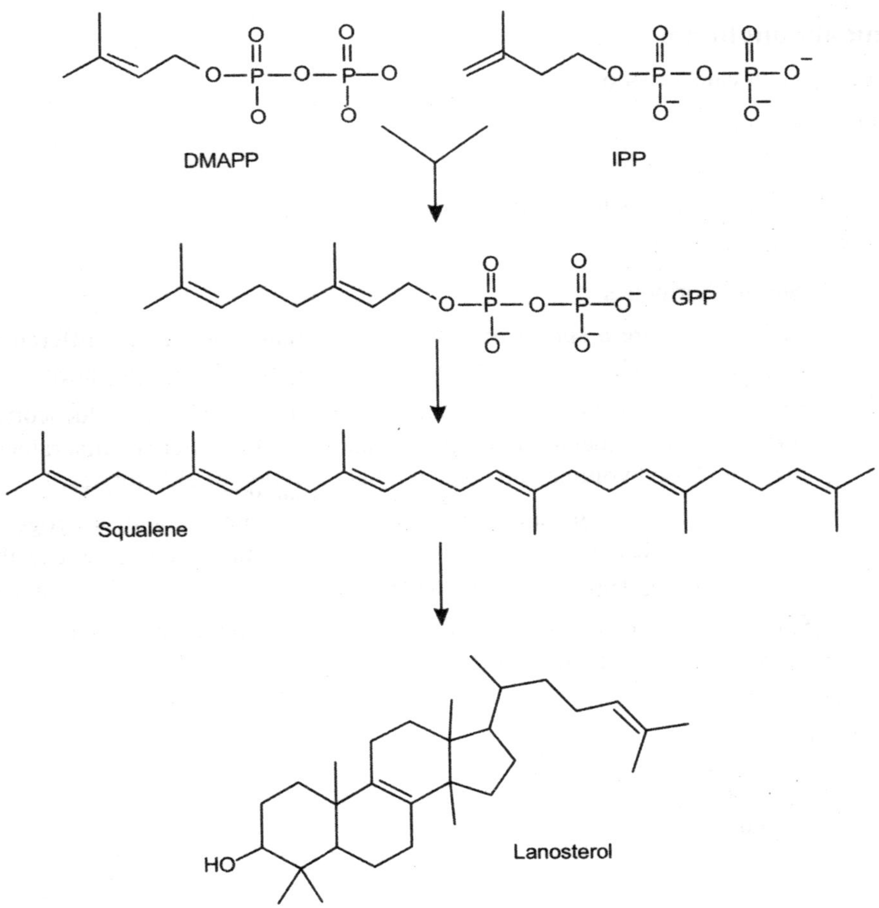

DMAPP

IPP

GPP

Squalene

Lanosterol

Simplified version of the steroid synthesis pathway with the intermediates isopentenyl pyrophosphate (IPP), dimethylallyl pyrophosphate (DMAPP), geranyl pyrophosphate (GPP) and squalene shown. Some intermediates are omitted.

Steroids include estrogen (US spelling) or oestrogen (UK/AUS spelling), progesterone and testosterone. Estrogen and progesterone are made primarily in the ovary and in the placenta during pregnancy and testosterone in the testes. Testosterone is also converted into estrogen to regulate the supply of each, in the bodies of both females and males. Certain neurons and glia in the central nervous system (CNS) express the enzymes that are required for the local synthesis of pregnane neurosteroids, either *de novo* or from peripherally derived sources. The rate limiting step of steroid synthesis is the conversion of cholesterol to pregnenolone which occurs inside the mitochondrion.

## Classification

### Taxonomical/Functional

Some of the common categories of steroids:

- Animal steroids
  - Insect steroids
    - Ecdysteroids such as ecdysterone
  - Vertebrate steroids
    - Steroid hormones
    - Sex steroids are a subset of sex hormones that produce sex differences or support reproduction. They include androgens, estrogens, and progestagens.
    - Corticosteroids include glucocorticoids and mineralocorticoids. Glucocorticoids regulate many aspects of metabolism and immune function, whereas mineralocorticoids help maintain blood volume and control renal excretion of electrolytes.
    - Anabolic steroids are a class of steroids that interact with androgen receptors to increase muscle and bone synthesis. There are natural and synthetic anabolic steroids. In popular language the word "steroids" usually refers to anabolic steroids.
    - Cholesterol which modulates the fluidity of cell membranes and is the principle constituent of the plaques implicated in atherosclerosis.
- Plant steroids
  - Phytosterols
  - Brassinosteroids
- Fungus steroids
  - Ergosterols

### Structural

It is also possible to classify steroids based upon their chemical composition. One example of how MeSH performs this classification is available at the Wikipedia MeSH catalog. Examples from this classification include:

| Class | Examples | Number of carbon atoms |
|---|---|---|
| Cholstanes | cholesterol | 27 |
| Cholanes | cholic acid | 22 |
| Pregnanes | progesterone | 21 |
| Androstanes | testosterone | 19 |
| Estranes | estradiol | 18 |

CHAPTER

41

# Organic Reactions and Mechanism

## Introduction

In this chapter we are going to discuss about the organic reaction and mechanism. The different methods and types of reactions used in the organic synthesis are also been discussed here. Different bonding and the methods of bonding are also being explained.

The students are requested to read this chapter and carefully use them while experimenting in organic chemistry and organic synthesis.

## Chemical Bonding

The chemical bonding in organic chemistry includes the carbon – carbon bonding, the carbon – hydrogen bonding and carbon – oxygen bonding mainly. We are going to discuss these aspects in detail. The bonding depends upon the electrons in motion, here when two atomic orbitals, bond together the shape of bonding becomes quite different. An *s* – orbital is spherical, the p-orbital looks like two touching spheres or spherical eight. These orbitals have negative positive or zero polarity, but this is not the type of the electrical charge since the both half of the electron clouds are of negative charge, these are the polarity of the wave function $\psi$ of electron.

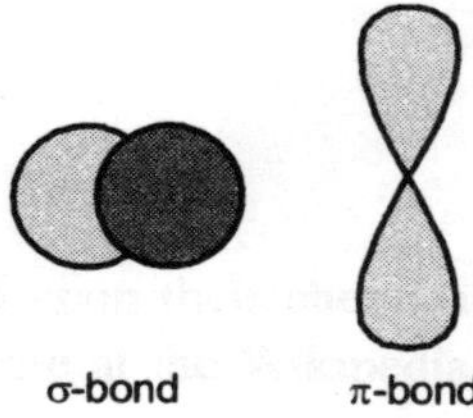

σ-bond π-bond

When the orbitals are separated by a node the have opposite sign on each half, and this increment of nodes increases with increment of the energy. Each orbital is occupied by two electrons of opposite spin

## Orbital Hybridization

### Historical Development

The hybridisation theory was promoted by chemist Linus Pauling in order to explain the structure of molecules such as methane ($CH_4$). Historically, this concept was developed for such simple chemical systems but the approach was later applied more widely, and today it is considered an effective heuristic for rationalizing the structures of organic compounds.

Hybridisation theory is not as practical for quantitative calculations as Molecular Orbital Theory. Problems with hybridisation are especially notable when the *d* orbitals are involved in bonding, as in coordination chemistry and organometallic chemistry. Although hybridisation schemes in transition metal chemistry can be used, they are not generally as accurate.

It is important to note that orbitals are a model representation of the behaviour of electrons within molecules. In the case of simple hybridisation, this approximation is based on the atomic orbitals of hydrogen. Hybridised orbitals are assumed to be mixtures of these atomic orbitals, superimposed on each other in various proportions. Hydrogen orbitals are used as a basis for simple schemes of hybridisation because it is one of the few examples of orbitals for which an exact analytic solution to its Schrödinger equation is known. These orbitals are then assumed to be slightly, but not significantly, distorted in heavier atoms, like carbon, nitrogen, and oxygen. Under these assumptions is the theory of hybridisation most applicable. It must be noted that one does not need hybridisation to describe molecules, but for molecules made up from carbon, nitrogen and oxygen (and to a lesser extent, sulfur and phosphorus) the hybridisation theory/model makes the description much easier.

The hybridisation theory finds its use mainly in organic chemistry, and mostly concerns C, N and O (and to a lesser extent P and S). Its explanation starts with the way bonding is organized in methane. In chemistry, **hybridisation** or **hybridization** (see also spelling differences) is the concept of mixing atomic orbitals to form new *hybrid orbitals* suitable for the qualitative description of atomic bonding properties. Hybridised orbitals are very useful in the explanation of the shape of molecular orbitals for molecules. It is an integral part of valence bond theory. Although sometimes taught together with the valence shell electron-pair repulsion (VSEPR) theory, valence bond and hybridization are in fact not related to the VSEPR model.

The hybridisation theory finds its use mainly in organic chemistry, and mostly concerns C, N and O (and to a lesser extent P and S). Its explanation starts with the way bonding is organized in methane.

### *sp*³ Hybrids

Hybridisation describes the bonding atoms from an atom's point of view. That is, for a tetrahedrally coordinated carbon (e.g. methane, $CH_4$), the carbon should have 4 orbitals with the correct symmetry

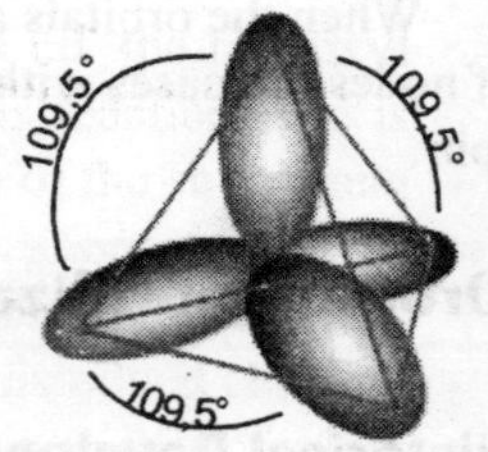


to bond to the 4 hydrogen atoms. The problem with the existence of methane is now this: Carbon's ground-state configuration is $1s^2\,2s^2\,2p_x^{\,1}\,2p_y^{\,1}$ or perhaps more easily read:

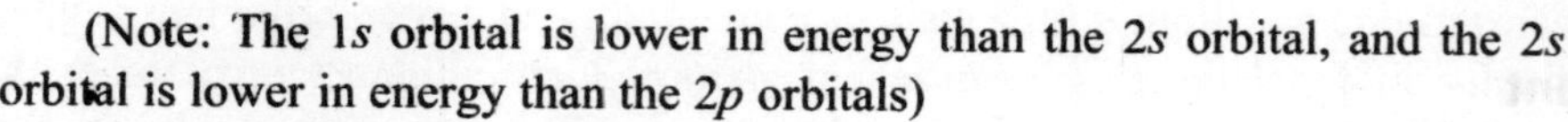


(Note: The 1*s* orbital is lower in energy than the 2*s* orbital, and the 2*s* orbital is lower in energy than the 2*p* orbitals)

The valence bond theory would predict, based on the existence of two half-filled *p*-type orbitals (the designations $p_x p_y$ or $p_z$ are meaningless at this point, as they do not fill in any particular order), that C forms two covalent bonds. $CH_2$. However, methylene is a very reactive molecule (see also: carbene) and cannot exist outside of a molecular system. Therefore, this theory alone cannot explain the existence of $CH_4$.

Furthermore, ground state orbitals cannot be used for bonding in $CH_4$. While exciting a 2*s* electron into a 2*p* orbital would theoretically allow for four bonds according to the valence bond theory, (which has been proved experimentally correct for systems like $O_2$) this would imply that the various bonds of $CH_4$ would have differing energies due to differing levels of orbital overlap. Once again, this has been experimentally disproved: any hydrogen can be removed from a carbon with equal ease.

To summarise, to explain the existence of $CH_4$ (and many other molecules) a method by which as many as 12 bonds (for transition metals) of equal strength (and therefore equal length) can be created was required.

The first step in hybridisation is the excitation of one (or more) electrons (we consider the carbon atom in methane, for simplicity of the discussion):

C* ↑↓ (1s) ↑ (2s) ↑ ($2p_x$) ↑ ($2p_y$) ↑ ($2p_z$)

The proton that forms the nucleus of a hydrogen atom attracts one of the valence electrons on carbon. This causes an excitation, moving a 2s electron into a 2p orbital. This, however, increases the influence of the carbon nucleus on the valence electrons by increasing the effective core potential (the amount of charge the nucleus exerts on a given electron = Charge of Core " Charge of all electrons closer to the nucleus).

The combination of these forces creates new mathematical functions known as hybridised orbitals. In the case of carbon attempting to bond with four hydrogens, four orbitals are required. Therefore, the 2*s* orbital (core orbitals are almost never involved in bonding) mixes with the three 2*p* orbitals to form four ***sp*³ hybrids** (read as *s-p-three*). See graphical summary below.

becomes

C* ↑↓ (1s) ↑ ($sp^3$) ↑ ($sp^3$) ↑ ($sp^3$) ↑ ($sp^3$)

In $CH_4$, four $sp^3$ hybridised orbitals are overlapped by hydrogen's 1*s* orbital, yielding four σ (sigma) bonds. The four bonds are of the same length and strength. This theory fits our requirements.

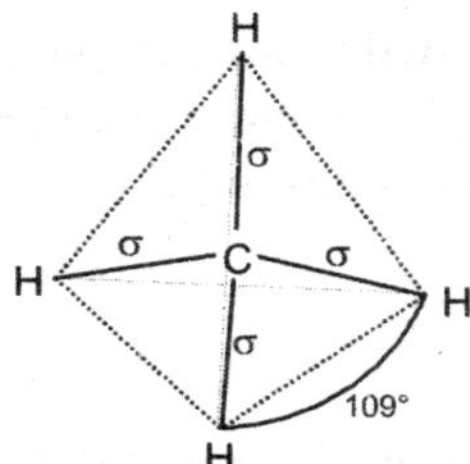


An alternative view is: View the carbon as the $C^{4''}$ anion. In this case all the orbitals on the carbon are filled:

$C^{4-}$ ↑↓ ↑↓ ↑↓ ↑↓ ↑↓
1s 2s $2p_x$ $2p_y$ $2p_z$

If we now recombine these orbitals with the empty *s*-orbitals of 4 hydrogens (4 protons, $H^+$) and allow maximum separation between the 4 hydrogens (i.e. tetrahedral surrounding of the carbon), we see that at any orientation of the *p*-orbitals, a single hydrogen has an overlap of 25% with the *s*-orbital of the C, and a total of 75% of overlap with the 3 *p*-orbitals (see that the relative percentages are the same as the character of the respective orbital in an $sp^3$-hybridisation model, 25% *s*- and 75% *p*-character).

According to the orbital hybridisation theory the valence electrons in methane should be equal in energy but its photoelectron spectrum [3] shows two bands, one at 12.7 eV (one electron pair) and one at 23 eV (three electron pairs). This apparent inconsistency can be explained when one considers additional orbital mixing taking place when the $sp^3$ orbitals mix with the 4 hydrogen orbitals.

## $sp^2$ Hybrids

Other carbon based compounds and other molecules may be explained in a similar way as methane. Take, for example, ethene ($C_2H_4$). Ethene has a double bond between the carbons. The Lewis structure looks like this:

H H
C=C
H H

Carbon will $sp^2$ hybridise, because hybrid orbitals will form only σ bonds and one π (pi) bond is required for the double bond between the carbons. The hydrogen-carbon bonds are all of equal strength and length, which agrees with experimental data.

In ***sp*² hybridisation** the 2*s* orbital is mixed with only two of the three available 2*p* orbitals:

C* ↑↓ ↑ ↑ ↑ ↑
1s $sp^2$ $sp^2$ $sp^2$ p

forming a total of 3 $sp^2$ orbitals with one p-orbital remaining. In ethylene the two carbon atoms form a σ bond by overlapping two $sp^2$ orbitals and each carbon atom forms two covalent bonds with hydrogen by $s$–$sp^2$ overlap all with 120° angles. The π bond between the carbon atoms perpendicular to the molecular plane is formed by 2*p*–2*p* overlap (however, the π bond may or may not occur).

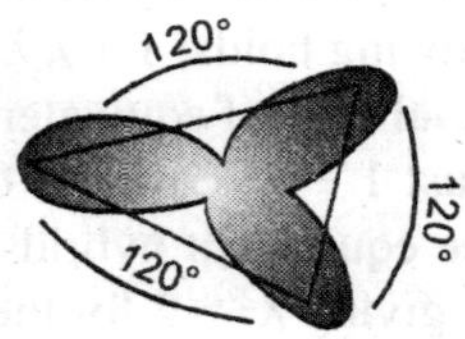

The amount of *p*-character is not restricted to integer values, i.e. hybridisations like $sp^{2.5}$ are also readily described. In this case the geometries are somewhat distorted from the ideally hybridised picture. For example, as stated in Bent's rule, a bond tends to have higher *p*-character when directed toward a more electronegative substituent.

## *sp* Hybrids

The chemical bonding in compounds such as alkynes with triple bonds is explained by ***sp* hybridization.**

C* ↑↓ (1s) ↑ (sp) ↑ (sp) ↑ (p) ↑ (p)

In this model the 2*s* orbital mixes with only one of the three *p*-orbitals resulting in two *sp* orbitals and two remaining unchanged *p* orbitals. The chemical bonding in acetylene (ethyne) ($C_2H_2$) consists of *sp*–*sp* overlap between the two carbon atoms forming a ó bond and two additional π bonds formed by *p*–*p* overlap. Each carbon also bonds to hydrogen in a sigma *s*–*sp* overlap at 180° angles.

**Hybridisation and molecule shape**

Hybridisation, along with the VSEPR theory, helps to explain molecule shape:

- $AX_1$ (e.g., LiH): no hybridisation; trivially linear shape
- $AX_2$ (e.g., $BeCl_2$): *sp* hybridisation; linear or **diagonal** shape; bond angles are $\cos^{-1}(-1) = 180°$
  - $AX_2E$ (e.g., $GeF_2$): bent/V shape, < 120°
- $AX_3$ (e.g., $BCl_3$): $sp^2$ hybridisation; trigonal planar shape; bond angles are $\cos^{-1}(-1/2) = 120°$
  - $AX_3E$ (e.g.,$NH_3$): trigonal pyramidal, 107°
- $AX_4$ (e.g., $CCl_4$): $sp^3$ hybridisation; tetrahedral shape; bond angles are $\cos^{-1}(-1/3) = 109.5°$
- $AX_5$ (e.g., $PCl_5$): $sp^3d$ hybridisation; trigonal bipyramidal shape
- $AX_6$ (e.g., $SF_6$): $sp^3d^2$ hybridisation; octahedral (or square bipyramidal) shape

This holds if there are no lone electron pairs on the central atom. If there are, they should be counted in the $X_i$ number, but bond angles become smaller due to increased repulsion. For example, in water ($H_2O$), the oxygen atom has two bonds with H and two lone electron pairs (as can be seen with the valence bond theory as well from the electronic configuration of oxygen), which means there are four such 'elements' on O. The model molecule is, then, $AX_4$: $sp^3$ hybridisation is utilized, and the electron arrangement of $H_2O$ is tetrahedral. This agrees with the experimentally-determined shape for water, a non-linear, bent structure, with a bond angle of 104.5 degrees (the two lone-pairs are not visible).

In general, for an atom with s and p orbitals forming hybrids $h_i$ and $h_j$ with included angle θ, the following holds: $1 + \lambda_i\lambda_j \cos(\theta) = 0$. The p-to-s ratio for hybrid i is $\lambda_i^2$, and for hybrid j it is $\lambda_j^2$. In the special case of equivalent hybrids on the same atom, again with included angle θ, the equation reduces to just $1 + \lambda^2 \cos(\theta) = 0$. For example, $BH_3$ has a trigonal planar geometry, three 120° bond angles, three equivalent hybrids about the boron atom, and thus $1 + \lambda^2 \cos(\theta) = 0$ becomes $1 + \lambda^2 \cos(120°) = 0$, giving $\lambda^2 = 2$ for the p-to-s ratio. In other words, $sp^2$ hybrids, just as expected from the list above.

## Controversy Regarding d-orbital Participation

Hybridisation theory has failed in a few aspects, notably in explaining the energy considerations for the involvement of d-orbitals in chemical bonding (See above for $sp^3d$ and $sp^3d^2$ hybridisation). This can be well-explained by means of an example. Consider, for instance, how the theory in question accounts for the bonding in phosphorus pentachloride ($PCl_5$). d-orbitals are large, comparatively distant from the nucleus and high in energy. Radial distances of orbitals from the nucleus seem to reveal that d-orbitals are far too high in energy to 'mix' with s- and p-orbitals. 3s - 0.47 , 3p - 0.55, 3d - 2.4 (in angstroms). Thus, at first sight, it seems improbable for $sp^3d$ hybridisation to occur.

However, a deeper look into the factors that affect orbital size(and energy) reveals more. Formal charge on the central atom is one such factor, and it's obvious that the P atom in $PCl_5$ has quite a large partial positive charge on itself. Thus the 3d orbital contracts in size to enough of an extent so that hybridisation may occur with s and p orbitals. Further, note the cases in which d-orbital participation was proposed in hybridisation: $SF_6$(sulfur hexafluoride), $IF_7$, $XeF_6$; in all these molecules, the central atom is surrounded by the highly electronegative fluorine atom, thus making hybridisation probable among s, p and d orbitals. A further study reveals that orbital size also depends on the number of electrons occupying it. And even further, coupling of d orbital electrons also results in contraction, albeit to a smaller extent.

The molecular orbital theory, however, offers a clearer insight into the bonding in these molecules.

## Hybridisation Theory vs. MO Theory

Hybridisation theory is an integral part of organic chemistry and in general discussed together with molecular orbital theory in advanced organic chemistry textbooks although for different reasons. One textbook notes that for drawing reaction mechanisms sometimes a classical bonding picture is needed with 2 atoms sharing two electrons. It also comments that predicting bond angles in methane with MO theory is not straightforward. Another textbook treats hybridisation theory when explaining bonding in alkenes and a third uses MO theory to explain bonding in hydrogen but hybridisation theory for methane.

Although the language and pictures arising from Hybridisation Theory, more widely known as Valence Bond Theory, remain widespread in synthetic organic chemistry, this qualitative analysis of bonding has been largely superseded by molecular orbital theory in other branches of chemistry. For example, inorganic chemistry texts have all but abandoned instruction of hybridisation, except as a historical footnote.[7][8] One specific problem with hybridisation is that it incorrectly predicts the photoelectron spectra of many molecules, including such fundamental species such as methane and water. From a pedagogical perspective, hybridisation approach tends to over-emphasize localisation of bonding electrons and does not effectively embrace molecular symmetry as does MO Theory.

# Ligands

In chemistry, a **ligand** is an atom, ion, or molecule (see also: functional group) that generally donates one or more of its electrons through a coordinate covalent bond to, or shares its electrons through a covalent bond with, one or more central atoms or ions (these ligands act as a Lewis base). Fewer examples exist where a molecule can be described as a ligand that accepts electrons from a Lewis base (hence, the ligand acts as a Lewis acid).

Most commonly the central atom is a metal or metalloid in inorganic chemistry. But in organic chemistry ligands are also used to protect functional groups (e.g. borane $BH_3$ as ligand for the protection of phosphine $PH_3$), or to stabilize reactive compounds (tetrahydrofuran, THF as a ligand for $BH_3$ to make $BH_3$ easier to handle). The molecule resulting from the coordination of a ligand (or an array of ligands) to a central atom is termed a complex.

Factors that characterize the ligands are their charge, size (bulk), and of course the nature of the constituent atoms.

The ligands in a complex:

- stabilize the central atom, and
- dictate the reactivity of the central atom.
- **Ligands in metal complexes**
- The constitution of metal complexes has been described by Alfred Werner, who developed the basis for modern coordination chemistry. The ligands that are directly bonded to the metal (that is, share electrons), are called "inner sphere" ligands. If the inner-sphere ligands do not balance the charge of the central atom (the oxidation number), this may be done by simple ionic bonding with another set of counter ions (the "outer-sphere" ligands). The complex of the metal with the inner sphere ligands is then called a complex ion (which can be either cationic or anionic). The complex, along with its counter ions, is called a coordination compound. The size of a ligand is indicated by its cone angle.
- **Donation and back-donation**
- In general, ligands donate electron density to the (electron deficient) central atom - that is, they overlap between the highest occupied molecular orbital (HOMO) of the ligand with the lowest unoccupied molecular orbital (LUMO) of the central atom. The ligand thus acts as a Lewis base by donating electron density (in general electron pairs) to the central atom, acting as a Lewis acid. In some cases ligands donate only one electron from a singly occupied orbital (the donating atom in these ligands is a radical).
- Some metal centers in combination with certain ligands (e.g. carbon monoxide (CO)) can be further stabilised by donating electron density back to the ligand in a process known as *back-bonding*. In this case a filled, central-atom-based orbital donates density into the LUMO of the (coordinated) ligand.
- **Strong field vs. weak field ligands**
- Ligands and metal ions can be ordered by their 'hardness' (see also hard soft acid base theory). Certain metal ions have a preference for certain ligands. In general, 'hard' metal ions prefer weak field ligands, whereas 'soft' metal ions prefer strong field ligands. From a MO point of view, the HOMO of the ligand should have an energy that makes overlap with the LUMO of the metal preferential. Metal ions bound to strong-field ligands follow the Aufbau principle, whereas complexes bound to weak-field ligands follow Hund's rule (see crystal field theory).

- Binding of the metal with the ligands results in a set of molecular orbitals, where the metal can be identified with a new HOMO and LUMO (the orbitals defining the properties and reactivity of the resulting complex) and a certain ordering of the 5 d-orbitals (which may be filled, or partially filled with electrons). In an octahedral environment, the 5 otherwise degenerate d-orbitals split in sets of 2 and 3 orbitals (for a more in depth explanation, see crystal field theory).
- 3 orbitals of low energy: $d_{xy}$, $d_{xz}$ and $d_{yz}$
- 2 of high energy: $d_{z^2}$ and $d_{x^2-y^2}$
- The energy difference between these 2 sets of d-orbitals is called the splitting parameter, $\Delta_o$. The magnitude of $\Delta_o$ is determined by the field-strength of the ligand: strong field ligands, by definition, increase $\Delta_o$ more than weak field ligands. Ligands can now be sorted according to the magnitude of $\Delta_o$ (see the table below). This ordering of ligands is almost invariable for all metal ions and is called spectrochemical series.
- For complexes with a tetrahedral surrounding, the d-orbitals again split into two sets, but this time in reverse order:
- 2 orbitals of low energy: $d_{z^2}$ and $d_{x^2-y^2}$
- 3 orbitals of high energy: $d_{xy}$, $d_{xz}$ and $d_{yz}$
- The energy difference between these 2 sets of d-orbitals is now called $\Delta_t$. The magnitude of $\Delta_t$ is smaller than for $\Delta_o$, because in a tetrahedral complex only 4 ligands influence the d-orbitals, whereas in an octahedral complex the d-orbitals are influenced by 6 ligands. When the coordination number is neither octahedral nor tetrahedral, the splitting becomes correspondingly more complex. For the purposes of ranking ligands, however, the properties of the octahedral complexes and the resulting $\Delta_o$ has been of primary interest.
- The arrangement of the d-orbitals on the central atom (as determined by the 'strength' of the ligand), has a strong effect on virtually all the properties of the resulting complexes. E.g. the energy differences in the d-orbitals has a strong effect in the optical absorption spectra of metal complexes. It turns out that valence electrons occupying orbitals with significant 3d-orbital character absorb in the 400-800 nm region of the spectrum (UV-visible range). The absorption of light (what we perceive as the color) by these electrons (that is, excitation of electrons from one orbital to another orbital under influence of light) can be correlated to the ground state of the metal complex, which reflects the bonding properties of the ligands. The relative change in (relative) energy of the d-orbitals as a function of the field-strength of the ligands is described in Tanabe-Sugano diagrams.
- **Denticity**
- Some ligand molecules are able to bind to the metal ion through multiple sites because they have free lone pairs on more than one atom. Ligands that bind to more than one site are termed *chelating* (from the Greek for *claw*). For example, a ligand binding through two sites is *bidentate*

and three sites is *tridentate*. The *bite angle* refers to the angle between the two bonds of a bidentate chelate. A classic example of a polydentate ligand is EDTA. It is able to bond through six sites, completely surrounding some metals. Polydentate ligands are commonly formed by joining organic molecules together. The number of atoms with which a polydentate ligand bind to the metal centre is called its denticity (symbol κ). κ indicates the number non-contiguous donor sites by which a ligand attaches to a metal. Hapticity (η) and denticity are often confused. Hapticity refers to contiguous atoms that are attached to a metal. Ethylene forms $\eta^2$ complexes. Ethylenediamine forms $\kappa^2$ complexes. Cyclopentadienyl is typically bonded in $\eta^5$ mode because each bonded carbon is connected directly. $EDTA^{4-}$ on the other hand, when it is sexidentate, is $\kappa^6$ mode, the amines and the carboxylate oxygen atoms are not connected directly. To simplify matters, $\eta^n$ tends to refer to unsaturated hydrocarbons and $\kappa^n$ tends to describe polydentate amine and carboxylate ligands.

- Complexes of polydentate ligands are called *chelate* complexes. They tend to be more stable than monodentate complexes, as it is necessary to break all of the bonds to the central atom for the ligand to be displaced. This increased stability or inertness is called the chelate effect. Entropy favors the displacement of many ligands by one polydentate ligand. The increase in the total number of molecules in solution is favourable.
- Related to the chelate effect is the macrocyclic effect. A macrocyclic ligand is any large cyclic ligand which at least partially surrounds the central atom and bonds to it, leaving the central atom at the centre of a large ring. The more rigid and the higher its denticity, the more inert will be the macrocyclic complex. Heme is a good example, the iron atom is at the centre of a porphyrin macrocycle, being bound to four nitrogen atoms of the tetrapyrrole macrocycle. The very stable dimethylglyoximate complex of nickel is a synthetic macrocycle derived from the anion of dimethylglyoxime.
- Unlike polydentate ligands, ambidentate ligands can attach to the central atom in two places but not both. A good example of this is thiocyanide, $SCN^-$, which can attach at either the sulfur atom or the nitrogen atom. Such compounds give rise to linkage isomerism.

**Common ligands**

- Virtually every molecule and every ion can serve as a ligand for (or "coordinate to") metals. Monodentate ligands include virtually all anions and all simple Lewis bases. Thus, the halides and pseudohalides are important anionic ligands whereas ammonia, carbon monoxide, and water are particularly important charge-neutral ligands. Simple organic compounds are also very important, be they anionic ($RO^-$ and $RCO_2^-$) or neutral ($R_2O$, $R_2S$, $R_{3-x}NH_x$, and $R_3P$). Beyond the classical Lewis bases and anions, all unsaturated molecules are also ligands, utilizing their π-electrons in forming the coordinate bond. Also, metals can bind to the ó bonds in for example silanes, hydrocarbons, and dihydrogen (see also: agostic interaction).

## Examples of Common Ligands (by Field Strength)

In the following table the ligands are sorted by field strength (weak field ligands first):

| Ligand | formula (bonding atom(s) in bold) | Charge | Most common denticity | Remark(s) |
|---|---|---|---|---|
| Iodide | $I^-$ | monoanionic | monodentate | |
| Bromide | $Br^-$ | monoanionic | monodentate | |
| Sulphide | $S^{2-}$ | dianionic | monodentate (M=S), or bidentate bridging (M-S-M') | |
| Thiocyanate | $S\text{-}CN^-$ | monoanionic | monodentate | ambidentate (see also isothiocyanate, *vide infra* |
| Chloride | $Cl^-$ | monoanionic | monodentate | also found bridging |
| Nitrate | $O\text{-}NO_2^-$ | monoanionic | monodentate | |
| Azide | $N\text{-}N_2^-$ | monoanionic | monodentate | |
| Fluoride | $F^-$ | monoanionic | monodentate | |
| Hydroxide | $O\text{-}H^-$ | monoanionic | monodentate | often found as a bridging ligand |
| Oxalate | $[O\text{-}C(=O)\text{-}C(=O)\text{-}O]^{2-}$ | dianionic | bidentate | |
| Water | H-O-H | neutral | monodentate | monodentate |
| Isothiocyanate | $N{=}C{=}S^-$ | monoanionic | monodentate | ambidentate (see also thiocyanate, *vide supra*) |
| Acetonitrile | $CH_3CN$ | neutral | monodentate | |
| Pyridine | $C_5H_5N$ | neutral | monodentate | |
| Ammonia | $NH_3$ | neutral | monodentate | |
| Ethylenediamine | en | neutral | bidentate | |
| 2,2'-Bipyridine | bipy | neutral | bidentate | easily reduced to its (radical) anion or even to its dianion |
| 1,10-Phenanthroline | phen | neutral | bidentate | |
| Nitrite | $O\text{-}N\text{-}O^-$ | monoanionic | monodentate | ambidentate |
| Triphenylphosphine | $PPh_3$ | neutral | monodentate | |
| Cyanide | $CN^-$ | monoanionic | monodentate | can bridge between metals (both metals bound to C, or one to C and one to N) |
| Carbon monoxide | CO | neutral | monodentate | can bridge between metals (both metals bound to C) |

**Note:** The entries in the table are sorted by field strength, binding through the stated atom (i.e. as a terminal ligand), the 'strength' of the ligand changes when the ligand binds in an alternative binding mode (e.g. when it bridges between metals) or when the conformation of the ligand gets distorted (e.g. a linear ligand that is forced through steric interactions to bind in a non-linear fashion).

## Other General Encountered Ligands (Alphabetical)

In this table other common ligands are listed in alphabetical order.

| Ligand | Formula (bonding atom(s) in bold) | Charge | Most common denticity | Remark(s) |
|---|---|---|---|---|
| Acetylacetonate (Acac) | $CH_3$-C(**O**)-CH-C(**O**)-$CH_3$ | monoanionic | bidentate | In general bidentate, bound through both oxygens, but sometimes bound through the central carbon only,see also analogous ketimine analogues |
| Alkenes | $R_2$**C=C**$R_2$ | neutral | | compounds with a C-C double bond |
| Benzene | **C**$_6H_6$ | neutral | | and other arenes |
| 1,2-Bis(diphenyl phosphino) ethane (dppe) | $Ph_2$**P**$C_2H_4$**P**$Ph_2$ | neutral | bidentate | |
| Corroles | | | tetradentate | |
| Crown ethers | | neutral | | primarily for alkali and alkaline earth metal cations |
| 2,2,2-crypt | | | hexadentate | primarily for alkali and alkaline earth metal cations |
| Cryptates | | neutral | | |
| Cyclopentadienyl | $[C_5H_5]^-$ | monoanionic | | |
| Diethylenetriamine (dien) | | neutral | tridentate | related to TACN, but not constrained to facial complexation |
| Dimethylglyoximate ($dmgH^-$) | | monoanionic | | |
| Ethylenediaminetetra-acetate (EDTA) | | tetra-anionic | hexadentate | actual ligand is the tetra anion |
| Ethylenediaminetria-cetate | | trianionic | pentadentate | actual ligand is the trianion |
| glycinate | | | bidentate | other α-amino acid anions are comparable (but chiral) |
| Heme | | dianionic | tetradentate | macrocyclic ligand |
| Nitrosyl | **NO**$^+$ | cationic | | bent (1e) and linear (3e) bonding mode |
| Scorpionate ligand | | | tridentate | |
| Sulfite | | monoanionic | monodentate | ambidentate |

| | | | | |
|---|---|---|---|---|
| 2,2',5',2-*Terpyridine* *(terpy)* | | neutral | tridentate | meridional bonding only |
| Thiocyanate | | monoanionic | monodentate | ambidentate, sometimes bridging |
| Triazacyclononane (tacn) | $(C_2H_4)_3(NR)_3$ | neutral | tridentate | macrocyclic ligand see also the N,N',N"-trimethylated analogue |
| Triethylenetetramine (trien) | | neutral | tetradentate | |
| Tris(2-aminoethyl) amine (trien) | | neutral | tetradentate | |
| Tris(2-diphenylpho-sphineethyl)amine $(np_3)$ | | neutral | tetradentate | |
| Terpyridine | | neutral | tridentate | |

## Lactam and Lactim

A **lactam** (the noun is a portmanteau of the words *lactone* + *amide*) is a cyclic amide. Prefixes may indicate the ring size: β-lactam (4-membered), γ-lactam (5-membered), δ-lactam (6-membered ring). That order in the nomenclature is because beta β, gamma γ and delta δ are the second, third and fourth letters in the alphabetical order of the Greek alphabet, respectively.

### Synthesis

General synthetic methods exist for the organic synthesis of lactams.

- Lactams form by the acid-catalyzed rearrangement of oximes in the Beckmann rearrangement.
- Lactams form from cyclic ketones and ammonia in the Schmidt reaction.
- lactams form from cyclisation of amino acids.
- In **iodolactamization** [1] an iminium ion reacts with an halonium ion formed in situ by reaction of an alkene with iodine.

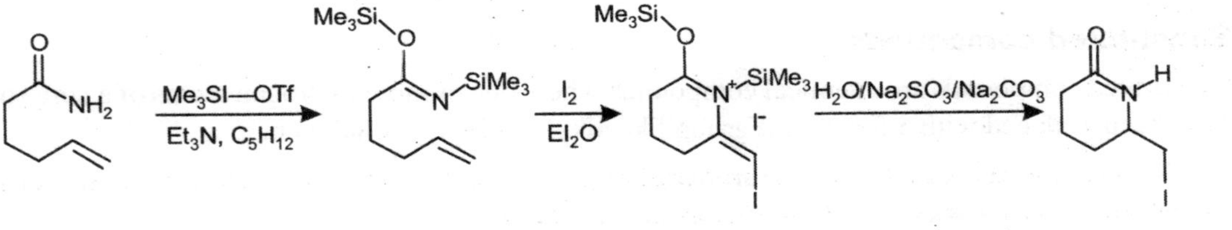

$Me_3SI—OTf$, $Et_3N$, $C_5H_{12}$; $I_2$, $EI_2O$; $H_2O/Na_2SO_3/Na_2CO_3$

- Lactams form by copper catalyzed 1,3-dipolar cycloaddition of alkynes and nitrones in the Kinugasa reaction

### Reactions

- Lactams can polymerize to polyamides.

**Lactim** is a hydroxy imide compound characterized by an enolic group in a ring configuration. It is tautomeric with lactam

## Substitution Reaction

In a **substitution reaction**, a functional group in a particular chemical compound is replaced by another group.

In organic chemistry, the electrophilic and nucleophilic substitution reactions are of prime importance. Organic substitution reactions are classified in several main organic reaction types depending on whether the reagent that brings about the substitution is considered an electrophile or a nucleophile, whether a reactive intermediate involved in the reaction is a carbocation, a carbanion or a free radical or whether the substrate is aliphatic or aromatic. Detailed understanding of a reaction type helps to predict the product outcome in a reaction. It also is helpful for optimizing a reaction with regard to variables such as temperature and choice of solvent.

### Nucleophilic substitutions

- A nucleophile reacts with an aliphatic substrate in a nucleophilic aliphatic substitution reaction.
- When the substrate is an aromatic compound the reaction type is nucleophilic aromatic substitution.
- Carboxylic acid derivatives react with nucleophiles in nucleophilic acyl substitution. This kind of reaction can be useful in preparing compounds

### Electrophilic substitutions

- Electrophiles are involved in electrophilic substitution reactions and particularly in electrophilic aromatic substitutions.
- Electrophilic reactions to other unsaturated compounds than arenes generally lead to electrophilic addition rather than substitution.

### Radical substitutions

A radical substitution reaction involves radicals.

### Substituted compounds

Substituted compounds are chemical compounds where one or more hydrogen atoms of a core structure have been replaced with a functional group like alkyl, hydroxy, or halogen.

For example benzene is a simple aromatic ring and substituted benzenes are a heterogeneous group of chemicals with a wide spectrum of uses and properties:

- Benzene $C_6H_6$
- Toluene $C_6H_5$-$CH_3$

- Xylene $C_6H_4(-CH_3)_2$
- Mesitylene $C_6H_3(-CH_3)_3$
- Phenol $C_6H_5$-OH
- Aniline $C_6H_5-NH_2$
- Chlorobenzene $C_6H_5$-Cl
- Nitrobenzene $C_6H_5-NO_2$
- Benzoic acid $C_6H_5$-COOH
- Picric acid $C_6H_2(-OH)(-NO_2)_3$
- Trinitrotoluene $C_6H_2(-CH_3)(-NO_2)_3$
- Salicylic acid $C_6H_4$(-OH)(-COOH)
- Acetylsalicylic acid $C_6H_4(-O-C(=O)-CH_3)$(-COOH)
- Paracetamol $C_6H_4(-NH-C(=O)-CH_3)$(-OH)
- Phenacetin $C_6H_4(-NH-C(=O)-CH_3)(-O-CH_2-CH_3)$

## $S_N1$ Reaction

The **$S_N1$ reaction** is a substitution reaction in organic chemistry. "$S_N$" stands for nucleophilic substitution and the "1" represents the fact that the rate-determining step is unimolecular. It involves a carbocation intermediate and is commonly seen in reactions of secondary or tertiary alkyl halides or, under strongly acidic conditions, with secondary or tertiary alcohols. With primary alkyl halides, the alternative $S_N2$ reaction occurs. Among inorganic chemists, $S_N1$ is referred to perhaps more accessibly as a *dissociative mechanism*. A reaction mechanism was first proposed by Christopher Ingold et al in 1940

## Mechanism

The $S_N1$ reaction between a molecule **A** and a nucleophile **B** takes place in three steps:

1. Formation of a carbocation from **A** by separation of a leaving group from the carbon; this step is slow and reversible [4].
2. Nucleophilic attack: **B** reacts with **A**. If the nucleophile is a neutral molecule (i.e. a solvent) a third step is required to complete the reaction. When the solvent is water, the intermediate is an oxonium ion.
3. Deprotonation: Removal of a proton on the protonated nucleophile by a nearby ion or molecule.

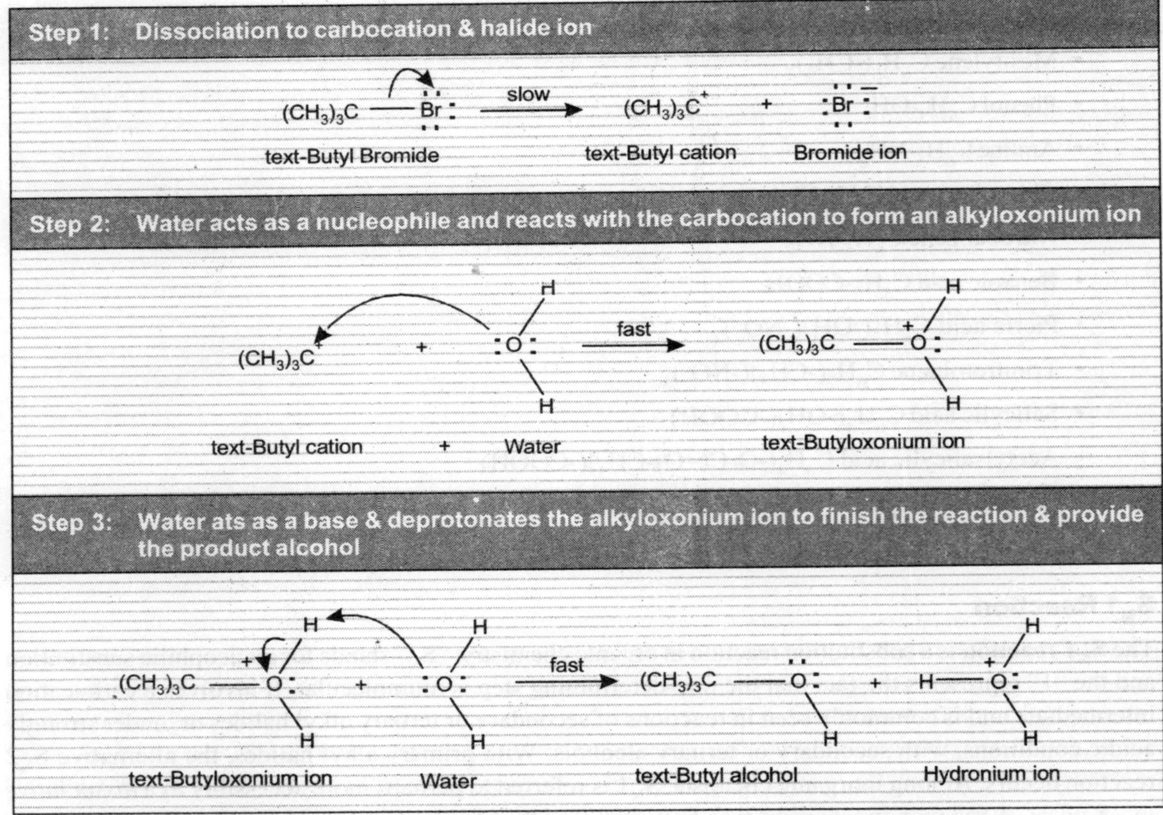


## Kinetics

In contrast to $S_N2$, $S_N1$ reactions take place in two steps (excluding any protonation or deprotonation). The rate determining step is the first step, so the rate of the overall reaction is essentially equal to that of carbocation formation and does not involve the attacking nucleophile. Thus nucleophilicity is irrelevant and the overall reaction rate depends on the concentration of the reactant only.

$$\text{rate} = k[\text{reactant}]$$

In 1954 it was found that addition of a small amount of lithium perchlorate to certain acetolysis reactions (for example that of the tosylate of cholesterol) led to a remarkable reaction rate increase [5]. Based on this **special salt effect** the general mechanism was refined to include a contact ion pair (**CIP**) with cation and anion together in a solvent cage which then dissociates to a so-called solvent-separated ion pair (SSIP) and then on to free ions (FI). All the interconversions are reversible and the added salt prevents the reformation of CIP from SSIP.

In some cases the $S_N1$ reaction will occur at an abnormally high rate due to neighbouring group participation (NGP). NGP often lowers the energy barrier required for the formation of the carbocation intermediate.

## Scope of the Reaction

The $S_N1$ mechanism tends to dominate when the central carbon atom is surrounded by bulky groups because such groups sterically hinder the $S_N2$ reaction. Additionally, bulky substituents on the central carbon increase the rate of carbocation formation because of the relief of steric strain that occurs. The resultant carbocation is also stabilized by both inductive stabilization and hyperconjugation from attached alkyl groups. The Hammond-Leffler postulate suggests that this too will increase the rate of carbocation formation. The $S_N1$ mechanism therefore dominates in reactions at tertiary alkyl centers and is further observed at secondary alkyl centers in the presence of weak nucleophiles.

## Stereochemistry

Because the intermediate carbocation is planar, the central carbon is *not* a stereocenter. Even if it were a stereocenter prior to becoming a carbocation, the original configuration at that atom is lost. Rather, the central carbon can be prochiral. Nucleophilic attack can occur from either side of the plane, so the product might consist of a mixture of two stereoisomers. In fact, if the central carbon is the only stereocenter in the reaction, racemization may occur. This stands in contrast to the $S_N2$ mechanism, where the chiral configuration of the substrate is inverted. However, an excess of inversion is usually observed, as the leaving group can remain in proximity to the carbocation intermediate for a short time and block nucleophilic attack. For example, in the reaction of *3S*-chloro-3-methylhexane with iodide ion, if the carbocation intermediate is free of the leaving group then it is achiral and stands an equal chance of attack on either side. This leads to a mixture of *3R*-iodo-3-methylhexane and *3S*-iodo-3-methylhexane:

S-enantiometer

Planar carbocation is achiral

$I^-$ attacks from left

$I^-$ attacks from right

Enantiomers

S-enantiometer

## Side Reactions

Two common side reactions are elimination reactions and carbocation rearrangement. If the reaction is performed under warm or hot conditions (which favour an increase in entropy), E1 elimination is likely to predominate, leading to formation of an alkene. Even if the reaction is performed cold, some alkene may be formed. If an attempt is made to perform an $S_N1$ reaction using a strongly basic nucleophile such as hydroxide or methoxide ion, the alkene will again be formed, this time via an E2 elimination.

This will be especially true if the reaction is heated. Finally, if the carbocation intermediate can rearrange to a more stable carbocation, it will give a product derived from the more stable carbocation rather than the simple substitution product.

## Solvent Effects

Since the $S_N1$ reaction involves formation of an unstable carbocation intermediate in the rate-determining step, anything that can facilitate this will speed up the reaction. The normal solvents of choice are both *polar* (to stabilise ionic intermediates in general) and *protic* (to solvate the leaving group in particular). Typical polar protic solvents include water and alcohols, which will also act as nucleophiles.

The **Y scale** correlates solvolysis reaction rates of any solvent (**k**) with that of a standard solvent (80% v/v ethanol/water) (**$k_0$**) through

$$\log\left(\frac{k}{k_0}\right) = mY$$

with **m** a reactant constant (m = 1 for tert-butyl chloride) and **Y** a solvent parameter [6]For example 100% ethanol gives Y = - 2.3, 50% ethanol in water Y = +1.65 and 15% concentration Y = +3.2

## $S_N2$ Reaction

The $S_N2$ **reaction** (also known as **bimolecular nucleophilic substitution**) is a type of nucleophilic substitution, where a lone pair from a nucleophile attacks an electron deficient electrophilic center and bonds to it, expelling another group called a leaving group. Thus the incoming group replaces the leaving group in one step. Since two reacting species are involved in the slow, rate-determining step of the reaction, this leads to the name ***bimolecular nucleophilic substitution***, or $S_N2$. The somewhat more transparently named analog to $S_N2$ among inorganic chemists is the *interchange mechanism*.

## Reaction Mechanism

The reaction most often occurs at an aliphatic $sp^3$ carbon center with an electronegative, stable leaving group attached to it - 'X' - frequently a halide atom. The breaking of the C-X bond and the formation of the new C-Nu bond occur simultaneously to form a transition state in which the carbon under nucleophilic attack is pentacoordinate, and approximately $sp^2$ hybridised. The nucleophile attacks the carbon at 180° to the leaving group, since this provides the best overlap between the nucleophile's lone pair and the C-X $\sigma^*$ antibonding orbital. The leaving group is then pushed off the opposite side and the product is formed.

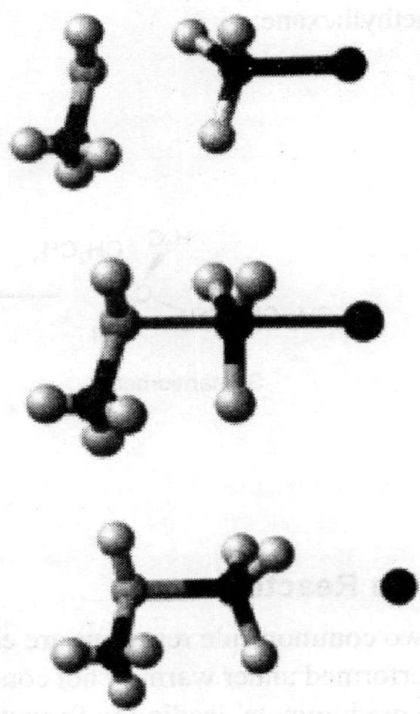

Ball-and-stick represenation of the $S_N2$ reaction of MeSH with MeI

If the substrate under nucleophilic attack is chiral, this leads to an inversion of stereochemistry, called the Walden inversion.

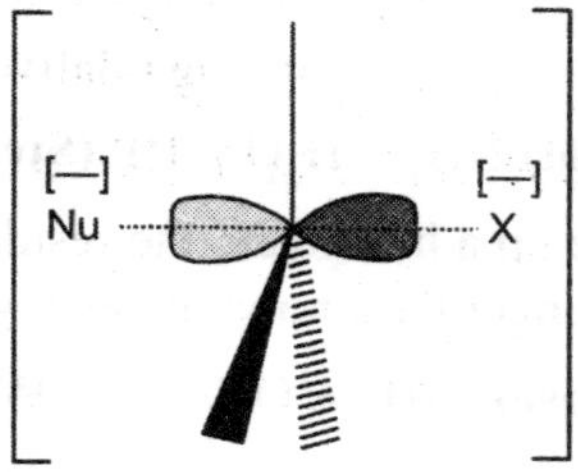

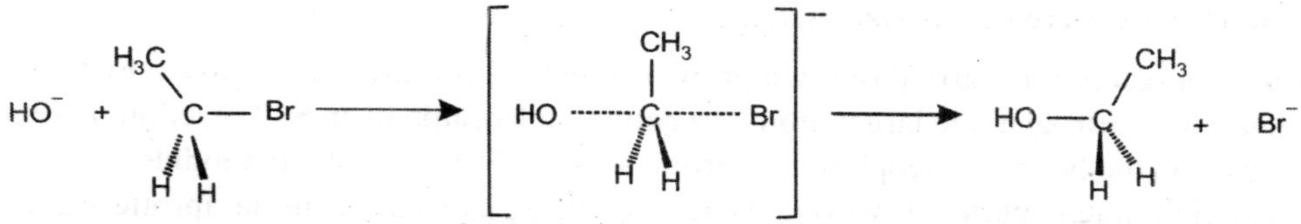

$S_N2$ reaction of bromoethane with hydroxide ion. The products are ethanol and a bromine ion.

In an example of the $S_N2$ reaction, the attack of $OH^-$ (the nucleophile) on a bromoethane (the electrophile) results in ethanol, with bromide ejected as the leaving group.

$S_N2$ attack occurs if the backside route of attack is not sterically hindered by substituents on the substrate. Therefore this mechanism usually occurs at an unhindered primary carbon centre. If there is steric crowding on the substrate near the leaving group, such as at a tertiary carbon centre, the substitution will involve an $S_N1$ rather than an $S_N2$ mechanism, (an $S_N1$ would also be more likely in this case because a sufficiently stable carbocation intermediary could be formed.)

## Factors Affecting Reaction

**The Basicity of the Leaving Group**. By comparing the relative $S_N2$ reaction rates of compounds with atoms in the same periodic group (the halides, for example), results show that the ability as a leaving group during an $S_N2$ reaction depends on its basicity. In general, *the weaker the basicity of a group, the greater its leaving ability*. For example, the iodide ion is a very weak base and because it is so, it is the most reactive. Weak bases do not share their electrons well because their electrons are farther away from the nucleus, making it easier for their bonds to be broken. In contrast, the fluoride ion is a strong base and, therefore, the least reactive. In fact, the fluoride ion is such a strong base that compounds involving them essentially do not undergo $S_N2$ reaction. Looking at the periodic table, relative basicity decreases down a group.

**(Stronger Base)** $F^- > Cl^- > Br^- > I^-$ **(Weaker Base)**

**The Size of the Nucleophile**. How readily a compound attacks an electron-deficient atom also affects an $S_N2$ reaction. As a rule, a negatively charged species (e.g. $OH^-$) are better nucleophiles than neutral species (e.g. $H_2O$, water). There is a direct relationship between basicity and nucleophilicity: *stronger bases are better nucleophiles*. Acidity, the ability of an atom to give up a proton ($H^+$), is comparatively relative when molecules whose *attacking atoms are approximately the same in size*, the weakest going toward the left side of the periodic table. Elements are approximately the same size across a row in the periodic table. If hydrogen were attached to second-row elements of the periodic table, the resulting compounds would have the following relative acidities:

**(Weaker Acid)** $NH_3 < H_2O < HF$ **(Stronger Acid)**

If each of these acids were to give up a hydrogen, the result would be its conjugate base, and the relative strengths will reverse. The stronger base now moves toward the left side of the periodic table.

**(Stronger Base)** $^-NH_2 > OH^- > F^-$ **(Weaker Base)**

Elements increase in size down the periodic table. Although basicity decreases down the periodic table, nucleophilicity increases as size increases *depending on the solvent used.*

**Solvent**. If a reaction is carried out in a protic solvent, whose molecules have a hydrogen bonded to an oxygen or to a nitrogen, the larger atom is a better nucleophile in an $S_N2$ reaction. In other words, the weaker base is the better nucleophile in a protic solvent. For example, the iodide ion is better than a fluoride ion as a nucleophile. However, if the reaction is carried out in an aprotic solvent, whose molecules do not have hydrogen bonded to an oxygen or to a nitrogen, then the stronger base is the better nucleophile. In this case, the fluoride ion is better than the iodide ion as a nucleophile.

**Sterics**. Steric hindrance is any effect of a compound due to the size and/or arrangement of its substituent groups. Steric effects affect nucleophilicity but *does not affect base strength*. A bulky nucleophile, such as a tert-butoxide ion with its specific arrangement of methyl groups, is a poorer nucleophile than an ethoxide ion with a straighter chain of carbons, even though *tert*-butoxide is a stronger base.

## Reaction Kinetics

The rate of an $S_N2$ reaction is second order, as the rate-determining step depends on the nucleophile concentration, $[Nu^-]$ as well as the concentration of substrate, [RX].

$$r = k[RX][Nu^-]$$

This is a key difference between the $S_N1$ and $S_N2$ mechanisms. In the $S_N1$ reaction the nucleophile attacks after the rate-limiting step is over, whereas in $S_N2$ the nucleophile forces off the leaving group in the limiting step. In cases where both mechanisms are possible (for example at a secondary carbon centre), the mechanism depends on solvent, temperature, concentration of the nucleophile or on the leaving group.

$S_N2$ reactions are generally favoured in primary alkyl halides or secondary alkyl halides with an aprotic solvent. They occur at a negligible rate in tertiary alkyl halides due to steric hindrance.

It is important to understand that $S_N2$ and $S_N1$ are two extremes of a sliding scale of reactions, it is possible to find many reactions which exhibit both $S_N2$ and $S_N1$ character in their mechanisms. For instance, it is possible to get a contact ion pairs formed from an alkyl halide in which the ions are not fully separated. When these undergo substitution the stereochemistry will be inverted (as in $S_N2$) for many of the reacting molecules but a few may show retention of configuration.

## E2 Competition

A common side reaction taking place with $S_N2$ reactions is E2 elimination: the incoming anion can act as a base rather than as a nucleophile, abstracting a proton and leading to formation of the alkene. This effect can be demonstrated in the gas-phase reaction between a sulfonate and a simple alkyl bromide taking place inside a mass spectrometer:

R' — O⁻, gas phase, –Br

R = H 90% 10%
R = $CH_3$ 15% 85%

With ethyl bromide, the reaction product is predominantly the substitution product. As steric hindrance around the electrophilic center increases, as with isobutyl bromide, substitution is disfavoured and elimination is the predominant reaction. Other factors favoring elimination are the strength of the base. With the less basic benzoate substrate, isopropyl bromide reacts with 55% substitution. In general, gas phase reactions and solution phase reactions of this type follow the same trends, even though in the first, solvent effects are eliminated.

### Roundabout Mechanism

A development attracting attention in 2008 concerns a $S_N2$ **roundabout mechanism** observed in a gas-phase reaction between chloride ions and methyl iodide with a special technique called *crossed molecular beam imaging*. When the chloride ions have sufficient velocity the energy of the resulting iodine ions after the collision is much lower than expected and it is theorized that energy is lost as a result of a full roundabout of the methyl group around the iodine atom before the actual displacement takes place

## $S_N$i or Substitution Nucleophilic Internal

**$S_N$i** or **Substitution Nucleophilic internal** stands for a specific but not often encountered nucleophilic aliphatic substitution reaction mechanism. A typical representative organic reaction displaying this mechanism is the chlorination of alcohols with thionyl chloride and the main feature is retention of stereochemical configuration. Thionyl chloride first reacts with the alcohol to form an alkyl chloro sulfite. In the second step the sulfite group is lost and just like in an $S_N1$ reaction an alkyl carbocation is created. The actual nucleophile in the third reaction step is the chlorine atom attached to the sulfite group which recombines with this carbocation. The crucial difference with the standard SN1 mechanism is that since the nucleophile resides at the same side as the original leaving group i.d. the hydroxyl group, the stereochemistry is *retention* of configuration and not racemization. This reaction type is linked to many forms of Neighbouring group participation, for instance the reaction of the sulfur lone pair in sulfur mustard to form the cationic intermediate. This reaction mechanism is supported by the observation that addition of pyridine to the reaction leads to inversion. The reasoning behind this finding is that pyridine reacts with the intermediate sulfite replacing chlorine. The dislodged chloride has to resort to nucleophilic attack from the rear as in a regular nucleophilic substitution.In the complete picture for this reaction the sulfite reacts with a chlorine ion in a standard $S_N2$ reaction with *inversion* of configuration. When the solvent is also a nucleophile such as dioxane two successive $S_N2$ reactions take place and the stereochemistry is again *retention*. With standard SN1 reaction conditions the reaction outcome is *retention* via a competing SNi mechanism and not racemization and with pyridine added the result is again *inversion*.

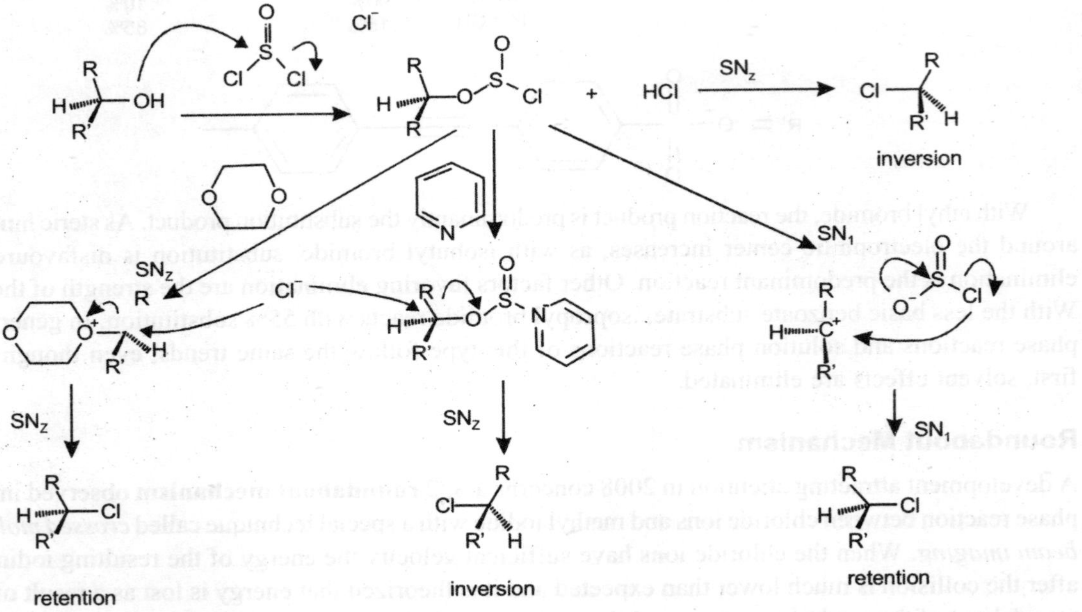

SN2 inversion SN1 SN2 SN2 SN2 SN1 retention inversion retention

## Nulceophile

In chemistry, a **nucleophile** (literally *nucleus lover* as in nucleus and phile) is a reagent that forms a chemical bond to its reaction partner (the electrophile) by donating both bonding electrons.[1] Because nucleophiles donate electrons, they are by definition Lewis bases (see *acid-base reaction theories*). All molecules or ions with a free pair of electrons can act as nucleophiles, although negative ions (anions) are more potent than neutral reagents. Neutral nucleophilic reactions with solvents such as alcohols and water are named solvolysis.

Nucleophiles may take part in nucleophilic substitution, whereby a nucleophile becomes attracted to a full or partial positive charge on an element and displaces the group it is bonded to.

**Nucleophilic** is an adjective that describes the affinity of a nucleophile to the nuclei, while **nucleophilicity** or *nucleophile strength* refers to the nucleophilic character. Nucleophilicity is often used to compare an atom's relative affinity to another's.

In general, in a row across the periodic table, the more basic the ion (the higher the $pK_a$ of the conjugate acid), the more reactive it is as a nucleophile. In a given group, polarizability is more important in the determination of the nucleophilicity: the easier it is to distort the electron cloud around an atom or molecule, the more readily it will react. *e.g.*, the iodide ion ($I^-$) is more nucleophilic than the fluoride ion ($F^-$).

An **ambident nucleophile** is one that can attack from two or more places, resulting in two or more products. For example, the thiocyanate ion ($SCN^-$) may attack from either the S or the N. For this reason, the $S_N2$ reaction of an alkyl halide with $SCN^-$ often leads to a mixture of RSCN (an alkyl thiocyanate) and RNCS (an alkyl isothiocyanate). Similar considerations apply in the Kolbe nitrile synthesis.

The terms *nucleophile* and *electrophile* were introduced by Christopher Kelk Ingold in 1929, replacing the terms *cationoid* and *anionoid* proposed earlier by A. J. Lapworth in 1925

## Common Examples

In the example below, the oxygen of the hydroxide ion donates an electron to bond with the carbon at the end of the bromopropane molecule. The bond between the carbon and the bromine then undergoes heterolytic fission, with the bromine atom taking the donated electron and becoming the bromide ion ($Br^-$): The picture is incorrect, because a Sn2 reaction occurs by backside attack. This means that the hydroxide ion attacks the carbon atom from the other side, exactly opposite the bromine ion.

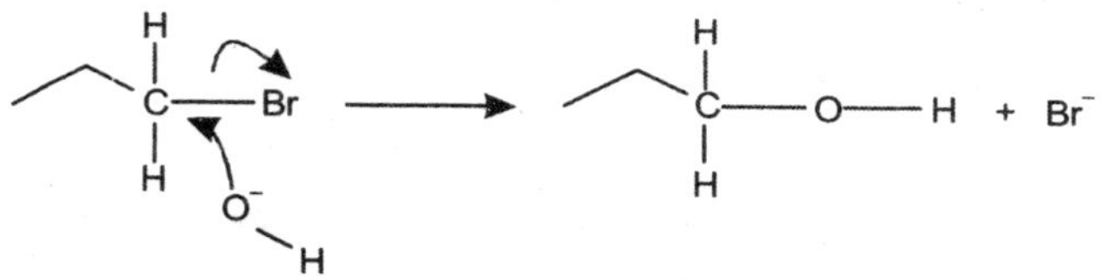

## Carbon Nucleophiles

Carbon nucleophiles are alkyl metal halides found in the Grignard reaction, Blaise reaction, Reformatsky reaction, and Barbier reaction, organolithium reagents and anions of a terminal alkyne.

Enols are also carbon nucleophiles. The formation of an enol is catalyzed by acid or base. Enols are ambident nucleophiles, but generally nucleophilic at the alpha carbon atom. Enols are commonly used in condensation reactions, including the Claisen condensation and the aldol condensation reactions.

## Oxygen Nucleophiles

Examples of oxygen nucleophiles are Water ($H_2O$) and Alcohols.

## Sulphur Nucleophiles

Sulfur nucleophiles are Thiols ($HS^-$).

Sulfur is generally very nucleophilic because of its large size, which makes it easily polarizable, and its lone pairs of electrons (in some cases).

## Nitrogen Nucleophiles

Nitrogen nucleophiles are Ammonia, Azide and Amines.

## Nucleophilicity Scales

Many schemes have been devised attempting to quantify relative nucleophilic strength. The following empirical data have been obtained by measuring reaction rates for a large number of reactions involving a large number of nucleophiles and electrophiles and linear regression. Nucleophiles displaying the so-called alpha effect are usually omitted in this type of treatment.

## Swain-Scott Equation

The first such attempt is found in the so-called **Swain-Scott equation** derived in 1953. This free-energy relationship relates the pseudo first order reaction rate constant (in water at 25°C), **k**, of a reaction, normalized to the reaction rate, $\mathbf{k_0}$, of a standard reaction with water as the nucleophile, to a **nucleophilic constant n** for a given nucleophile and a **substrate constant s** that depends on the sensitivity of a substrate to nucleophilic attack (defined as 1 for methyl bromide).

This treatment results in the following values for typical nucleophilic anions: acetate 2.7, chloride 3.0, azide 4.0, hydroxide 4.2, aniline 4.5, iodide 5.0 and thiosulfate 6.4. Typical substrate constants are 0.66 for ethyl tosylate, 0.77 for α-propiolactone, 1.00 for 2,3-epoxypropanol, 0.87 for benzyl chloride and 1.43 for benzoyl chloride.

The equation predicts that in a nucleophilic displacement on benzyl chloride, the azide anion reacts 3000 times faster than water.

## Richie Equation

The **Richie equation** named after its creator and derived in 1972 is another free-energy relationship:[6][7][8]

$$\log (k/k_0) = N^+$$

or

$$\log (k) = N^+ + \log (k_0)$$

where $N^+$ is the nucleophile dependent parameter and $k_0$ the reaction rate constant for water. In this equation a substrate dependent parameter like s in the Swain-Scott equation is absent. The equation states that two nucleophiles react with the same relative reactivity regardless of the nature of the electrophile which is in violation of the Reactivity–selectivity principle. For this reason this equation is also called the **constant selectivity relationship**.

In the original publication the data were obtained by reactions of selected nucleophiles with selected electrophilic carbocations such as tropylium cations:

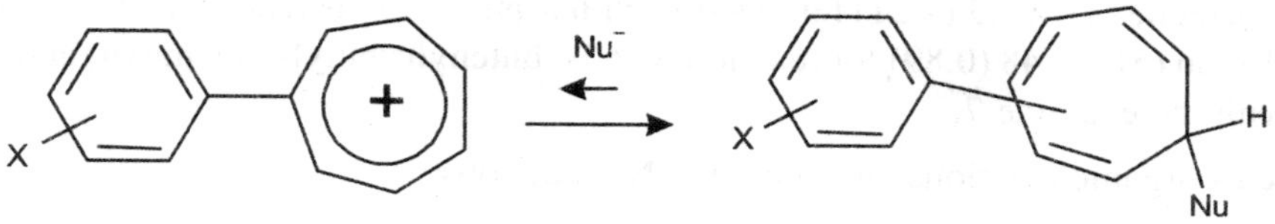

or diazonium cations:

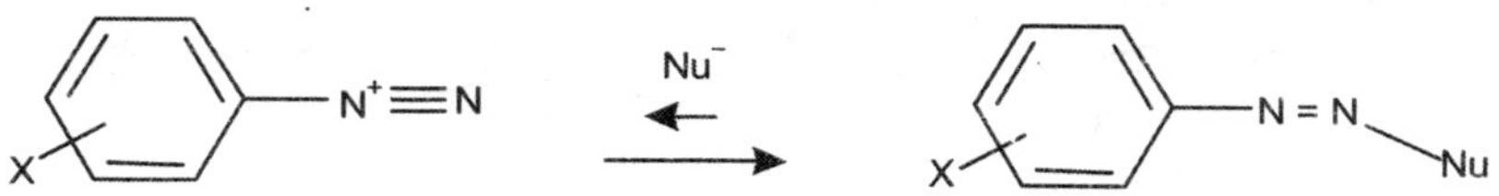

or (not displayed) ions based on Malachite green. Subsequently many other reaction types were described.

Typical Richie $N^+$ values (in methanol) are: 0.5 for methanol, 5.9 for the cyanide anion, 7.5 for the methoxide anion , 8.5 for the azide anion and 10.7 for the thiophenol anion. The values for the relative cation reactivities are -0.4 for the malachite green cation, +2.6 for the benzenediazonium cation and +4.5 for the tropylium cation.

## Mayr-Patz Equation

$$log(k) = S(N + E)$$

In the Mayr-Patz equation (1994):

The second order reaction rate constant **k** at 20°C for a reaction is related to a **nucleophilicity parameter N**, an **electrophilicity parameter E** and a **nucleophile-dependent slope parameter s**. The constant s is defined as 1 with *2-methyl-1-pentene* as the nucleophile.

Many of the constants have been derived from reaction of so-called benzhydrylium ions as the electrophiles:[10] 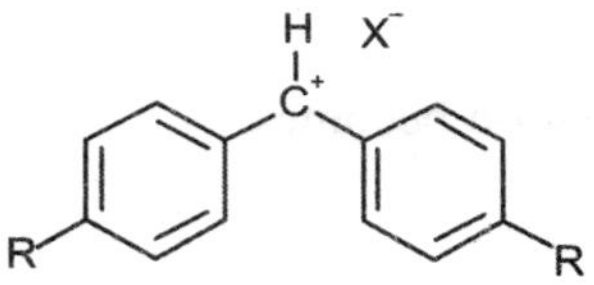 and a diverse collection of π-nucleophiles:

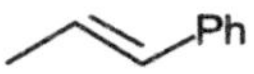

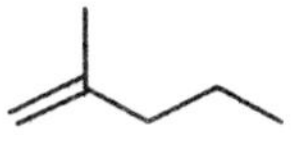

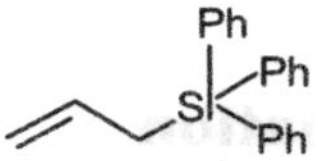

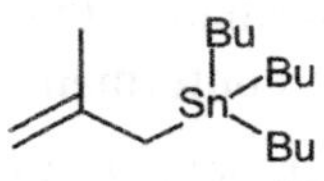

Typical E values are +6.2 for R = chlorine, +5.90 for R = hydrogen, 0 for R = methoxy and -7.02 for R = dimethylamine.

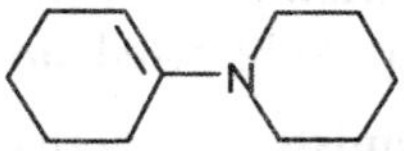

Typical N values with s in parenthesis are -4.47 (1.32) for electrophilic aromatic substitution to toluene (1), -1.41 (1.12) for electrophilic addition to 1-phenyl-2-propene (2) and 0.96 (1) for addition to 2-methyl-1-pentene (3), -0.13 (1.21) for reaction with triphenylallylsilane (4), 3.61 (1.11) for reaction with 2-methylfuran (5), +7.48 (0.89) for reaction with isobutenyltributylstannane (6) and +13.36 (0.81) for reaction with the enamine 7.

The range of organic reactions also include SN2 reactions:

Nu

+ Nu — $CH_3$

S⁺ X⁻ $CH_3$ S

With E = -9.15 for the *S-methyldibenzothiophenium ion*, typical nucleophile values N (s) are 15.63 (0.64) for piperidine, 10.49 (0.68) for methoxide and 5.20 (0.89) for water. In short: nucleophilicities towards $sp_2$ or $sp_3$ centers follow the same pattern.

## Unified Equation

In an effort to unify the above described equations the Mayr equation is rewritten as:

$$\log(k) = s_E s_N (N + E)$$

with $s_E$ the electrophile-dependent slope parameter and $s_N$ the nucleophile-dependent slop parameter. This equation can be rewritten in several ways:

- with $s_E$ = 1 for carbocations this equation is equal to the original Mayr-Patz equation of 1994,
- with $s_N$ = 0.6 for most n nucleophiles the equation becomes
  $\log(k) = 0.6 s_E N + 0.6 s_E E$
  or the original Scott-Swain equation written as:
  $\log(k) = \log(k_0) + s_E N$
- with $s_E$ = 1 for carbocations and $s_N$ = 0.6 the equation becomes:
  $\log(k) = 0.6N + 0.6E$
  or the original Ritchie equation written as:
  $\log(k)$ " $\log(k_0) = N^+$

# Staggered Conformation

A **staggered conformation** is a chemical conformation that exists in any open chain single chemical bond connecting two $sp_3$ hybridised atoms as a conformational *energy minimum*.

**Alkane stereochemistry** concerns the stereochemistry of linear alkanes and the linear alkane conformers. The existence of more than one conformation is due to hindered rotation around $sp^3$ hybridised carbon carbon bonds. The smallest molecule with such a chemical bond, ethane, is found to exist as two conformers, **staggered** and **eclipsed**.

- A staggered conformation is a chemical conformation that exists in any open chain single chemical bond connecting two $sp^3$ hybridised atoms as a conformational *energy minimum*.

- An eclipsed conformation is a chemical conformation that exists in any open chain single chemical bond connecting two $sp^3$ hybridised atoms as a conformational *energy maximum*. This is believed to be due to steric hindrance, although a role for hyperconjugation is proposed by a com.

## Conformations

Alkane stereochemistry concerns the stereochemistry of linear alkanes and the linear alkane conformers. The existence of more than one conformation is due to hindered rotation around $sp^3$ hybridised carbon carbon bonds. The smallest molecule with such a chemical bond, ethane, exists as an infinite number of conformations with respect to rotation around the C-C bond; two of these are recognised as energy minimum (staggered) and energy maximum (eclipsed) forms. The importance of these is seen by extension of these concepts to more complex molecules for which stable conformations may be predicted as minimum energy forms.

In the example of staggered ethane in Newman projection, a hydrogen atom on one carbon atom has a 60° **torsional angle** or **torsion angle** with respect to the nearest hydrogen atom on the other carbon so that steric hindrance is minimised. The staggered conformation is more stable by 12.5 kJ/mol than the eclipsed conformation, which is the energy maximum for ethane. In the eclipsed conformation the torsional angle is minimized.

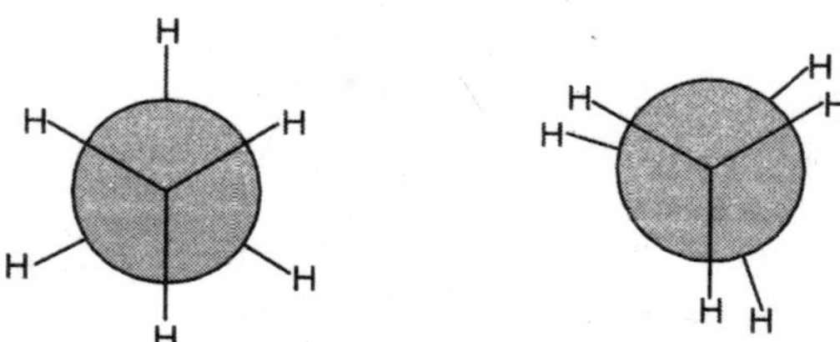

staggered conformation left, eclipsed conformation right in Newman projection

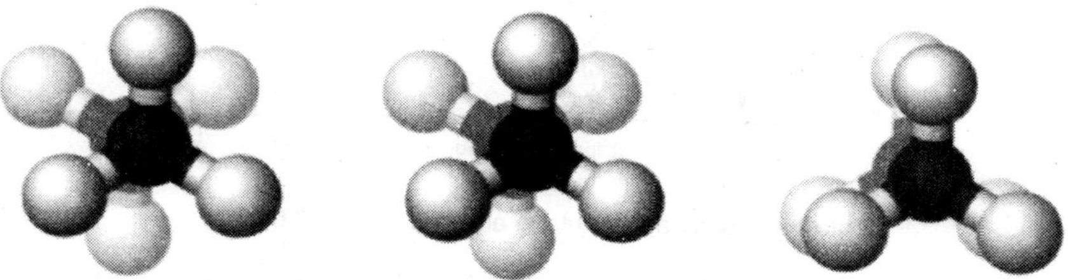

In butane, the two staggered conformations are no longer equivalent and represent two distinct conformers:the **anti-conformation** (left-most, below) and the **gauche conformation** (right-most, below).

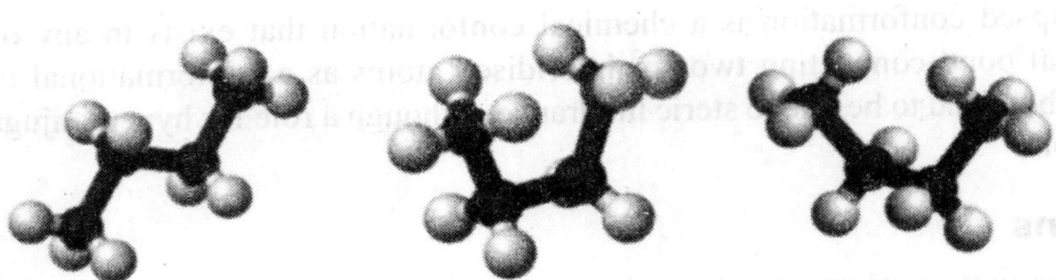

Both conformations are free of torsional strain, but, in the gauche conformation, the two methyl groups are in closer proximity than the sum of their van der Waals radii. The interaction between the two methyl groups is repulsive (van der Waals strain), and an energy barrier results.

A measure of the potential energy stored in butane conformers with greater steric hindrance than the 'anti'-conformer ground state is given by these values:

- Gauche, conformer - 3.8 kJ/mol
- Eclipsed H and $CH_3$ - 16 kJ/mol
- Eclipsed $CH_3$ and $CH_3$ - 19 kJ/mol.

The eclipsed methyl groups exert a greater steric strain because of their greater electron density compared to lone hydrogen atoms.

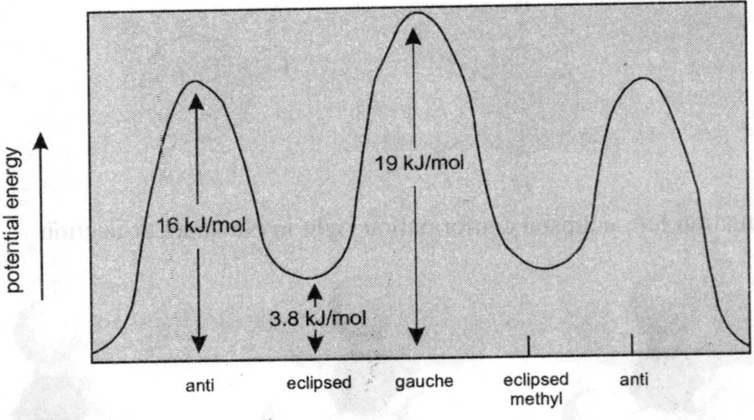


conformer potentials of butane about central C-C bond

The textbook explanation for the existence of the energy maximum for an eclipsed conformation in ethane is steric hindrance, but, with a C-C bond length of 154 pm and a Van der Waals radius for hydrogen of 120 pm, the hydrogen atoms in ethane are never in each other's way. The question of whether steric hindrance is responsible for the eclipsed energy maximum is a topic of debate to this day. One alternative to the steric hindrance explanation is based on hyperconjugation. In terms of molecular orbital theory in the staggered conformation, one C-H sigma bonding orbital donates electron density to the antibonding orbital of the other C-H bond. The energetic stabilization of this effect is maximized when the two orbitals have maximal overlap, occurring in the staggered conformation. There is no overlap in the eclipsed conformation, leading to a disfavored energy maximum.

### Definitions

Many definitions that describe a specific conformation (IUPAC Gold Book) exist:

- a torsion angle of ±60° is called **gauche**
- a torsion angle between 0° and ± 90° is called **syn** (s)
- a torsion angle between ± 90° and 180° is called **anti** (a)
- a torsion angle between 30° and 150° or between –30° and –150° is called **clinal**
- a torsion angle between 0° and 30° or 150° and 180° is called **periplanar** (p)
- a torsion angle between 0° to 30° is called **synperiplanar** or **syn-** or **cis-conformation** (sp)
- a torsion angle between 30° to 90° and –30° to –90° is called **synclinal** or **gauche** or **skew** (sc)
- a torsion angle between 90° to 150°, and –90° to –150° is called **anticlinal** (ac)
- a torsion angle between ± 150° to 180° is called **antiperiplanar** or **anti** or **trans** (ap).

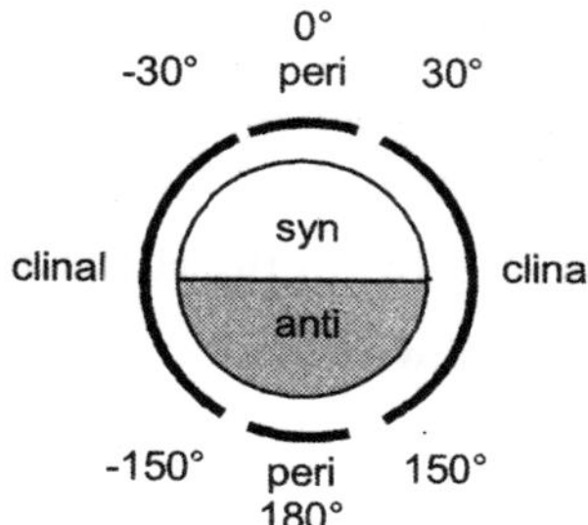


Any strain resulting from torsion is also called **Pitzer Strain** or **eclipsing strain.**

### Special Cases

In n-pentane, the terminal methyl groups experience additional pentane interference.

Replacing hydrogen by fluorine in polytetrafluoroethylene changes the stereochemistry from the zigzag geometry to that of a helix due to electrostatic repulsion of the fluorine atoms in the 1,3 positions. Evidence for the helix structure in the crystalline state is derived from X-ray crystallography and from NMR spectroscopy and circular dichroism in solution.

## Electrophile

In chemistry, an **electrophile** (literally *electron-lover*) is a reagent attracted to electrons that participates in a chemical reaction by accepting an electron pair in order to bond to a nucleophile. Because electrophiles accept electrons, they are Lewis acids (see acid-base reaction theories). Most electrophiles are positively charged, have an atom which carries a partial positive charge, or have an atom which does not have an octet of electrons.

The electrophiles attack the most electron-populated part of a nucleophile. The electrophiles frequently seen in the organic syntheses are cations such as $H^+$ and $NO^+$, polarized neutral molecules such as HCl, alkyl halides, acyl halides, and carbonyl compounds, polarizable neutral molecules such as $Cl_2$ and $Br_2$, oxidizing agents such as organic peracids, chemical species that do not satisfy the octet rule such as carbenes and radicals, and some of lewis acids such as $BH_3$

## Electrophiles in Organic Chemistry

### Alkenes

Electrophilic addition is one of the three main forms of reaction concerning alkenes. They consist of:

- Hydrogenation by the addition of hydrogen over the double bond.
- Electrophilic addition reactions with halogens and sulfuric acid.
- Hydration to form alcohols.

### Addition of Halogens

These occur between alkenes and electrophiles, often halogens as in halogen addition reactions. Common reactions include use of bromine water to titrate against a sample to deduce the number of double bonds present. For example, ethylene + bromine → 1,2-dibromoethane:

$$C_2H_4 + Br_2 \rightarrow BrCH_2CH_2Br$$

This takes the form of 3 main steps shown below:

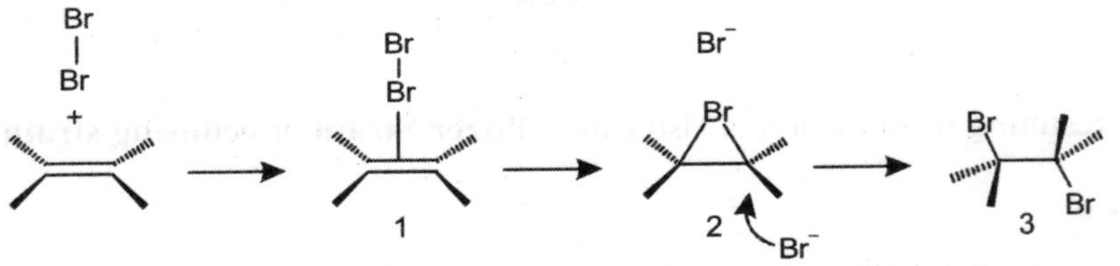

Br–Br + alkene → 1 → 2 ($Br^-$) → 3

1. **Forming of a π-complex**
   The electrophilic Br-Br molecule interacts with electron-rich alkene molecure to form a π-complex **1**.
2. **Forming of a three-membered bromonium ion**
   The alkene is working as an electron donor and bromine as an electrophile. The three-membered bromonium ion **2** consisted with two carbon atoms and a bromine atom forms with a release of $Br^-$.
3. **Attacking of bromide ion**
   The bromonium ion is opened by the attack of $Br^-$ from the back side. This yields the vicinal dibromide with an antiperiplanar configuration. When other nucleophiles such as water or alcohol are existing, these may attack **2** to give an alcohol or an ether.

This process is called **$Ad_E2$ mechanism**. Iodine ($I_2$), chlorine ($Cl_2$), sulfenyl ion ($RS^+$), mercury cation ($Hg^{2+}$), and dichlorocarbene ($:CCl_2$) also react through similar pathways. The direct conversion of **1** to **3** will appear when the Br" is large excess in the reaction medium. A β-bromo carbenium ion intermediate may be predominant instead of **3** if the alkene has a cation-stabilizing substituent like phenyl group. There is an example of the isolation of the bromonium ion **2**.

### Addition of Hydrogen Halides

Hydrogen halides such as hydrogen chloride (HCl) adds to alkenes to give alkyl halide in hydrohalogenation. For example, the reaction of HCl with ethylene furnishes chloroethane. The reaction proceeds with a cation intermediate, being different from the above halogen addition. An example is shown below:

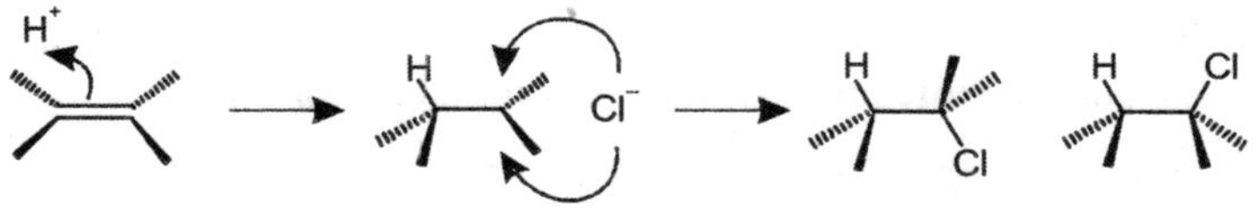

H⁺ → H, Cl⁻ → H, Cl; H, Cl

1 2 3

1. Proton ($H^+$) adds (by working as an electrophile) to one of the carbon atoms on the alkene to form cation **1**.
2. Chloride ion ($Cl^-$) combines with the cation **1** to form the adducts **2** and **3**.

In this manner, the stereoselectivity of the product, that is, from which side $Cl^-$ will attack relies on the types of alkenes applied and conditions of the reaction. At least, which of the two carbon atoms will be attacked by $H^+$ is usually decided by Markovnikov's rule. Thus, $H^+$ attacks the carbon atom which carries the less number of substituents so as to the more stabilized carbocation (with the more stabilizing substituents) will form.

This process is called **A-S$_E$2 mechanism**. Hydrogen fluoride (HF) and hydrogen iodide (HI) react with alkenes similarly and Markovnikov-type products will be given. Hydrogen bromide (HBr) also takes this pathway, but sometimes a radical process competes and a mixture of isomers may form.

### Hydration

One of the more complex hydration reactions utilises sulfuric acid as a catalyst. This reaction occurs in a similar way to the addition reaction but has an extra step in which the $OSO_3H$ group is replaced by an OH group, forming an alcohol:

$$C_2H_4 + H_2O \rightarrow C_2H_5OH$$

As you can see the $H_2SO_4$ does not take part in the overall reaction, however it does take part but remains unchanged so is classified as a catalyst.

This is the reaction in more detail:

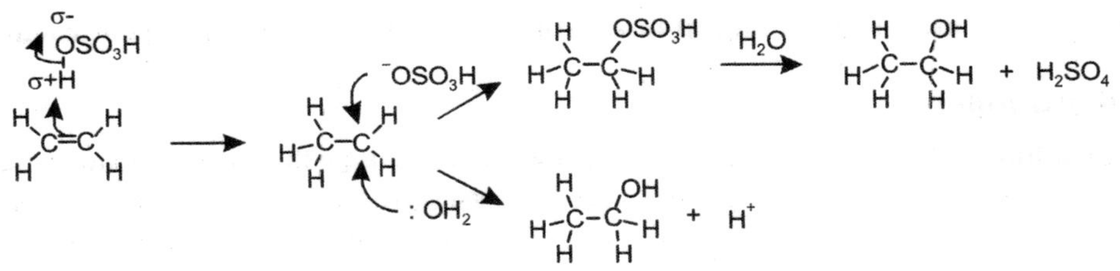

σ- $OSO_3H$ σ+H → $^-OSO_3H$ → H–C(H)(H)–C(OSO$_3$H)(H)–H $\xrightarrow{H_2O}$ H–C(H)(H)–C(OH)(H)–H + $H_2SO_4$

: $OH_2$ → H–C(H)(H)–C(OH)(H)–H + $H^+$

1. The H-$OSO_3H$ molecule has a δ+ charge on the initial H atom, this is attracted to and reacts with the double bond in the same way as before.
2. The remaining (negatively charged) $^-OSO_3H$ ion then attaches to the carbocation. Forming ethyl hydrogensulphate (upper way on the above scheme).
3. When water ($H_2O$) is added and the mixture headed ethanol ($C_2H_5OH$) is produced, the "spare" hydrogen atom from the water goes into "replacing" the "lost" hydrogen and thus reproduces sulfuric acid. Another pathway in which water molecule combines directly to the intermediate carbocation (lower way) is also possible. This pathway become predominant when aqueous sulfuric acid is used.

Overall this process adds a molecule of water to a molecule of ethene.

This is an important reaction in industry as it produces ethanol, which is the alcohol having various purposes including fuels and starting material for other chemicals.

## Electrophilicity Scale

Several methods exist to rank electrophiles in order of reactivity and one of them is devised by Robert Parr with the **electrophilicity index** ω given as:

$$\omega = \frac{x^2}{2\eta}$$

with chi the electronegativity and eta chemical hardness. This equation is related to classical equation for electrical power:

$$P = \frac{V^2}{R}$$

where **R** is the resistance (Ohm or Ω) and **V** is voltage. In this sense the electrophilicity index is a kind of electrophilic power. Correlations have been found between electrophilicity of various chemical compounds and reaction rates in biochemical systems and such phenomena as allergic contact dermititis.

| Electrophilicity index | |
|---|---|
| Fluorine | 3.86 |
| Chlorine | 3.67 |
| Bromine | 3.40 |
| Iodine | 3.09 |
| Hypochlorite | 2.52 |
| sulfur dioxide | 2.01 |
| Carbon disulfide | 1.64 |
| Benzene | 1.45 |
| Sodium | 0.88 |
| Some selected values (no dimensions) | |

A electrophilicity index also exists for free radicals. Strongly electrophilic radicals such as the halogens react with electron-rich reaction sites and strongly nucleophilic radicals such as the 2-hydroxypropyl-2-yl and tert-butyl radical react with a preference for electron-poor reaction sites.

## Superelectrophiles

**Superelectrophiles** are defined as cationic electrophilic reagents with greatly enhanced reactivities in the presence of superacids. These compounds were first described by George A. Olah. Superelectrophiles form as a doubly electron deficient superelectrophile by protosolvation of a cationic electrophile. As observed by Olah, a mixture of acetic acid and boron trifluoride is able to deprotonate isobutane when combined with hydrofluoric acid via the formation of a superacid from $BF_3$ and HF. The responsible reactive intermediate is the $CH_3CO_H$ dication. Likewise methane can be nitrated to nitromethane with nitronium tetrafluoroborate $NO_2^+BF_4^-$ only in presence of a strong acid like fluorosulfuric acid.

In **gitionic** superelectrophiles charged centers are separated by no more than one atom, for example the protonitronium ion $O=N^+=O^+$-H (a protonated nitronium ion) and in **distonic** superelectrophiles they are separated by 2 or more atoms for example in the fluorination reagent F-TEDA-$BF_4$.

## Electrophilic Addition

In Organic chemistry, an **Electrophilic addition** reaction is an addition reaction where, in a chemical compound, a pi bond is removed by the creation of two new covalent bonds. In Electrophilic addition reactions, common substrates have a carbon-carbon double bond or triple bond.

$$\text{Y-Z} + \text{C=C} \rightarrow \text{Y-C-C-Z}$$

The driving force for this reaction is the formation of an electrophile $Y^+$ that forms a covalent bond with an electron-rich unsaturated C=C bond. The positive charge on Y is transferred to the carbon-carbon bond.

Step (1) $Y^+ + \text{C=C} \rightarrow \text{Y-C-C}^+\text{-}$

In step 2 of an Electrophilic addition, the positively charged intermediate combines with (Z) that is electron-rich to form the second covalent bond.

Step (2) $\text{Y-C-C}^+\text{-} + Z \rightarrow \text{Y-C-C-Z}$

Step 2 is also found in a SN1 reaction. The exact nature of the electrophile and the nature of the positively charged intermediate is not always clear and depends on reactants and reaction conditions.

In all asymmetric addition reactions to Carbon Regioselectivity is important and often determined by Markovnikov's rule. Organoborane compounds give anti-Markovnikov additions. Electrophilic attack to an aromatic system results in electrophilic aromatic substitution rather than an addition reaction.

### Typical Electrophilic Additions

Typical electrophilic additions to alkenes with reagents are:

- dihalo addition reactions: $X_2$
- Hydrohalogenations:HX
- Hydration reactions: $H_2O$
- Hydrogenations $H_2$
- Oxymercuration reactions: mercuric acetate, water
- Hydroboration-oxidation reactions : diborane
- the Prins reaction : formaldehyde, water

## Electrophilic Aromatic Substitution

**Electrophilic aromatic substitution** or **EAS** is an organic reaction in which an atom, usually hydrogen, appended to an aromatic system is replaced by an electrophile. The most important reactions of this type that take place are aromatic nitration, aromatic halogenation, aromatic sulfonation, and acylation and alkylating Friedel-Crafts reactions.

### Basic Reactions

Aromatic nitrations to form nitro compounds take place by generating a nitronium ion from nitric acid and sulphuric acid.

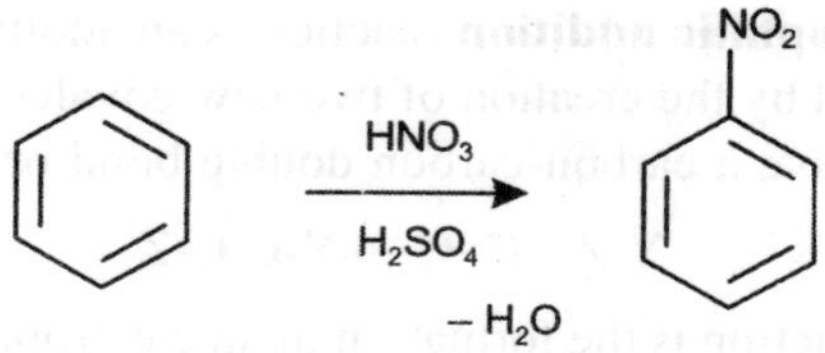

$NO_2$

$HNO_3$

$H_2SO_4$

$- H_2O$

Aromatic sulfonation of benzene with fuming sulfuric acid gives benzenesulfonic acid.

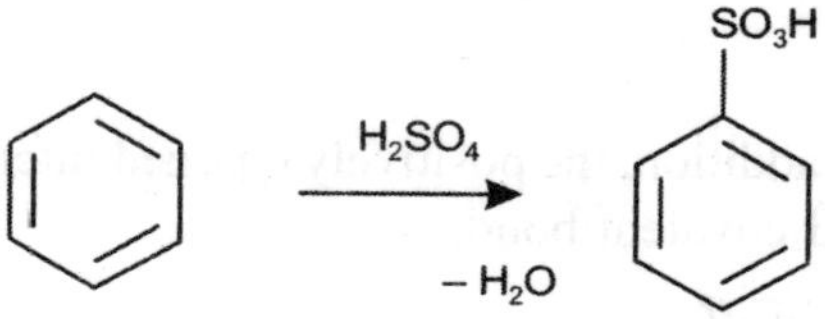

$SO_3H$

$H_2SO_4$

$- H_2O$

Aromatic halogenation of benzene with bromine, chlorine or iodine gives the corresponding aryl halogen compounds catalyzed by iron tribromide.

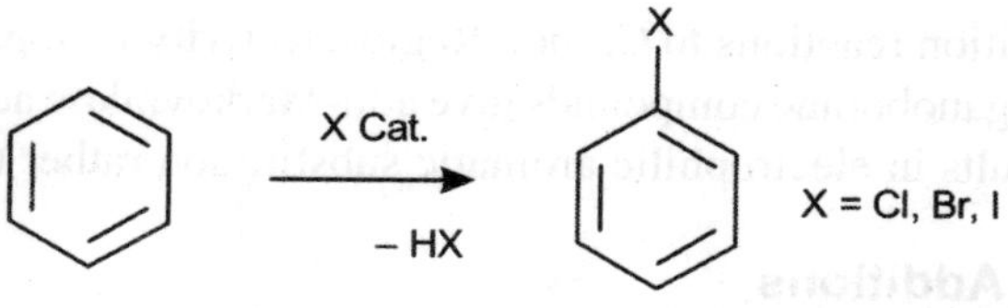

X

X Cat.

– HX

X = Cl, Br, I

The Friedel-Crafts reaction exists as an acylation and an alkylation with as reactants acyl halides or alkyl halides.

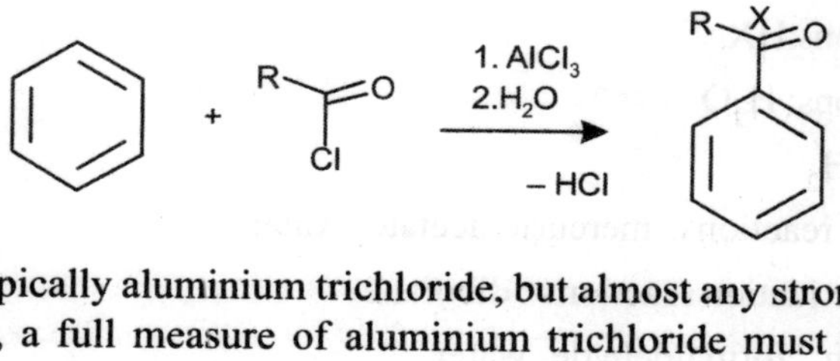

R

O

Cl

+

1. $AlCl_3$

2. $H_2O$

$- HCl$

R X O

The catalyst is most typically aluminium trichloride, but almost any strong Lewis acid can be used. In Fridel-Crafts acylation, a full measure of aluminium trichloride must be used, as opposed to a catalytic amount.

$$C_6H_6 + R\text{–}Cl \xrightarrow[-\,HCl]{AlCl_3} C_6H_5R$$

## Other Reactions

- Other reactions that follow an electrophilic aromatic substitution pattern are a group of aromatic formylation reactions including the Vilsmeier-Haack reaction, the Gattermann Koch reaction and the Reimer-Tiemann reaction.
- Other electrophiles are aromatic diazonium salts in diazonium couplings, carbon dioxide in the Kolbe-Schmitt reaction and activated carbonyl groups in the Pechmann condensation.
- In the multistep Lehmstedt-Tanasescu reaction, one of the electrophiles is a N-nitroso intermediate.
- In the **Tscherniac-Einhorn reaction** (named after Alfred Einhorn) the electrophile is a N-methanol derivative of an amide

## Basic Reaction Mechanism

In the first step of the reaction mechanism for this reaction, the electron-rich aromatic ring which in the simplest case is benzene is attacked by the electrophile **A**. This leads to the formation of a positively-charged cyclohexadienyl cation, also known as an arenium ion. This carbocation is unstable, owing both to the positive charge on the molecule and to the temporary loss of aromaticity. However, the cyclohexadienyl cation is partially stabilized by resonance, which allows the positive charge to be distributed over three carbon atoms.

In the second stage of the reaction, a Lewis base **B** donates electrons to the hydrogen atom at the point of electrophilic attack, and the electrons shared by the hydrogen return to the *pi* system, restoring aromaticity.

An electrophilic substitution reaction on benzene does not always result in monosubstitution. While electrophilic substituents usually withdraw electrons from the aromatic ring and thus deactivate it against further reaction, a sufficiently strong electrophile can perform a second or even a third substitution. This is especially the case with the use of catalysts.

## Substituted Aromatic Rings

Electrophiles may attack aromatic rings with functional groups. Performing an electrophilic substitution on an already substituted benzene compound raises the problem of regioselectivity. In case of a monosubstituted benzene, there are 4 different reactive positions. For a monosubstituted benzene, the ring carbon atom bearing the substituent is position 1 or ipso, the next ring atom is position 2 or ortho, position 3 is meta and position 4 is para. Positions 5 and 6 are respectively equal to 3 and 2.

Substituents can generally be divided into two classes regarding electrophilic substitution: activating and deactivating towards the aromatic ring. **Activating substituents** or activating groups stabilize the cationic intermediate formed during the substitution by donating electrons into the ring system, by either inductive effect or resonance effects. Examples of activated aromatic rings are toluene, aniline and phenol.

The extra electron density delivered into the ring by the substituent is not equally divided over the entire ring, but is concentrated on atoms 2, 4 and 6 (the ortho and para positions). These positions are thus the most reactive towards an electron-poor electrophile. The highest electron density is located on both ortho positions, though this increased reactivity might be offset by steric hindrance between substituent and electrophile. The final result of the elecrophilic aromatic substitution might thus be hard to predict, and it is usually only established by doing the reaction and determining the ratio of ortho versus para substitution.

On the other hand, deactivating substituents destabilize the intermediate cation and thus decrease the reaction rate. They do so by withdrawing electron density from the aromatic ring, though the positions most affected are again the ortho and para ones. This means that the most reactive positions (or, least unreactive) are the meta ones (atoms 3 and 5). Examples of deactivated aromatic rings are nitrobenzene, benzaldehyde and trifluoromethylbenzene. The deactivation of the aromatic system also means that generally harsher conditions are required to drive the reaction to completion. An example of this is the nitration of toluene during the production of trinitrotoluene (TNT). While the first nitration, on the activated toluene ring, can be done at room temperature and with dilute acid, the second one, on the deactivated nitrotoluene ring, already needs prolonged heating and more concentrated acid, and the third one, on very strongly deactivated dinitrotoluene, has to be done in boiling concentrated sulfuric acid.

Functional groups thus usually tend to favour one or two of these positions above the others; that is, they *direct* the electrophile to specific positions. A functional group that tends to direct attacking electrophiles to the *meta* position, for example, is said to be **meta-directing**.

### *Ortho/para* Directors

Groups with unshared pairs of electrons, such as the amino group of aniline, are strongly *activating* and *ortho/para*-directing. Such activating groups donate those unshared electrons to the *pi* system.

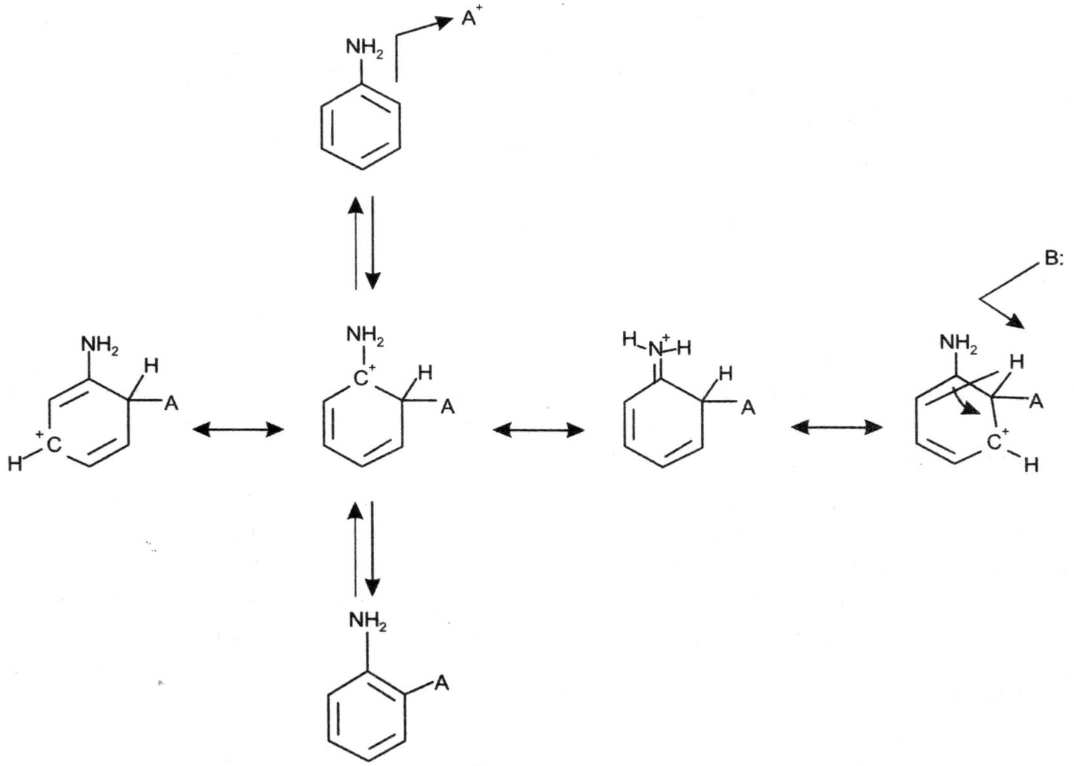

When the electrophile attacks the *ortho* and *para* positions of aniline, the nitrogen atom can donate electron density to the *pi* system, giving four resonance structures (as opposed to three in the basic reaction). This substantially enhances the stability of the cationic intermediate.

Compare this with the case when the electrophile attacks the *meta* position. In that case, the nitrogen atom cannot donate electron density to the *pi* system, giving only three resonance contributors. For this reason, the *meta*-substituted product is produced in much smaller proportion to the *ortho* and *para* products.

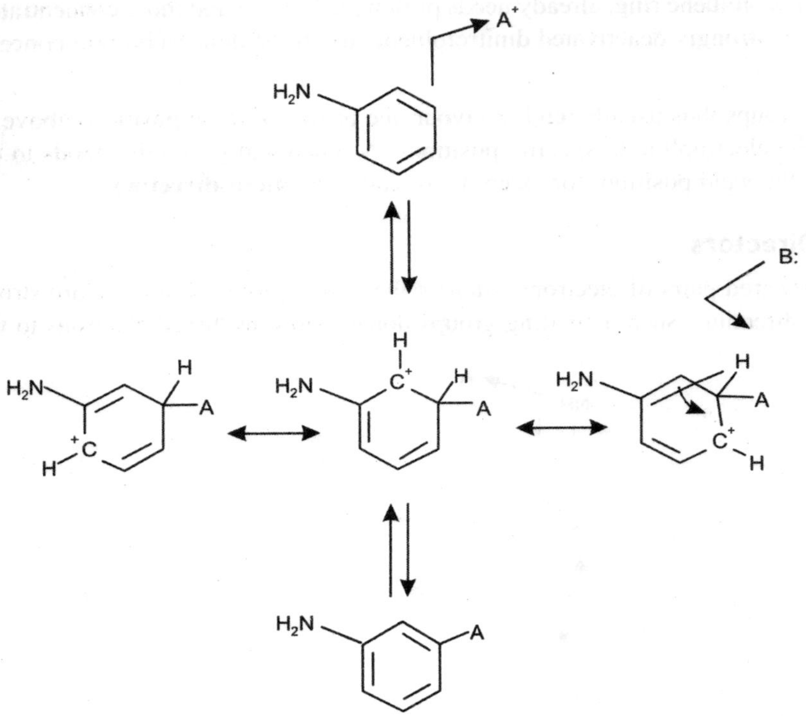

Other substituents, such as the alkyl and aryl substituents, may also donate electron density to the *pi* system; however, since they lack an available unshared pair of electrons, their ability to do this is rather limited. Thus they only weakly activate the ring and do not strongly disfavour the *meta* position.

Halogens are *ortho/para* directors, since they possess an unshared pair of electrons just as nitrogen does. However, the stability this provides is offset by the fact that halogens are substantially more electronegative than carbon, and thus draw electron density away from the *pi* system. This destabilizes the cationic intermediate, and EAS occurs less readily. Halogens are therefore deactivating groups.

Directed ortho metalation is a special type of EAS with special **ortho directors**.

## *Meta* Directors

Non-halogen groups with atoms that are more electronegative than carbon, such as the nitro group ($NO_2$) draw substantial electron density from the *pi* system. These groups are strongly deactivating groups. Additionally, since the substituted carbon is already electron-poor, the resonance contributor with a positive charge on this carbon (produced by *ortho/para* attack) is less stable than the others. Therefore, these electron-withdrawing groups are *meta* directors.

## Ipso Substitution

Ipso substitution is a special case of electrophilic aromatic substitution where the leaving group is not hydrogen.

A classic example is the reaction of salicylic acid with a mixture of nitric and sulfuric acid to form picric acid. The nitration of the 2 position involves the loss of $CO_2$ as the leaving group.

Desulfonation in which a sulfonyl group is substituted by a proton is a common example. See also Hayashi rearrangement

## Five Membered Heterocyclic Compounds

Furan, Thiophene, Pyrrole and their derivatives are all highly activated compared to benzene. These compounds all contain an atom with an unshared pair of electrons (oxygen, sulfur, or nitrogen) as a member of the aromatic ring, which substantially increases the stability of the cationic intermediate. Examples of electrophilic substitutions to pyrrole are the Pictet-Spengler reaction and the Bischler-Napieralski reaction.

## Asymmetric Electrophilic Aromatic Substitution

Electrophilic aromatic substitutions with prochiral carbon electrophiles have been adapted for asymmetric synthesis by switching to chiral lewis acid catalysts especially in friedel-Crafts type reactions. An early example concerns the addition of chloral to phenols catalyzed by aluminium chloride modified with (-)-menthol A glyoxylate compound has been added to N,N-dimethylaniline with a chiral bisoxazoline ligand - copper(II) triflate catalyst system also in a Friedel-Crafts hydroxyalkylation.

1.5 eq. O H OEt O; $Et_2O$, rt; 0.1 eq. O N t-Bu O N t-Bu; $Cu(OTf)_2$; HO O OEt N; 78% 89% ee

In another alkylation N-methylpryrrole reacts with crotonaldehyde catalyzed by trifluoroacetic acid modified with a chiral imidazolidinone

O; $THF/H_2O$-60°C, 72 hrs.; N; 0.2 eq. O N TFA bn N H; N O; 83% 91% EE

Indole reacts with an enamide catalyzed by a chiral BINOL derived phosphoric acid

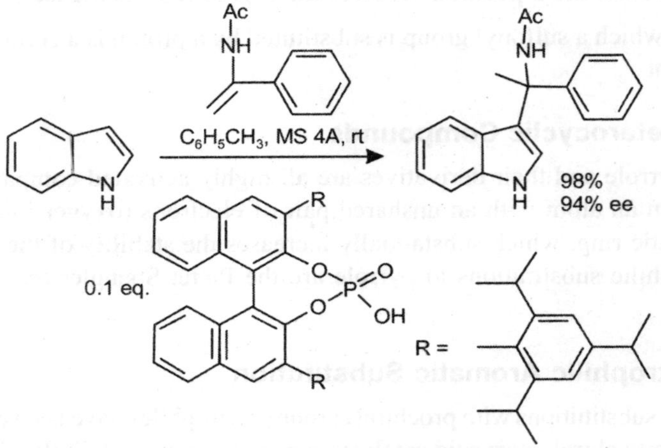

$C_6H_5CH_3$, MS 4A, rt

0.1 eq.

98%
94% ee

R =

In all these reactions the chiral catalyst load is between 10 to 20% and a new chiral carbon center is formed with 80-90 ee.

## Electrophilic Aliphatic Substitution

In electrophilic substitution in aliphatic compounds, an electrophile displaces a functional group. This reaction is similar to nucleophilic aliphatic substitution where the reactant is a nucleophile rather than an electrophile. The two electrophilic reaction mechanisms, **$S_E$1** and **$S_E$2** (**Substitution Electrophilic**), are also similar to the nucleophile counterparts [[$S_N$1|SN1]] and [[$S_N$2|SN2]]. In the $S_E$1 course of action the substrate first ionizes into a carbanion and a positively charged organic residue. The carbanion then quickly recombines with the electrophile. The $S_E$2 reaction mechanism has a single transition state in which the old bond and the newly formed bond are both present.

Electrophilic aliphatic substitution reactions are:

- Nitrosation
- Ketone halogenation
- Keto-enol tautomerism
- aliphatic diazonium coupling
- carbene insertion into C-H bonds

**Nitrosation** is a process of converting organic compounds into *nitroso* derivatives, i.e. compounds containing the R-NO functionality.

## C-Nitroso Compounds

C-Nitroso compounds, such as nitrosobenzene, are typically prepared by *oxidation* of *hydroxylamines*:

$$RNHOH + [O] \rightarrow RNO + H_2O$$

### *N*-nitrosamines

*N*-nitrosamines, including the carcinogenic variety, arise from the reaction of *nitrite* sources with *amino compounds*. Typically this reaction proceeds via the attack of the *nitrosonium* electrophile on an amine:

$$NO_2^- + 2\,H^+ \rightarrow NO^+ + H_2O$$

$$R_2NH + NO^+ \rightarrow R_2N\text{-}NO + H^+$$

### Reaction Mechanism

Formation of an N-nitrosamine

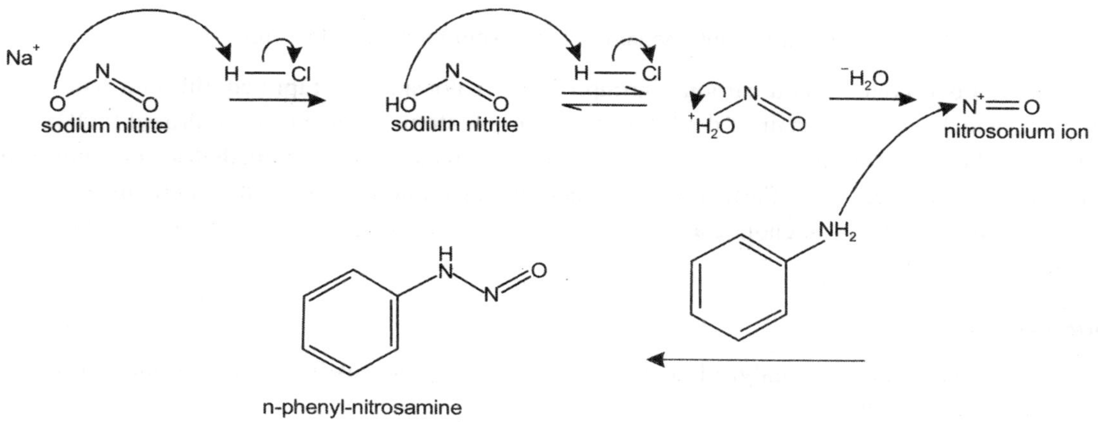

Na⁺ sodium nitrite; H—Cl; HO sodium nitrite; H—Cl; $^+H_2O$; $^-H_2O$; $N^+{=}O$ nitrosonium ion; $NH_2$

n-phenyl-nitrosamine

## Ketone Halogenation

In organic chemistry **ketone halogenation** is a special type of halogenation.

The position alpha (next) to the carbonyl group in a ketone is easily halogenated, due to the ability to form an enolate in basic solution, or an enol in acidic solution. An example is the bromination of acetone in basic solution:

$$CH_3\text{-}CO\text{-}CH_3 + OH^- \rightarrow CH_3\text{-}CO\text{-}CH_2^- + H_2O$$
$$CH_3\text{-}CO\text{-}CH_2^- + Br_2 \rightarrow CH_3\text{-}CO\text{-}CH_2Br + Br^-$$

In acidic solution, usually only one alpha hydrogen is replaced by a halogen, because each successive halogenation is slower that the first. The halogen decreases the basicity of the carbonyl oxygen, thus making protonation less likely. However, in basic solution successive halogenations are *more rapid*, because the halogen withdraws electrons by induction and makes remaining hydrogens more acidic. In the case of methyl ketones, this results in what is called the haloform reaction.

## Keto-enol Tautomerism

In organic chemistry, **keto-enol tautomerism** refers to a chemical equilibrium between a **keto** form (a ketone or an aldehyde) and an **enol**. The enol and keto forms are said to be tautomers of each other. The interconversion of the two forms involves the movement of a proton and the shifting of bonding electrons; hence, the isomerism qualifies as tautomerism.

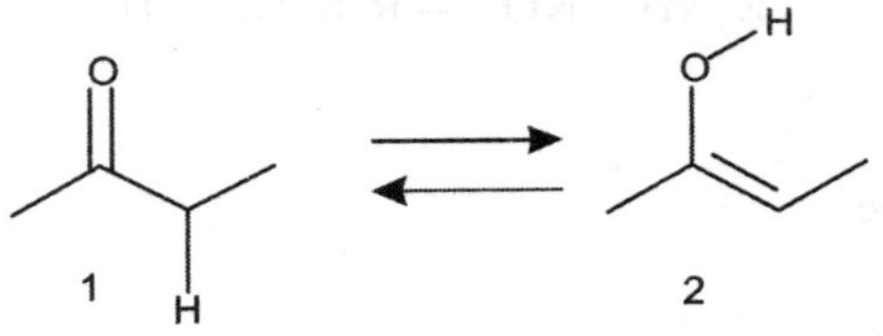

Keto-enol tautomerism. 1 is the keto form; 2 is the enol.

A compound containing a carbonyl group (C=O) is normally in rapid equilibrium with an enol tautomer, which contains a pair of doubly bonded carbon atoms adjacent to a hydroxyl (–OH) group, C=C-OH. The keto form predominates at equilibrium for most ketones. Nonetheless, the enol form is important for some reactions. Furthermore, the deprotonated intermediate in the interconversion of the two forms, referred to as an enolate anion, is important in carbonyl chemistry, in large part because it is a strong nucleophile.

### Mechanism

The conversion of an acid catalyzed enol to the keto form proceeds by a two step mechanism in an aqueous acidic solution.

First, the exposed electrons of the C=C double bond of the enol are donated to a hydronium ion ($H_3O^+$). This addition follows Markovnikov's rule, thus the proton is added to the carbon with more hydrogens. This is a concerted step with the oxygen in the hydroxyl group donating electrons to produce the eventual carbonyl group.

Second, the oxygen in a water molecule donates electrons to the hydrogen in the hydroxyl group, thus relieving the positive charge on the electronegative oxygen atom.

Overall Reaction :

Reaction Mechanism

Acid catalyzed Enolization

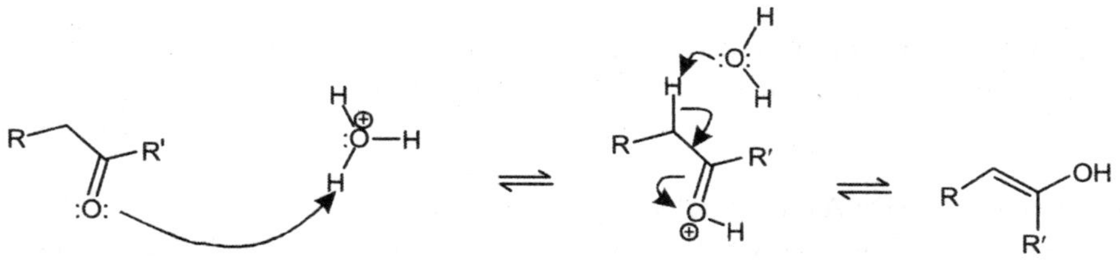

Base catalyzed Enolization

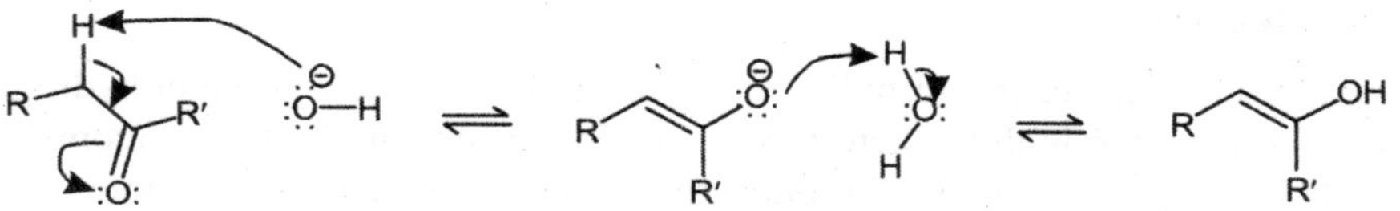

## Erlenmeyer Rule

One of the early investigators into keto-enol tatomerism was Richard August Carl Emil Erlenmeyer and his **Erlenmeyer rule** (developed in 1880) states that all alcohols in which the hydroxyl group is attached directly to a double-bonded carbon atom become aldehydes or ketones. This occurs because the keto form is generally more stable than its enol tautomer. As the lower energy form, the keto form is favored at equilibrium.

## Significance in Biochemistry

Keto-enol tautomerism is important in several areas of biochemistry. The high phosphate-transfer potential of phosphoenolpyruvate results from the fact that the phosphorylated compound is "trapped" in the less stable enol form, whereas after dephosphorylation it can assume the keto form. Rare enol tautomers of the bases guanine and thymine can lead to mutation because of their altered base-pairing properties.

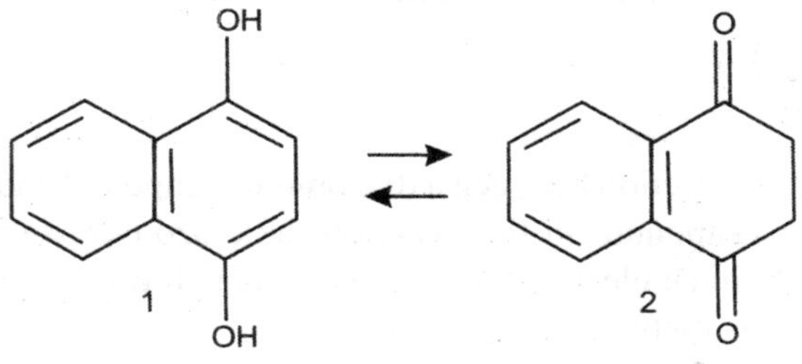

In certain aromatic compounds such as phenols the enol is important due to the aromatic character of the enol but not the keto form. Melting the naphthalene derivative 1,4-dihydroxynaphthalene **1** at 200 °C results in a 2:1 mixture with the keto form **2**. Heating the keto form in benzene at 120°C for three days also affords a mixture (1:1 with first order reaction kinetics) The keto product is kinetically stable and reverts back to the enol in presence of a base. The keto form can be obtained in a pure form by stirring the keto form in triflic acid and toluene (1:9 ratio) followed recrystallisation from isopropyl ether.

When the enol form is complexed with chromium tricarbonyl, complete conversion to the keto form accelerated and occurs even at room temperature in benzene.

## DNA

In deoxyribonucleic acids(DNA), the nucleotide bases are in keto form. However, James Watson and Francis Crick first believed them to be in the enol tautomeric form, delaying the solution of the structure for several months.

## Hydration of Alkynes

Hydration of alkynes (of the general form RC$\equiv$CR, where R is an alkyl group or hydrogen), produces an enol that is in equilibrium with the keto form. If R, R', or both are hydrogen atoms, the keto form is an aldehyde. If both R and R' are alkyl groups, the keto form is a ketone. The most commonly used set of reagents is sulfuric acid ($H_2SO_4$) and mercury(II) sulfate ($HgSO_4$).

**General Reaction**

R—C≡C—R' (Alkyne) $\xrightarrow{H^+, Hg^{2+}}$ enol $\rightarrow$ keto form

Example:

acetylene $\xrightarrow{H^+, Hg^{2+}}$ 1-ethern-1-ol $\rightarrow$ acetayldehyde

Here, acetylene (ethyne) is reacted with $H_2SO_4$ and $HgSO_4$, adding H to one carbon and OH to the other; this forms the intermediate enol. In this reaction, the carbons are equivalent and there is no stereoselectivity. The reaction immediately continues with keto-enol tautomerization.

In general, the equilibrium lies far toward the keto side; in fact, the enol intermediate cannot be isolated as a product. Hydration of alkynes is unlike hydration of alkenes, where the product is an alcohol rather than an enol; therefore, no such equilibrium occurs.

# Azo Coupling

An **azo coupling** is an organic reaction between a diazonium compound and an aniline or a phenol. The reaction product is an azo compound. In this reaction the diazonium salt is an electrophile and the activated arene a nucleophile in an electrophilic aromatic substitution. Azo couplings are important in the production of dyes such as methyl red and pigment red 170.

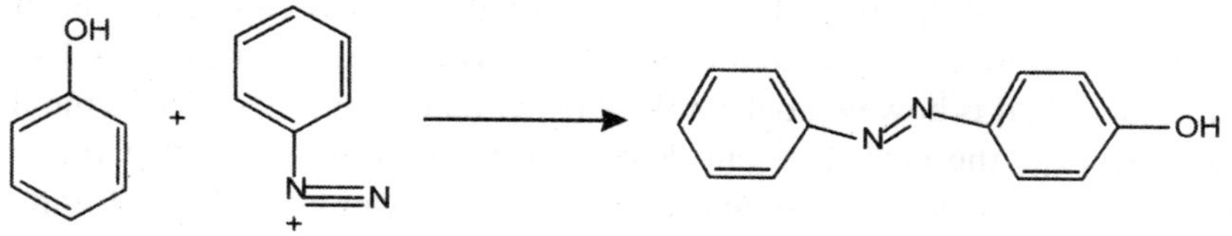

## Carbene

In chemistry, a **carbene** is a highly reactive organic molecule with a divalent carbon atom with only six valence electrons and the general formula: $R^1R^2C$: (two substituents and two electrons).[1] The carbene comes in two varieties: a singlet and triplet. The singlet type has its carbon atom $sp^2$ hybridised with an empty p-orbital extending above and below a plane containing $R^1$ and $R^2$ and the free electron pair. Typically these molecules are very short lived, although persistent carbenes are now known.

The parent carbene is $H_2C$: also called methylene. An often encountered carbene is $Cl_2C$: or dichlorocarbene which can be generated in situ from chloroform and a strong base.

### Structure

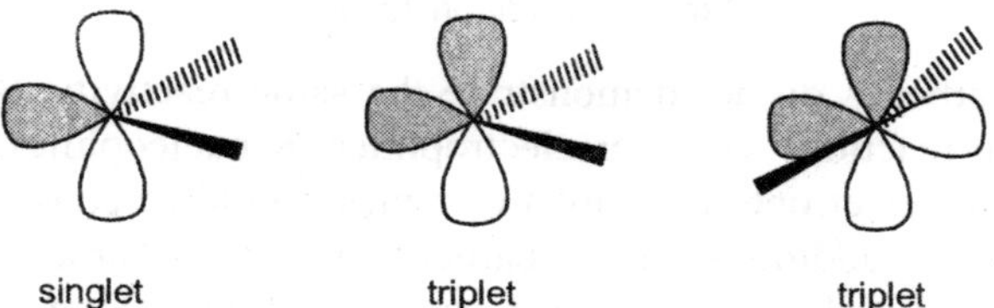


Singlet and triplet carbenes

Generally there are two types of carbenes; singlet or triplet carbenes. Singlet carbenes have a pair of electrons and an $sp^2$ hybrid structure. Triplet carbenes have two unpaired electrons. They may be either $sp^2$ hybrid or linear sp hybrid. Most carbenes have a nonlinear triplet ground state with the exception of carbenes with nitrogen, oxygen, sulfur atoms, and dihalocarbenes.

Singlet and triplet carbenes are named so because of the electronic spins they possess. Triplet carbenes are paramagnetic and may be observed by electron spin resonance spectroscopy if they can exist long enough without undergoing further reactions. The total spin of singlet carbenes is zero while that of triplet carbenes is one (in units of *h*). Bond angles are 125-140° for triplet methylene and 102° for singlet methylene (determined by EPR). Triplet carbenes are generally stable in gaseous state while singlet carbenes are often found in aqueous media.

For simple hydrocarbons, triplet carbenes usually have energies 8 kcal/mol (33 kJ/mol) lower than singlet carbenes (see also Hund's rule of Maximum Multiplicity), thus, in general, triplet is the more stable state (the ground state) and singlet is the excited state species. Substituents that can donate electron pairs may stabilize the singlet state by delocalizing the pair into an empty p-orbital. If the energy of the singlet state is sufficiently reduced it will actually become the ground state. No viable strategies exist for triplet stabilization. The carbene called 9-fluorenylidene has been shown to be a rapidly equilibrating mixture of singlet and triplet states with an approximately 1.1 kcal/mol (4.6 kJ/mol) energy difference.[2]. It is however debatable whether diaryl carbenes such as the fluorene carbene are true carbenes because the electrons can delocalize to such an extent that they become in fact biradicals. In silico experiments suggest that triplet carbenes can be stabilized with electropositive groups such as trifluorosilyl groups.

## Reactivity

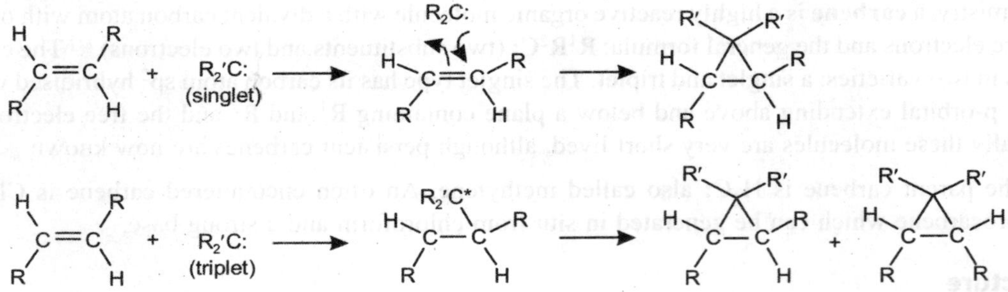

Carbene addition to alkenes

Singlet and triplet carbenes do not demonstrate the same reactivity. Singlet carbenes generally participate in cheletropic reactions as either electrophiles or nucleophiles. Singlet carbene with its unfilled p-orbital should be electrophilic. Triplet carbenes should be considered to be diradicals, and participate in stepwise radical additions. Triplet carbenes have to go through an intermediate with two unpaired electrons whereas singlet carbene can react in a single concerted step. Addition of singlet carbenes to olefinic double bonds is more stereoselective than that of triplet carbenes. Addition reactions with alkenes can be used to determine whether the singlet or triplet carbene is involved.

Reactions of singlet methylene are stereospecific while those of triplet methylene are not. For instance the reaction of methylene generated from photolysis of diazomethane with cis-2-butene and trans-2-butene is stereospecific which proves that in this reaction methylene is a singlet.[4]

Reactivity of a particular carbene depends on the substituent groups, preparation method, reaction conditions such as presence or absence of metals. Some of the reactions carbenes can do are insertions into C-H bonds, skeletal rearrangements, and additions to double bonds. Carbenes can be classified as nucleophilic, electrophilic, or ambiphilic. Reactivity is especially strongly influenced by substituents. For example, if a substituent is able to donate a pair of electrons, most likely carbene will not be electrophilic. Alkyl carbenes insert much more selectively than methylene, which does not differentiate between primary, secondary, and tertiary C-H bonds.

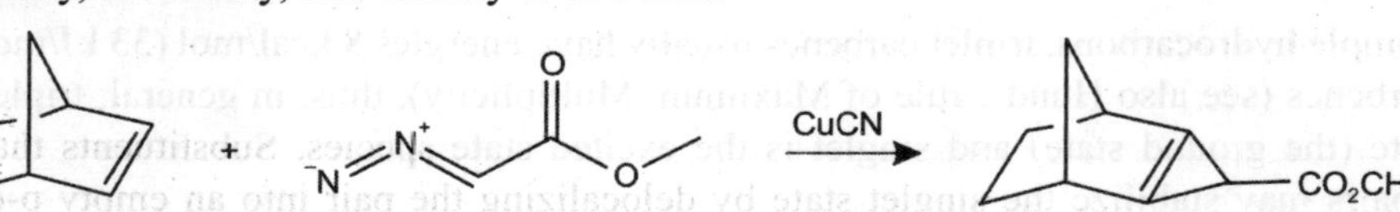

Carbene cyclopropanation

Carbenes add to double bonds to form cyclopropanes. A concerted mechanism is available for singlet carbenes. Triplet carbenes do not retain stereochemistry in the product molecule. Addition reactions are commonly very fast and exothermic. The slow step in most instances is generation of carbene. A well-known reagent employed for alkene-to-cyclopropane reactions is Simmons-Smith reagent. This reagents is a system of copper, zinc, and iodine, where the active reagent is believed to be iodomethylzinc iodide. Reagent is complexed by hydroxy groups such that addition commonly happens syn to such group.

Insertions are another common type of carbene reactions. The carbene basically interposes itself into an existing bond. The order of preference is commonly: 1. X-H bonds where X is not carbon 2. C-H bond 3. C-C bond. Insertions may or may not occur in single step.

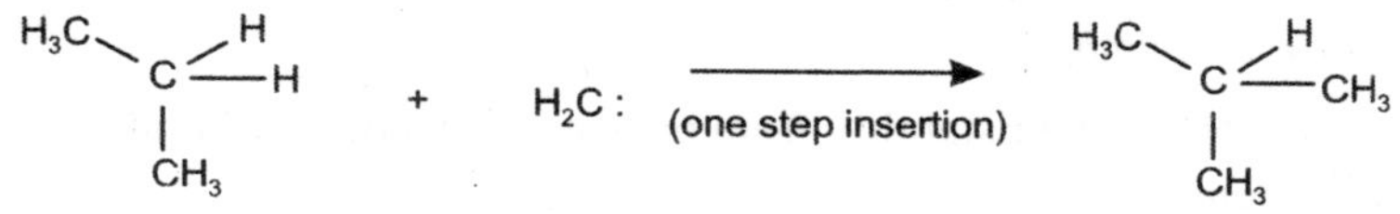

$H_3C$, H, H, $CH_3$ + $H_2C:$ → (one step insertion) $H_3C$, H, $CH_3$, $CH_3$

Carbene insertion

Intramolecular insertion reactions present new synthetic solutions. Generally, rigid structures favour such insertions to happen. When an intramolecular insertion is possible, no intermolecular insertions are seen. In flexible structures, five-membered ring formation is preferred to six-membered ring formation. Both inter- and intramolecular insertions are amendable to asymmetric induction by choosing chiral ligands on metal centers.

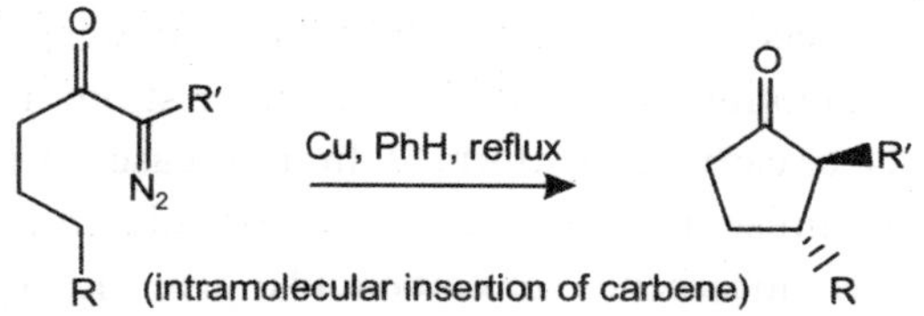

O, R′, $N_2$, R — Cu, PhH, reflux → O, R′, R

(intramolecular insertion of carbene)

Carbene intramolecular reaction

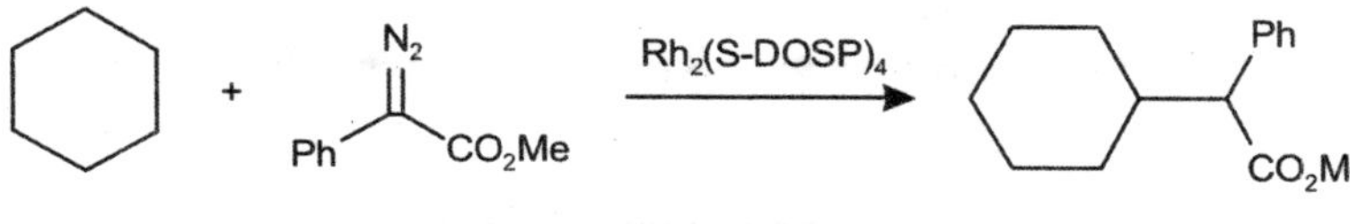

+ $N_2$, Ph, $CO_2Me$ — $Rh_2$(S-DOSP)$_4$ → Ph, $CO_2Me$

(intermolecular insertion of carbene)

Carbene intermolecular reaction

Alkylidene carbenes are alluring in that they offer formation of cyclopentene moieties. To generate an alkylidene carbene a ketone can be exposed to trimethylsilyl diazomethane.

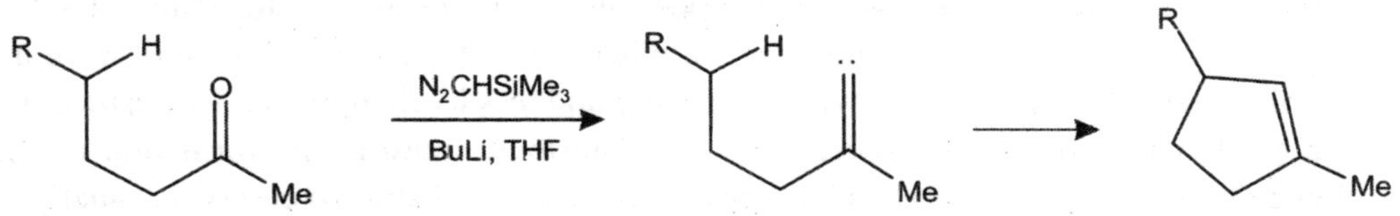

R, H, O, Me — $N_2CHSiMe_3$, BuLi, THF → R, H, Me → R, Me

Alkylidene carbene

## Carbenes and Carbene Ligands in Organometallic Chemistry

Carbenes can be stabilized as organometallic species. These transition metal carbene complexes fall into three categories, with the first two being the most clearly defined:

- Fischer carbenes, in which the carbene is tethered to a metal that bears an electron-withdrawing group (usually a carbonyl).
- Schrock carbenes, in which the carbene is tethered to a metal that bears an electron-donating group.
- Persistent carbenes, also known as stable carbenes or Arduengo carbenes. These include the class of *N*-heterocyclic carbenes (NHCs) and are often are used as ancillary ligands in organometallic chemistry.
- Foiled carbenes derive their stability from proximity of a double bond (i.e their ability to form conjugated systems).

## Generation of Carbenes

- Most commonly, photolytic, thermal, or transition metal catalyzed decomposition of diazoalkanes is used to create carbene molecules. A variation on catalyzed decomposition of diazoalkanes is the Bamford-Stevens reaction, which gives carbenes in aprotic solvents and carbenium ions in protic solvents.
- Another method is induced elimination of halogen from gem-dihalides or HX from $CHX_3$ moiety, employing organolithium reagents (or another strong base). It is not certain that in these reactions actual free carbenes are formed. In some cases there is evidence that completely free carbene is never present. It is likely that instead a metal-carbene complex forms. Nevertheless, these metallocarbenes (or carbenoids) give the expected products.

$$R-C(Br)(R)-Br \xrightarrow{\text{butyllithium}} R-C^{-}(Li^{+})(R)-Br \longrightarrow R_2C: \quad BrLi$$

Carbene preparation

- Photolysis of diazarines and epoxides can also be employed. Diazarines contain 3-membered rings and are cyclic forms or diazoalkanes. The strain of the small ring makes photoexcitation easy. Photolysis of epoxides gives carbonyl compounds as side products. With asymmetric epoxides, two different carbonyl compounds can potentially form. The nature of substituents usually favours formation of one over the other. One of the C-O bonds will have a greater double bond character and thus will be stronger and less likely to break. Resonance structures can be drawn to determine which part will contribute more to the formation of carbonyl. When one substituent is alkyl and another aryl, the aryl-substituted carbon is usually released as a carbene fragment.

- Thermolysis of alpha-halomercury compounds is another method to generate carbenes.
- Rhodium and copper complexes promote carbene formation.
- Carbenes are intermediates in the Wolff rearrangement.

## Carbocations

### History

The history of carbocations dates back to 1891 when G. Merling [6] reported that he added bromine to tropylidene (cycloheptatriene) and then heated the product to obtain a crystalline, water soluble material, $C_7H_7Br$. He did not suggest a structure for it; however Doering and Knox [7] convincingly showed that it was tropylium (cycloheptatrienylium) bromide. This ion is predicted to be aromatic by the Hückel Rule.

In 1902 Norris and Kehrman independently discovered that colourless triphenylmethanol gave deep yellow solutions in concentrated sulfuric acid. Triphenylmethyl chloride similarly formed orange complexes with aluminium and tin chlorides. Adolf von Baeyer recognized in 1902 the salt like character of the compounds formed.

$$Ph_3C—OH + H_2SO_4 \rightarrow Ph_3C^{+}HSO_4^{-} + H_2O$$ (Ph stands for a phenyl substituent)

He dubbed the relationship between colour and salt formation **halochromy** of which malachite green is a prime example.

Carbocations are reactive intermediates in many organic reactions. This idea, first proposed by Julius Stieglitz in 1899 (On the Constitution of the Salts of Imido-Ethers and other Carbimide Derivatives; Am. Chem. J. 21, 101; ISSN: 0096-4085) was further developed by Hans Meerwein in his 1922 study of the Wagner-Meerwein rearrangement. Carbocations were also found to be involved in the SN1 reaction and E1 reaction and in rearrangement reactions such as the Whitmore 1,2 shift. The chemical establishment was reluctant to accept the notion of a carbocation and for a long time the Journal of the American Chemical Society refused articles that mentioned them.

The first NMR spectrum of a stable carbocation in solution was published by Doering et al. [9]. It was the heptamethylbenzenonium ion, made by treating hexamethylbenzene with methyl chloride and aluminium chloride. The stable 7-norbornadienyl cation was prepared by Story et al. by reacting norbornadienyl chloride with silver tetrafluoroborate in sulfur dioxide at –80°C . The NMR spectrum established that it was nonclassically bridged (the first stable non-classical ion observed).

In 1962 Olah directly observed the tert-butyl carbocation by Nuclear magnetic resonance as a stable species on dissolving tert-butyl fluoride in magic acid. The NMR of norbornyl cation was first reported by Schleyer et al. [11] and it was shown to undergo proton scrambling over a barrier by Saunders et al.

A **carbocation** (pronounced /□kɑrboʊ□kætaɪɒn/) is an ion with a positively-charged carbon atom. The charged carbon atom in a carbocation is a "sextet", i.e. it has only six electrons in its outer valence shell instead of the eight valence electrons that ensures maximum stability (octet rule). Therefore carbocations are often reactive, seeking to fill the octet of valence electrons as well as regain a neutral charge. One could reasonably assume a carbocation to have $sp^3$ hybridization with an empty $sp^3$ orbital giving positive charge. However, the reactivity of a carbocation more closely resembles $sp^2$ hybridization with a trigonal planar molecular geometry.

## Definitions

A carbocation was previously often called a *carbonium ion* but questions arose on the exact meaning. In present day chemistry a carbocation is any positively charged carbon atom. Two special types have been suggested: *carbenium ions* are trivalent and *carbonium ions* are pentavalent or hexavalent. University level textbooks only discuss carbocations as if they are carbenium ions, or discuss carbocations with a fleeting reference to the older phrase of carbonium ion or carbenium and carbonium ions. One textbook to this day clings on to the older name of carbonium ion for carbenium ion and reserves the phrase *hypervalent carbenium ion* for $CH_5^+$

## Properties

III R–C⁺(R′)(R″) > II R–C⁺(R′)(H) > I R–C⁺(H)(H) > R–C⁺(H)(H)

Order of stability of examples oftertiary (III), secondary (II), and primary (I) alkyl carbocations.

In organic chemistry, a carbocation is often the target of nucleophilic attack by nucleophiles like $OH^-$ ions or halogen ions.

Carbocations are classified as *primary*, *secondary*, or *tertiary* depending on the number of carbon atoms bonded to the ionized carbon. Primary carbocations have one or zero carbons attached to the ionized carbon, secondary carbocations have two carbons attached to the ionized carbon, and tertiary carbocations have three carbons attached to the ionized carbon.

Stability of the carbocation increases with the number of alkyl groups bonded to the charge-bearing carbon. Tertiary carbocations are more stable (and form more readily) than secondary carbocations; primary carbocations are highly unstable because, while ionized higher-order carbons are stabilized by Hyperconjugation, unsubstituted (primary) carbons are not. Therefore, reactions such as the $S_N1$ reaction and the E1 elimination reaction normally do not occur if a primary carbocation would be formed. An exception to this occurs when there is a carbon-carbon double bond next to the ionized carbon. Such cations as *allyl* cation $CH_2{=}CH\text{-}CH_2^+$ and *benzyl* cation $C_6H_5\text{-}CH_2^+$ are more stable than most other carbocations. Molecules which can form allyl or benzyl carbocations are especially reactive.

Carbocations undergo rearrangement reactions from less stable structures to equally stable or more stable ones with rate constants in excess of 1.0E9/sec. This fact complicates synthetic pathways to many compounds. For example, when 3-pentanol is heated with aqueous HCl, the initially formed 3-pentyl carbocation rearranges to a statistical mixture of the 3-pentyl and 2-pentyl. These cations react with chloride ion to produce about 1/3 3-chloropentane and 2/3 2-chloropentane.

Some carbocations such as the norbornyl cation exhibit more or less symmetrical three centre bonding. Cations of this sort have been referred to as non-classical ions. The energy difference between "classical" carbocations and "non-classical" isomers is often very small, and there is generally little, if any activation energy involved in the transition between "classical" and "non-classical" structures. The "non-classical" form of the 2-butyl carbocation is essentially 2-butene with a proton directly above the centre of what would be the carbon-carbon double bond. "Non-classical" carbocations were once the subject of great controversy. One of George Olah's greatest contributions to chemistry was resolving this controversy

## Deprotonation

Deprotonation is a chemistry term that refers to the removal of a proton (hydrogen cation $H^+$) from a molecule, forming the conjugate base. The relative ability for a molecule to give up a proton is measured by a $pK_a$ value. A low $pK_a$ value indicates that the compound is acidic and will easily give up its proton to a base. The $pK_a$ of a compound is determined by many things, but most significantly it is impacted by the conjugate base's ability (or inability) to stabilize the negative charge through resonance.

Bases used to deprotonate depend on the $pK_a$ of the compound. Where the proton is not particularly acidic, and as such, the molecule does not give up its proton easily, a base stronger than the commonly known hydroxides are required. Hydrides are one of the many types of powerful deprotonating agents. Common hydrides used are sodium hydride and potassium hydride. These bases are so powerful because the hydride forms hydrogen gas when the proton from the other molecule is removed. However, the production of hydrogen also means that deprotonation using agents that release hydrogen are dangerous and should be done in an inert atmosphere (e.g. nitrogen) as water is a source of protons that is present in the air around us all the time and may react with the hydride instead of the desired molecule or may set on fire.

## Organic Reactions

An **Aldol condensation** is an organic reaction in which an enolate ion reacts with a carbonyl compound to form a β-hydroxyaldehyde or β-hydroxyketone, followed by dehydration to give a conjugated enone.

$$RC(=O)CH_2R' + R''C(=O)R''' \xrightarrow[-H_2O]{B:} RC(=O)C(R')=C(R'')R'''$$

Aldol condensations are important in organic synthesis, providing a good way to form carbon–carbon bonds. The Robinson annulation reaction sequence features an aldol condensation; the Wieland-Miescher ketone product is an important starting material for many organic syntheses. Aldol condensations are also commonly discussed in university level organic chemistry classes as a good bond-forming reaction that demonstrates important reaction mechanisms.[1][2][3] In its usual form, it involves the nucleophilic addition of a ketone enolate to an aldehyde to form a β-hydroxy ketone, or "**aldol**" (**ald**ehyde + alcoh**ol**), a structural unit found in many naturally occurring molecules and pharmaceuticals.[4][5][6]

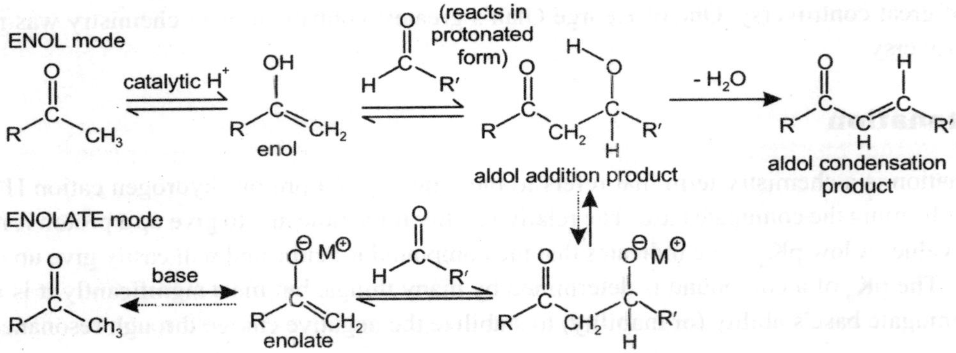

The name **aldol condensation** is also commonly used, especially in biochemistry, to refer to the aldol reaction itself, as catalyzed by aldolases. However, the aldol reaction is not formally a condensation reaction because it does not involve the loss of a small molecule. The reactions between a ketone and an aldehyde (crossed aldol condensation) or between two aldehydes also go by the name **Claisen-Schmidt condensation**. These reactions are named after two of its pioneering investigators Rainer Ludwig Claisen and J. G. Schmidt, who independently published on this topic in 1880 and 1881. An example is the synthesis of dibenzylideneacetone.

## Mechanism

The first part of this reaction is an aldol reaction, the second part a dehydration—an elimination reaction. Dehydration may be accompanied by decarboxylation when an activated carboxyl group is present. The aldol addition product can be dehydrated via two mechanisms; a strong base like potassium *t*-butoxide, potassium hydroxide or sodium hydride in an enolate mechanism,[10] or in an acid-catalyzed enol mechanism.

Base catalyzed aldol reaction (shown using $^-CH_3$ base)

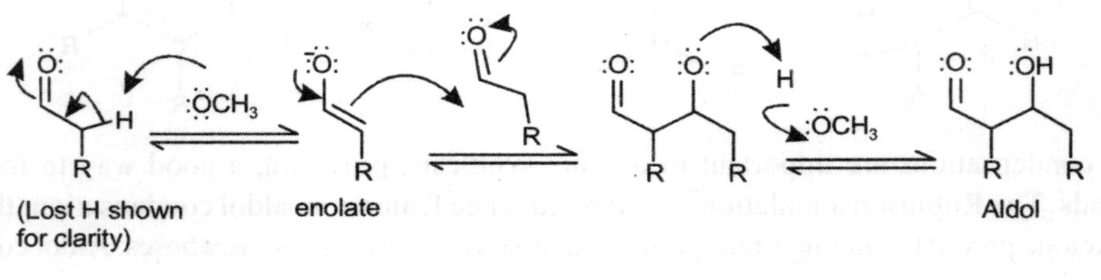

Base catalyzed dehydration (sometimes written as a single step)

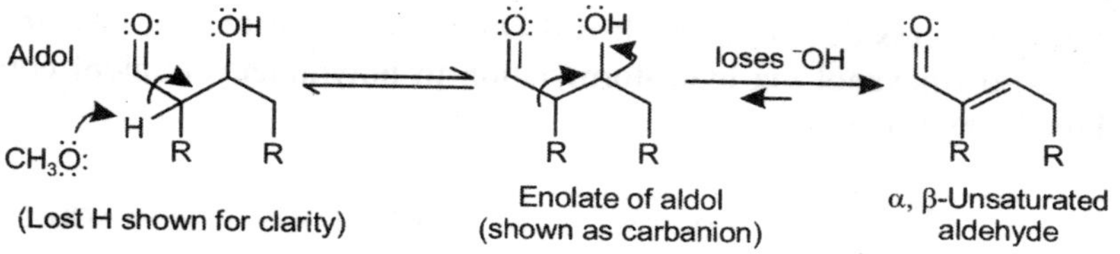

Aldol

$CH_3\ddot{O}$:

(Lost H shown for clarity)

Enolate of aldol (shown as carbanion)

loses $^-OH$

α, β-Unsaturated aldehyde

Acid catalyzed aldol reaction

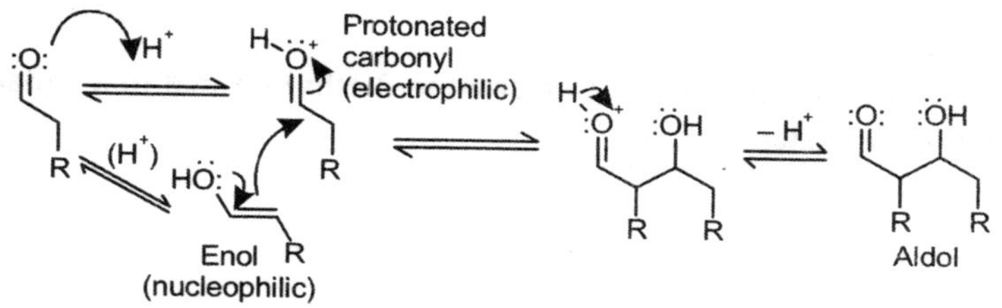

Protonated carbonyl (electrophilic)

Enol (nucleophilic)

$-H^+$

Aldol

Acid catalyzed dehydration

Aldol

(Lost H shown for clarity)

$H^+$

$H_2\ddot{O}$:

$-H_2O$

α, β–unsaturated aldehyde

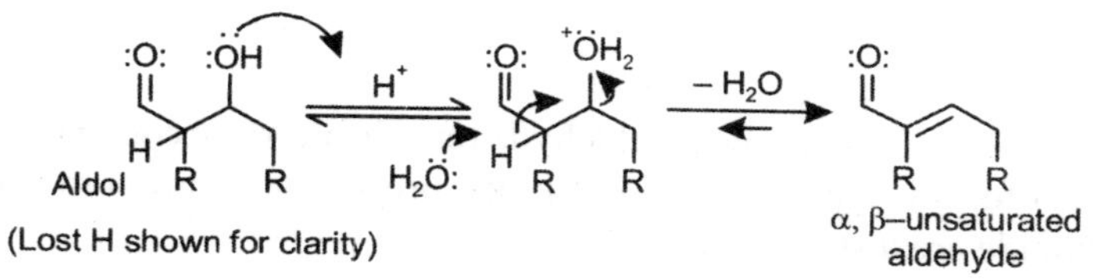

## Condensation Types

It is important to distinguish the Aldol condensation from other addition reactions to carbonyl compounds.

- When the base is an amine and the active hydrogen compound is sufficiently activated the reaction is called a Knoevenagel condensation.
- In a Perkin reaction the aldehyde is aromatic and the enolate generated from an anhydride.
- A Claisen condensation involves two ester compounds.
- A Dieckmann condensation involves two ester groups in the *same molecule* and yields a cyclic molecule
- A Henry reaction involves an aldehyde and an aliphatic nitro compound.
- A Robinson annulation involves a α, β-unsaturated ketone and a carbonyl group, which first engage in a Michael reaction prior to the aldol condensation.
- In the Guerbet reaction, an aldehyde, formed *in situ* from an alcohol, self-condenses to the dimerized alcohol.

## Aldox Process

In industry the **Aldox process** developed by Royal Dutch Shell and Exxon, converts propylene and syngas directly to 2-Ethylhexanol via hydroformylation to butyraldehyde, aldol condensation to 2-ethylhexenal and finally hydrogenation [11].

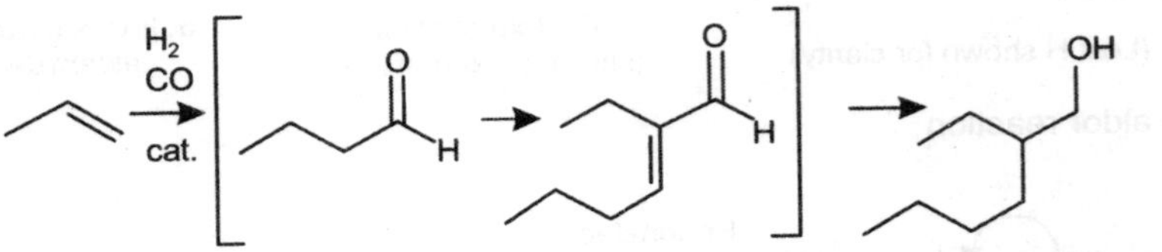

$H_2$ CO cat.

In one study crotonaldehyde is directly converted to 2-ethylhexanal in a palladium / Amberlyst / supercritical carbon dioxide system [12]:

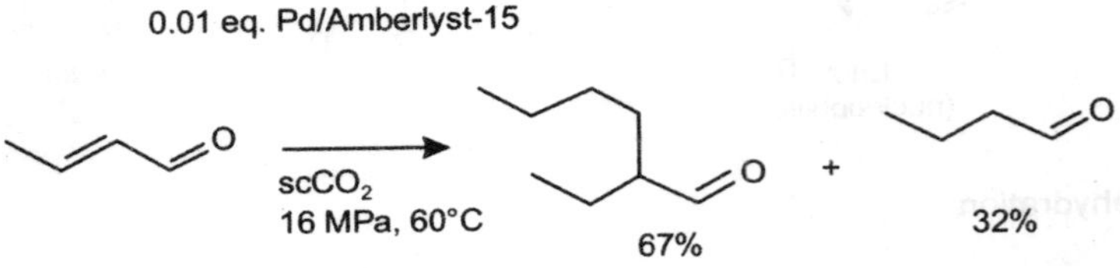

0.01 eq. Pd/Amberlyst-15

$scCO_2$ 16 MPa, 60°C

67% + 32%

## Scope

Ethyl 2-methylacetoacetate and campholenic aldehyde react in an Aldol condensation.[13] The synthetic procedure [14] is typical for this type of reactions. In the process in addition to water, an equivalent of ethanol and carbondioxide are lost in decarboxylation.

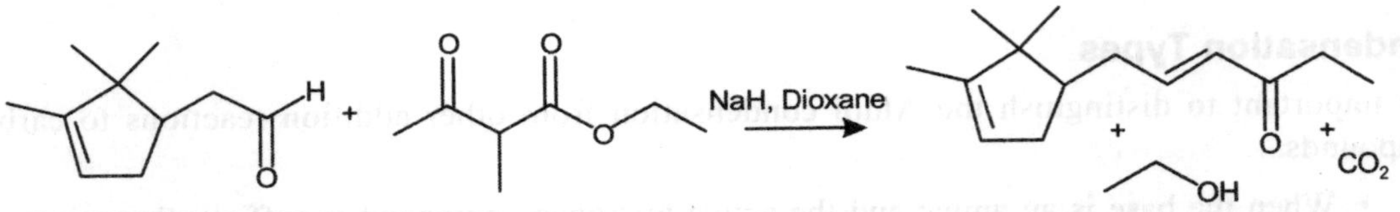

NaH, Dioxane

$CO_2$

Ethyl glyoxylate **2** and diethyl 2-methylglutaconate **1** react to *isoprenetricarboxylic acid* **3** (isoprene skeleton) with sodium ethoxide. This reaction product is very unstable with initial loss of carbon dioxide and followed by many secondary reactions. This is believed to be due to steric strain resulting from the methyl group and the carboxylic group in the *cis*-dienoid structure.[15]

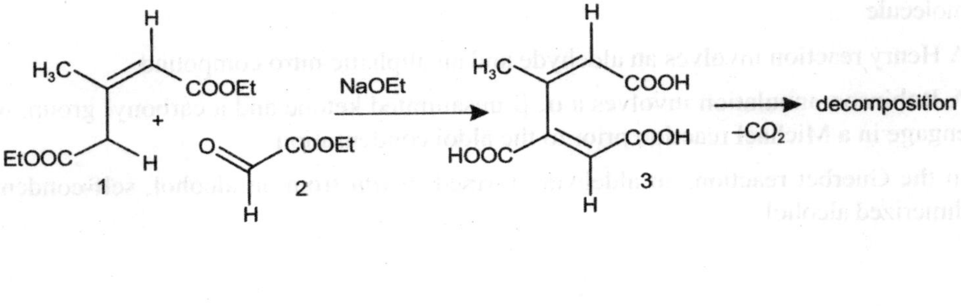

NaOEt

1 2 3

$-CO_2$ decomposition

Occasionally an aldol condensation is buried in a multistep reaction or in catalytic cycle such as the one sketched below:

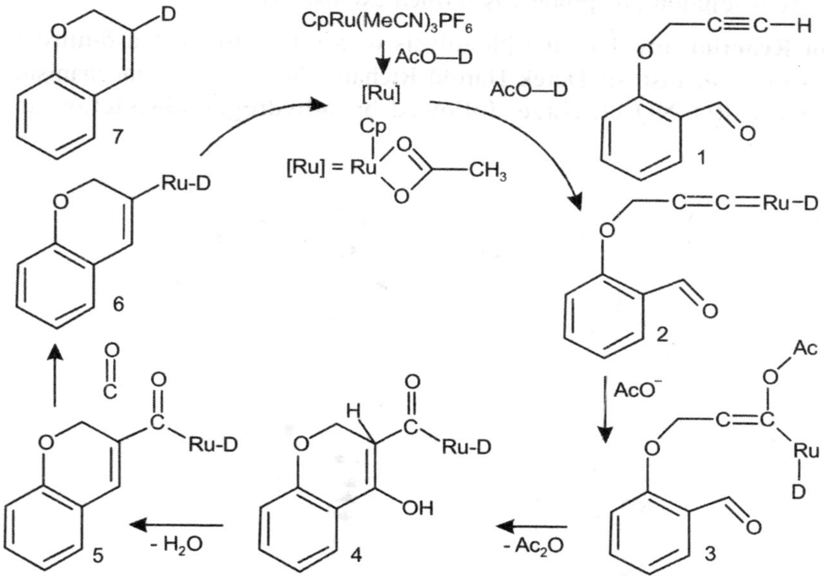

In this reaction an *alkynal* **1** is converted into a cycloalkene **7** with a ruthenium catalyst and the actual condensation takes place with intermediate **3** through **5**. Support for the reaction mechanism is based on isotope labelling.

The reaction between menthone and anisaldehyde is complicated due to steric shielding of the ketone group. The solution is use of a strong base such as potassium hydroxide and a very polar solvent such as DMSO in the reaction below:

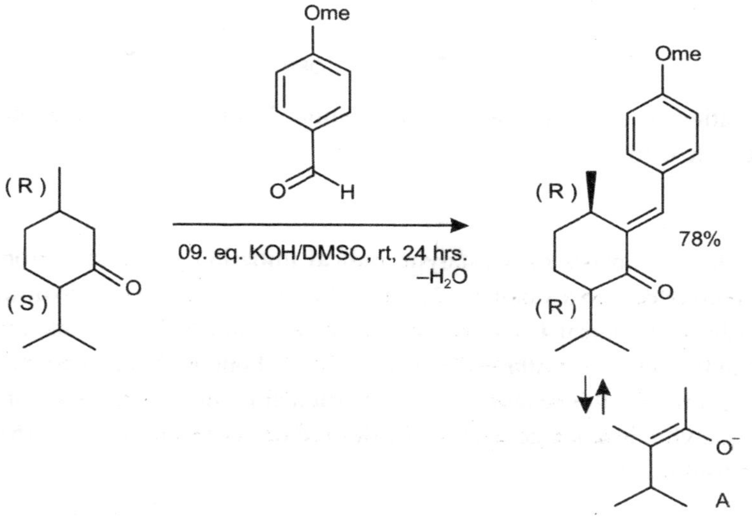

Due to epimerization through a common enolate ion (intermediate **A**) the reaction product has (R,R) cis configuration and not (R,S) trans as in the starting material. Because it is only the cis isomer that precipitates from solution this product is formed exclusively.

The **Barton Reaction** involves the photolysis of a nitrite to form a δ-nitroso alcohol. It is named for the British chemist Sir Derek Harold Richard Barton.[1] The mechanism is believed to involve a homolytic RO–NO cleavage, followed by δ-hydrogen abstraction and free radical recombination.[2]

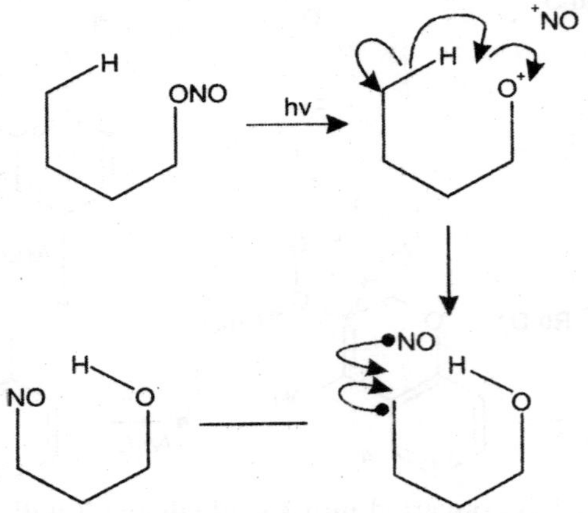

The **Barton-McCombie deoxygenation** is an organic reaction in which an hydroxy functional group in an organic compound is replaced by a hydride to give an alkyl group. It is named for the British chemists Sir Derek Harold Richard Barton (1918–1998) and Stuart W. McCombie.

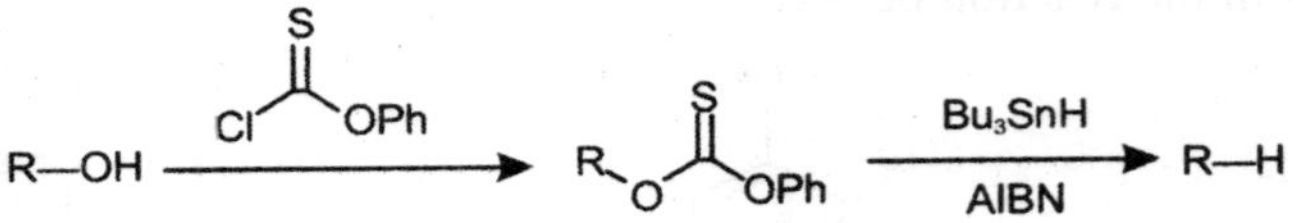

This deoxygenation reaction is a radical substitution. In the related Barton decarboxylation the reactant is a carboxylic acid.

## Mechanism

The reaction mechanism consists of a catalytic radical initiation step and a propagation step. The alcohol (**1**) is first converted into a xanthate (**2**). The other reactant tributyltin hydride **3** is decomposed by AIBN **8** into a tributyltin radical **4**. The tributyltin radical abstracts the xanthate group from **2** leaving an alkyl radical **5** and tributyltin xanthate (**7**). The sulfur tin bond in this compound is very stable and provides the driving force for this reaction. The alkyl radical in turn abstracts a hydrogen atom from a new molecule of tributyltin hydride generating the desired deoxygenated product (**6**) and a new radical species ready for propagation.

Bu3Sn⁻—S—C(=O)—R'
7
R—O—C(=S)—R'
2
R—OH
1
AIBN
8
9
CN
Bu3Sn
4
R
5
H
CN
10
Bu3Sn—H
3
R—H
6

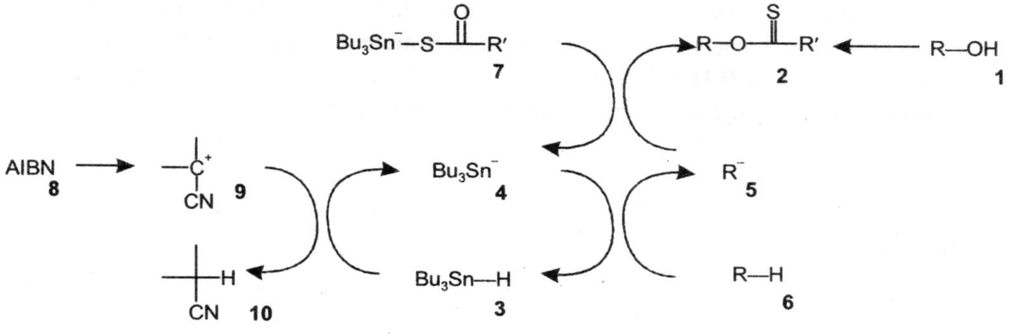

## Variations

### Alternative Hydride Sources

Main disadvantage of this reaction is the use of the tin hydride which is toxic, expensive and difficult to remove from the reaction mixture. One alternative is the use of tributyltin anhydride as the radical source and poly(methylhydridesiloxane) (PMHS) as the hydride source. Phenyl chlorothionoformate used as the starting material ultimately generates carbonyl sulfide.

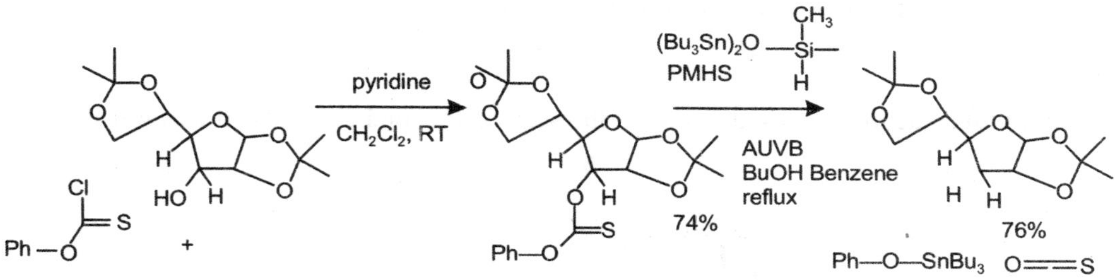

pyridine
CH2Cl2, RT
(Bu3Sn)2O
PMHS
CH3
Si
H
AUVB
BuOH Benzene
reflux
74%
76%
Ph—O
Ph—O—SnBu3
O==S
HO
Cl
S

### Trialkyl Boranes

An even more convenient hydrogen donor is provided by trialkylborane-water complexes [5] such as trimethylborane contaminated with small amounts of water.

MeO2C
CO2Me
C4H9
OH
X
S
CH3
Me3B-H2O
Benzene

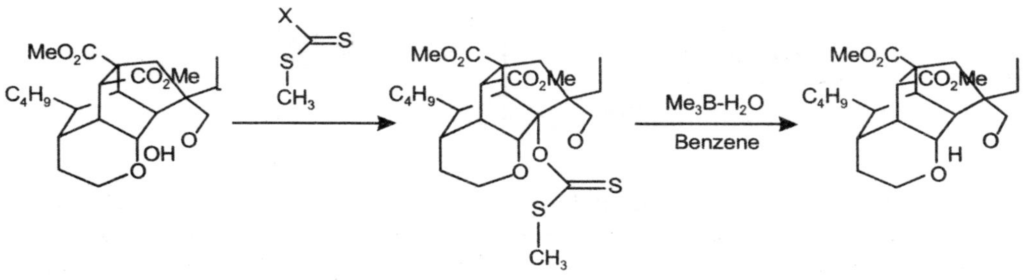

In this catalytic cycle the reaction is initiated by air oxidation of the trialkylborane **3** by air to the methyl radical **4**. This radical reacts with the xanthate **2** to S-methyl-S-methyl dithiocarbonate **7** and the radical intermediate **5**. The $CH_3B.H_2O$ complex **3** provides a hydrogen for recombining with this radical to the alkane **6** leaving behind diethyl borinic acid and a new methyl radical.

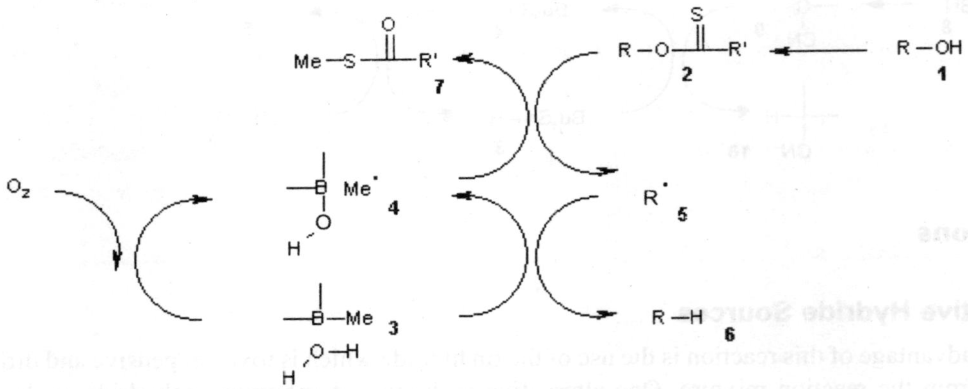

It is found by theoretical calculations that that a O-H homolysis reaction in the borane-water complex is endothermic with an energy similar to that of the homolysis reaction in tributyltin hydride but much lower than the homolysis reaction of pure water.

## Scope

A variation of this reaction was used as one of the steps in the total synthesis of azadirachtin

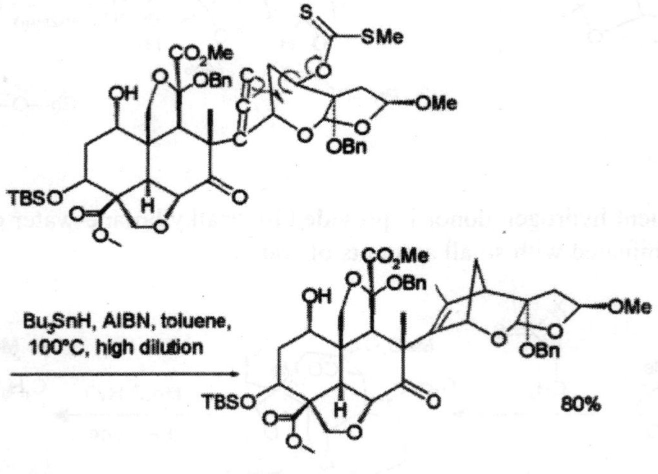

$Bu_3SnH$, AIBN, toluene, 100°C, high dilution

80%

In another variation the reagent is the imidazole 1,1'-thiocarbonyldiimidazole (TCDI), for example in the total synthesis of pallescensin B. TCDI is especially good to primary alcohols because there is no resonance stabilization of the xanthate because the nitrogen lonepair is involved in the aromatic sextet.

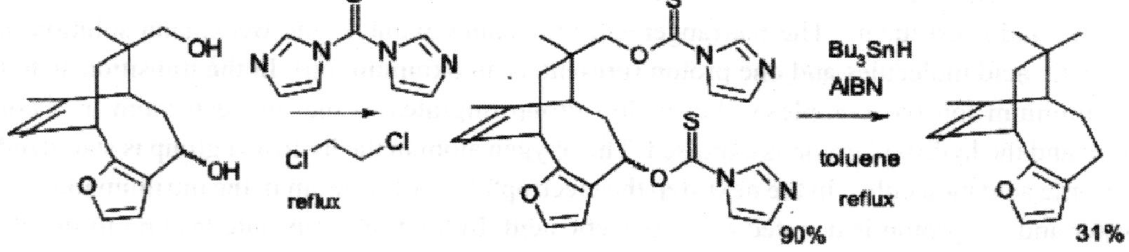

The reaction also applies to S-alkylxanthates. With triethylborane as a novel metal-free reagent, the required hydrogen atoms are abstracted from protic solvents, the reactor wall or even (in strictly anhydrous conditions) the borane itself.

**The Beckmann rearrangement**, named after the German chemist Ernst Otto Beckmann (1853-1923), is an acid-catalyzed rearrangement of an oxime to an amide. [1][2][3] Cyclic oximes yield lactams.

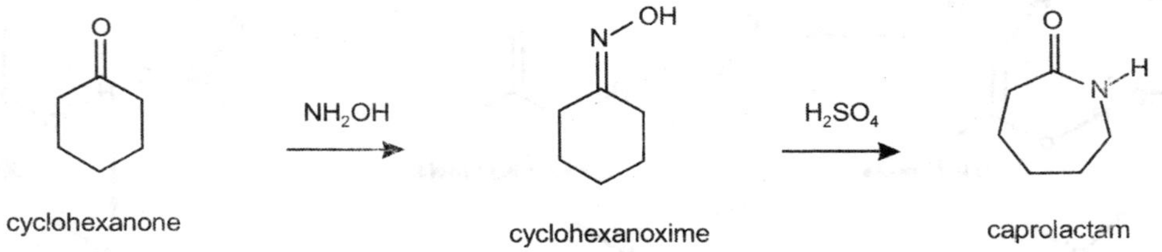

cyclohexanone $\xrightarrow{NH_2OH}$ cyclohexanoxime $\xrightarrow{H_2SO_4}$ caprolactam

This example reaction starting with cyclohexanone and forming caprolactam is one of the most important applications of the Beckmann rearrangement, as caprolactam is the feedstock in the production of Nylon 6.

The Beckmann solution consists of acetic acid, hydrochloric acid and acetic anhydride, and was widely used to catalyze the rearrangement. Other acids, such as sulfuric acid or polyphosphoric acid, can also be used. Sulphuric acid is the most commonly used acid for commercial lactam production due to its formation of an ammonium sulfate by-product when neutralized with ammonia. Ammonium sulfate is a common agricultural fertilizer providing nitrogen and sulfur.

## Reaction Mechanism

The reaction mechanism of the Beckmann rearrangement is generally believed to consist of an alkyl migration with expulsion of the hydroxyl group to form a nitrilium ion followed by hydrolysis:

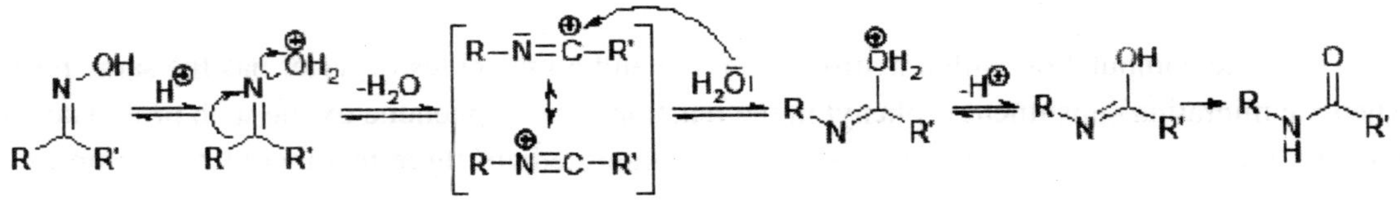

In one study, [5] the mechanism is established in silico taking into account the presence of solvent molecules and substituents. The rearrangement of acetone oxime in the Beckmann solution involves three acetic acid molecules and one proton (present as an oxonium ion). In the transition state leading to the iminium ion (σ- complex), the methyl group migrates to the nitrogen atom in a concerted reaction and the hydroxyl group is expulsed. The oxygen atom in the hydroxyl group is stabilized by the three acetic acid molecules. In the next step the electrophilic carbon atom in the nitrilium ion is attacked by water and the proton is donated back to acetic acid. In the transition state leading to the N-methyl acetimidic acid, the water oxygen atom is coordinated to 4 other atoms. In the third step, an isomerization step protonates the nitrogen atom leading to the amide.

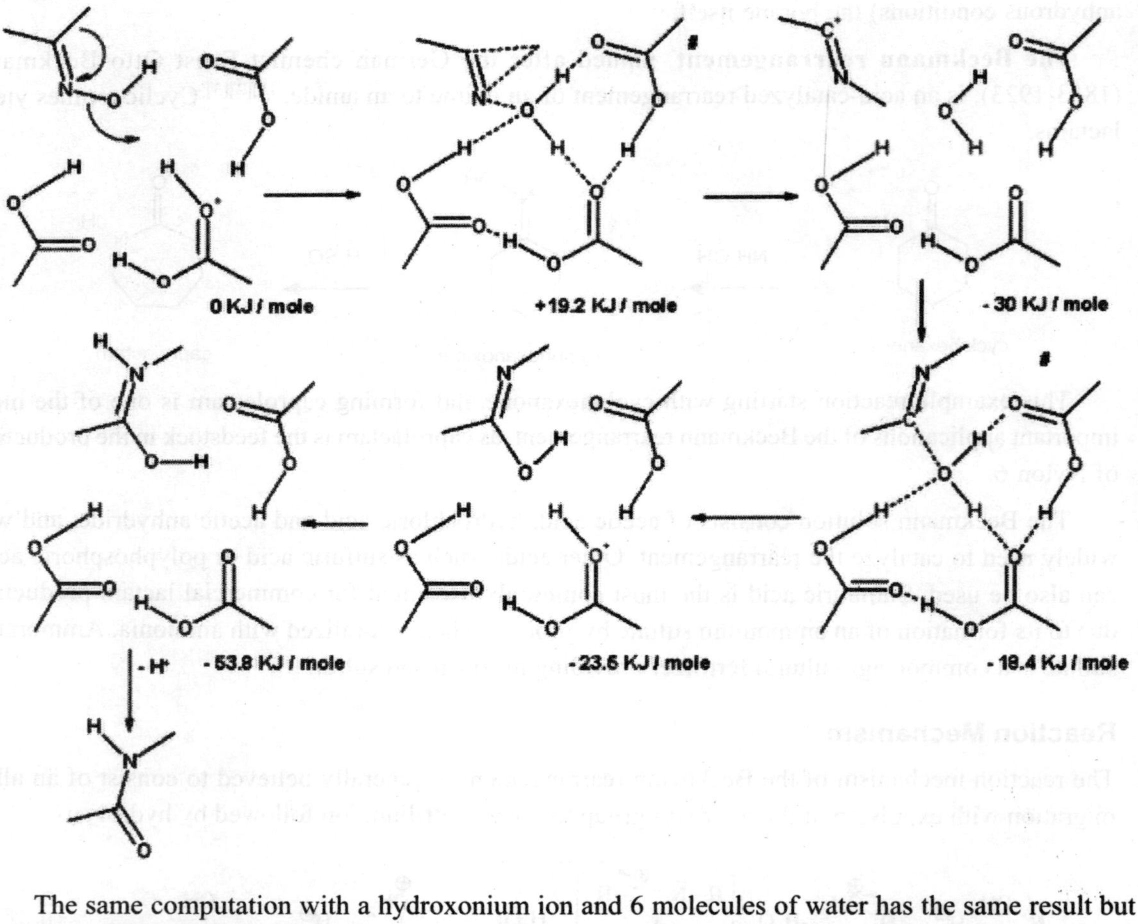

The same computation with a hydroxonium ion and 6 molecules of water has the same result but when the migrating substituent is phenyl in the reaction of acetophenone oxime with protonated acetic acid the mechanism favours the formation of an intermediate three-membered π - complex. This π - complex is again not found in the $H_3O^+(H_2O)_6$.

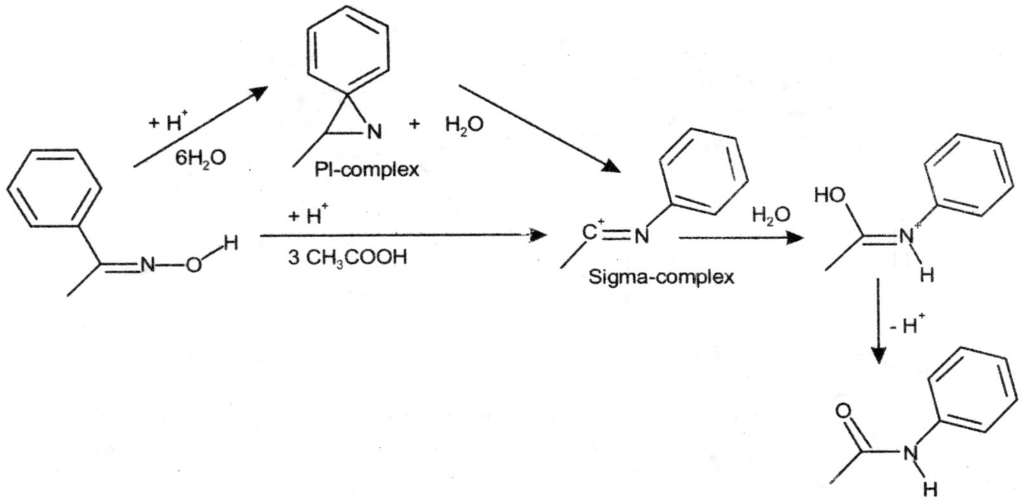

With the cyclohexanone-oxime, the relief of ring strain results in a third reaction mechanism leading directly to the protonated caprolactam in a single concerted step without the intermediate formation of a $\pi$ - complex or $\sigma$ - complex.

## Cyanuric Chloride Assisted Beckmann Reaction

The Beckmann reaction is known to be catalyzed by cyanuric chloride and zinc chloride co-catalyst. For example, cyclododecanone can be smoothly convered to the corresponding lactam, a monomer for the production of nylon 12.

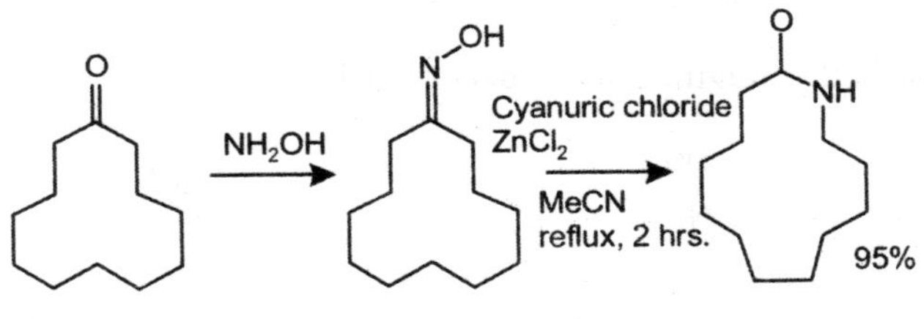

The reaction mechanism for this reaction is based on a catalytic cycle with cyanuric chloride activating the hydroxyl group via a nucleophilic aromatic substitution. The reaction product is dislodged and replaced by new reactant via an intermediate Meisenheimer complex.

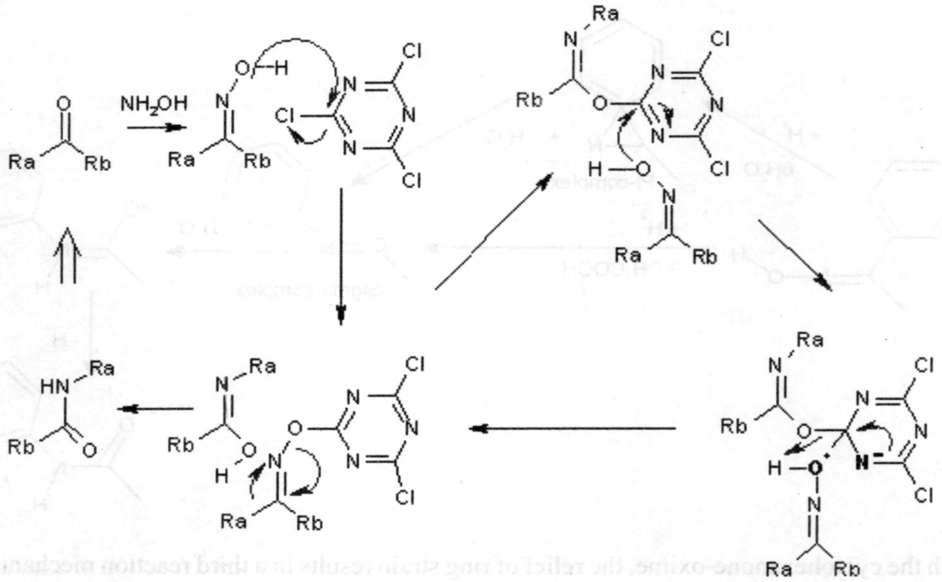

## Beckmann Fragmentation

When the oxime has a quaternary carbon atom in a anti position to the hydroxyl group a fragmentation occurs forming a nitrile:

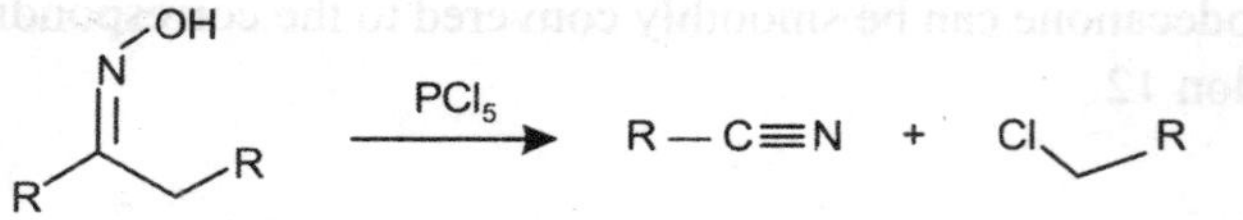

The fluorine donor in this fragmentation reaction is DAST[7]:

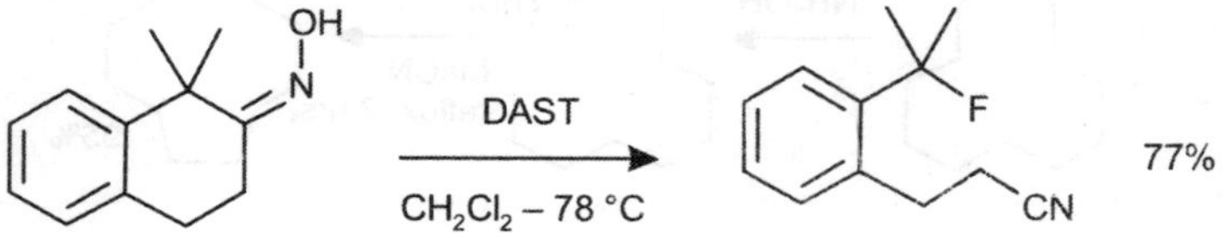

The **Cannizzaro reaction**, named after its discoverer Stanislao Cannizzaro, is a chemical reaction that involves the base-induced disproportionation of an aldehyde lacking a hydrogen atom in the alpha position. Cannizzaro first accomplished this transformation in 1853, when he obtained benzyl alcohol and benzoic acid from the treatment of benzaldehyde with potash (potassium carbonate).

The oxidation product is a carboxylic acid and the reduction product is an alcohol. For aldehydes with a hydrogen atom alpha to the carbonyl, i.e. RC*H*R'CHO, the preferred reaction is an aldol condensation, originating from deprotonation of this hydrogen. Reviews have been published.

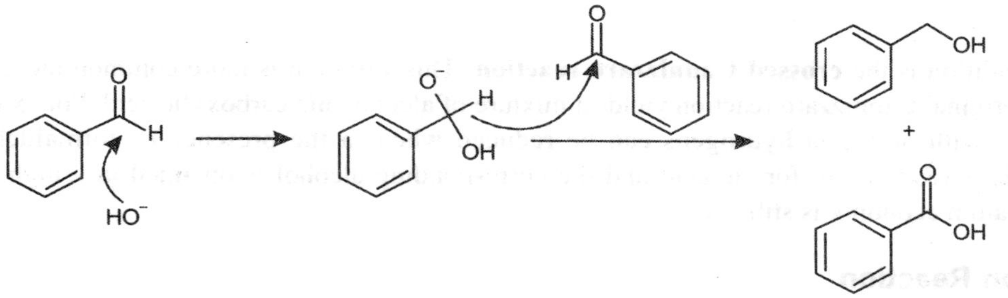

## Reaction Mechanism

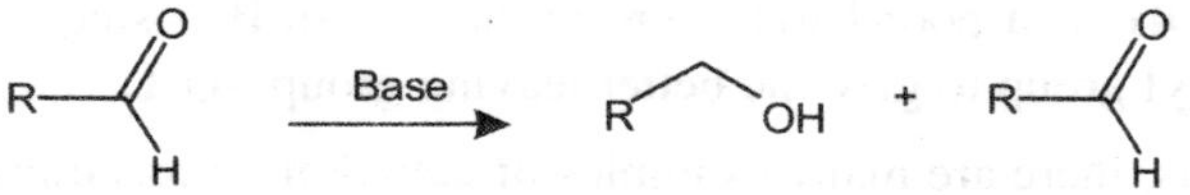

The first reaction step is nucleophilic addition of the base (for instance the hydroxy anion) to the carbonyl carbon of the aldehyde. The resulting alkoxide is deprotonated to give a di-anion, known as the Cannizzaro intermediate. Formation of this intermediate requires a strongly basic environment.

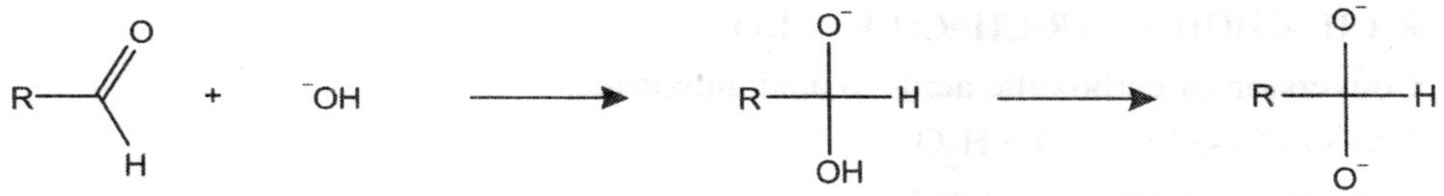

Both intermediates can react further with aldehyde to transfer a hydride, "H"". The hydridic character of the C-*H* is enhanced by the electron-donating character of the alpha oxygen anion. This hydride transfer simultaneously generates a hydroxyl anion and a carboxylate. Further evidence for the hydridic character of the Cannizzaro intermediate is provided by the formation of $H_2$ by its reaction with water.

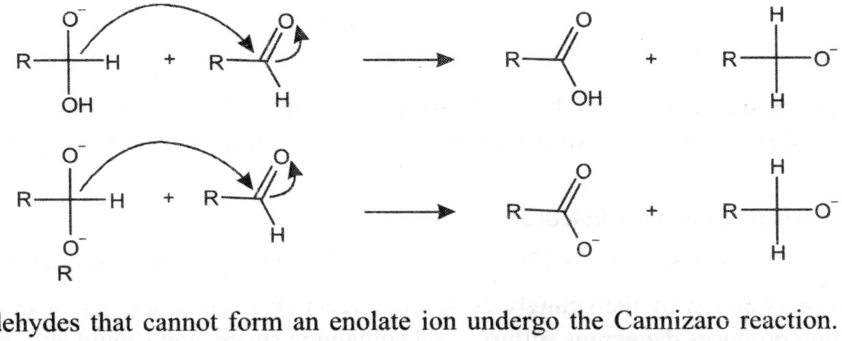

Only aldehydes that cannot form an enolate ion undergo the Cannizaro reaction. The aldehyde cannot have an enolizable proton. Under the basic conditions that facilitate the reaction, aldehydes that can form an enolate instead undergo aldol condensation. Examples of aldehydes that can undergo a Cannizaro reaction include formaldehyde and aromatic aldehydes such as benzaldehyde.

## Variations

A special condition is the **crossed Cannizzaro reaction**. This variation is more common these days because the original Cannizzaro reaction yields a mixture of alcohol and carboxylic acid. For example any aldehyde with no alpha hydrogens can be reduced when in the presence of formaldehyde. Formaldehyde is oxidized to formic acid and the corresponding alcohol is obtained in a high yield although the atom economy is still low.

## Dehydration Reaction

n chemistry, a **dehydration reaction** is usually defined as a chemical reaction that involves the loss of water from the reacting molecule. Dehydration reactions are a subset of elimination reactions. Because the hydroxyl group (-OH) is a poor leaving group, having an Brønsted acid catalyst often helps by protonating the hydroxyl group to give the better leaving group, $-OH_2^+$.

In organic synthesis, there are many examples of dehydration reactions:

- Conversion of alcohols to ethers
  $2\ R\text{-}OH \rightarrow R\text{-}O\text{-}R + H_2O$
- Conversion of alcohols to alkenes
  $R\text{-}CH_2\text{-}CHOH\text{-}R \rightarrow R\text{-}CH{=}CH\text{-}R + H_2O$
- Conversion of carboxylic acids to acid anhydrides
  $2\ RCO_2H \rightarrow (RCO)_2O + H_2O$
- Conversion of amides to nitriles
  $RCONH_2 \rightarrow R\text{-}CN + H_2O$
- in this rearrangement reaction called the **dienol benzene rearrangement**

$H_3C$, $CCl_3$ ... HO, H — HCl, $-H_2O$ → $CH_3$, $CCl_3$

Some dehydration reactions can be mechanistically complex, for instance the reaction of a sugar with concentrated sulphuric acid (experiment with video) to form carbon involves formation of carbon carbon bonds.[1]

- Sugar (sucrose) is dehydrated[2]:

$C_{12}H_{22}O_{11}$ + 98% Sulfuric acid → $12C_{\text{(graphitic foam)}} + 11H_2O_{\text{steam}}$ + Sulfuric acid/water mixture

The reaction is driven by the strongly exothermic reaction sulfuric acid has with water. (Beware that this reaction produces dangerous sulfuric-acid containing steam, and should only be performed in a fume-hood or well ventilated area.)

Common dehydrating agents; concentrated sulfuric acid, concentrated phosphoric acid, hot aluminium oxide, hot ceramic.

## Halogen Addition Reaction

A **halogen addition reaction** is a simple organic reaction where a halogen molecule is added to the carbon-carbon double bond of an alkene functional group.

Here is the general chemical formula of the halogen addition reaction:

$$C{=}C + X_2 \rightarrow X\text{-}C\text{-}C\text{-}X$$

(X represents the halogens bromine or chlorine, and in this case, a solvent could be $CH_2Cl_2$ or $CCl_4$). The product is a vicinal dihalide.

This type of reaction is a halogenation and an electrophilic addition.

## Reaction Mechanism

The reaction mechanism for an alkene bromination can be described as follows (*scheme 1*). In the first step of the reaction, a bromine molecule approaches the electron-rich alkene carbon-carbon double bond. The bromine atom closer to the bond takes on a partial positive charge as its electrons are repelled by the electrons of the double bond.

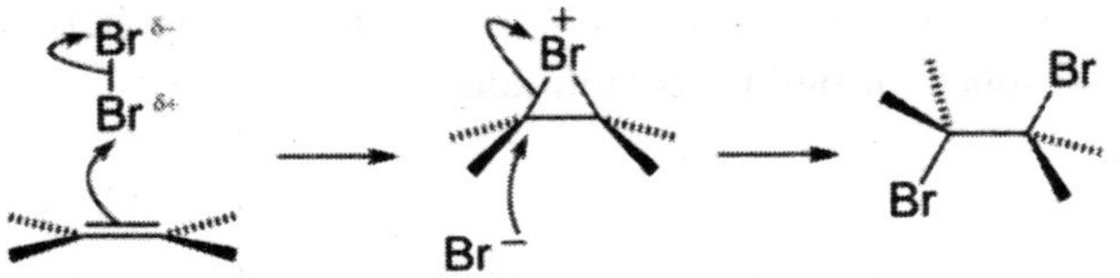

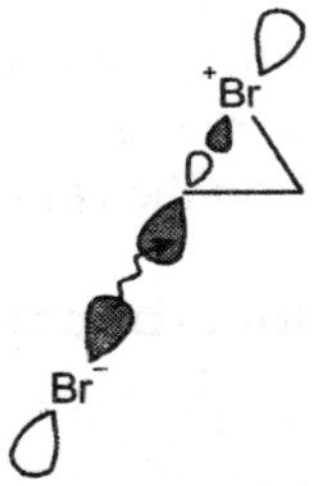

A bromide ion attacks the C-Br σ* antibonding molecular orbital of a bromonium ion.

The atom is electrophilic at this time and attacks the negatively charged, high energy pi bond portion of the alkene's carbon-carbon double bond. It forms for an instant a single sigma bond to *both* of the carbon atoms involved. The bonding of bromine is special in this intermediate, due to its relatively large size compared to carbon, the bromide ion is capable to latching onto both carbons which once formed the π-bond, making a three-membered ring. The bromide ion acquires a positive charge, which it shares slightly with the two carbon atoms. At this moment the halogen ion is called a "bromonium ion" or "chloronium ion", respectively. When the first bromine atom attacks the carbon-carbon pi-bond, it leaves behind one of its electrons with the other bromine that it was bonded to in $Br_2$. That other atom is now a negative bromide anion and is attracted to the slight positive charge on the carbon atoms. It is blocked from nucleophilic attack on one side of the carbon chain by the first bromine atom and can only attack on the other side. As it attacks and forms a bond with one of the carbons, the bond between the first bromine atom and the other carbon atoms breaks, leaving each carbon atom with a halogen substituent.

In this way the two halogens add in an anti addition fashion, and when the alkene is part of a cycle the dibromide adopts the trans configuration. For maximum overlap of the C-Br σ* antibonding molecular orbital (the LUMO, shown to the right in red) and the nucleophile ($X^-$) lone pair (the HOMO, shown to the right below in green), $X^-$ must attack the bromonium ion from behind, at carbon.

This reaction mechanism was proposed by Roberts and Kimball in 1937 [2]. With it they explained the observed trans-additions in brominations of maleic acid and fumaric acid. Maleic acid with a cis-double bond forms the dibromide as a mixture of enantiomers:

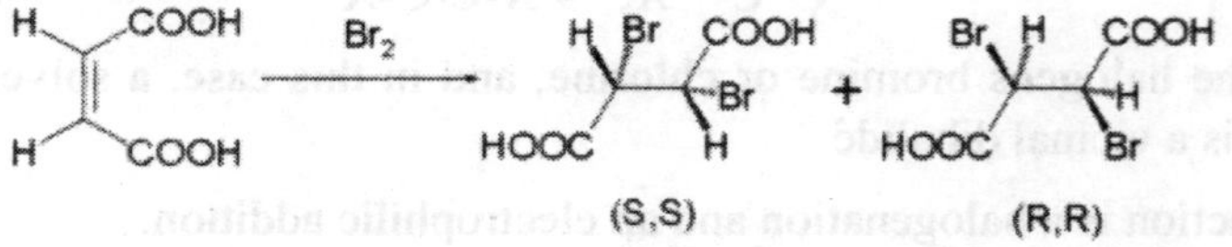

(S,S) (R,R)

while the trans-isomer fumaric acid reaction is completely stereoselective and stereospecific forming a single meso compound:

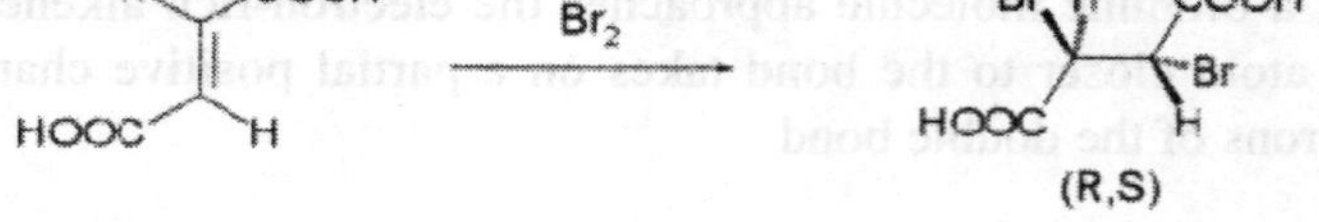

(R,S)

but both by anti-addition. The reaction is even stereoselective in alkenes with two bulky tert-butyl groups in a cis-position as in the compound *cis-di-tert-butylethylene* [3]. Despite the steric repulsion present in the chloronium ion, the only product formed is the anti-adduct.

## β-Halocarbocations

In an alternative reaction scheme depicted below the reactive intermediate is a β-bromocarbocation or β-bromocarbonium ion with one of the carbon atoms a genuine carbocation.

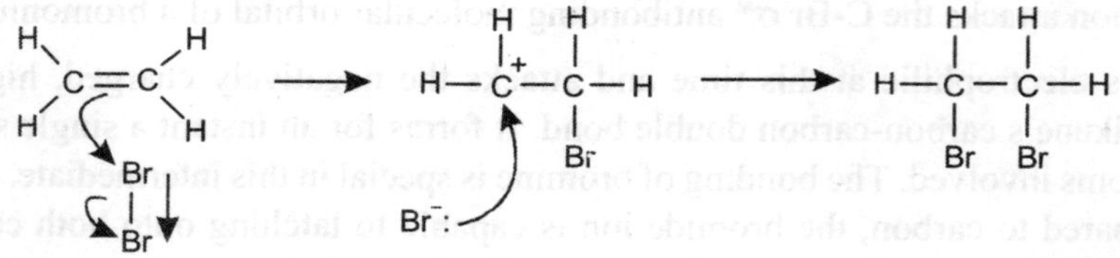

For reactions taking place through this mechanism no stereoselectivity is expected and indeed not found.

Roberts and Kimball in 1937 already accounted for the fact that brominations with the maleate ion resulted in cis-addition driven by repulsion between the negatively charged carboxylic acid anions being stronger than halonium ion formation. In alkenes such as anetholes and stilbenes the substituents are able to stabilize the carbocation by donating electrons at the expense of the halonium ion [4].

Halonium ions can be identified by means of NMR spectroscopy. In 1967 the group of George A. Olah obtained NMR spectra of tetramethylethylenebromonium ions by dissolving 2,3-dibromo-2,3-dimethylbutane in magic acid at -60°C [5]. The spectrum for the corresponding fluorine compound on the other hand was consistent with a rapidly equilibrating pair of β-fluorocarbocations.

## Dowd-Beckwith Ring Expansion Reaction

The **Dowd-Beckwith Ring Expansion Reaction** is an organic reaction in which a cyclic β-keto ester is expanded by up to 4 carbons in a free radical ring expansion reaction through an α-alkylhalo substituent The radical initiator system is based on AIBN and tributyltin hydride. The cyclic β-keto ester can be obtained through a Dieckmann condensation. The original reaction consisted of a nucleophilic aliphatic substitution of the enolate of ethyl cyclohexanone-2-carboxylate with 1,4-diiodobutane and sodium hydride followed by ring expansion to ethyl cyclodecanone-6-carboxylate. A side-reaction is organic reduction of the iodoalkane.

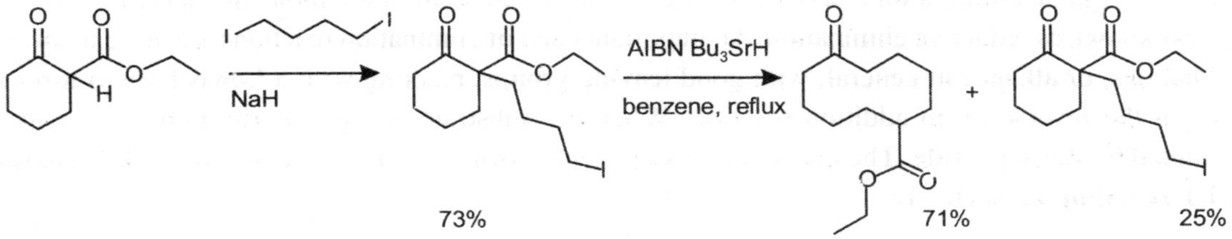

NaH

AIBN $Bu_3SrH$

benzene, reflux

+

73% 71% 25%

## Reaction Mechanism

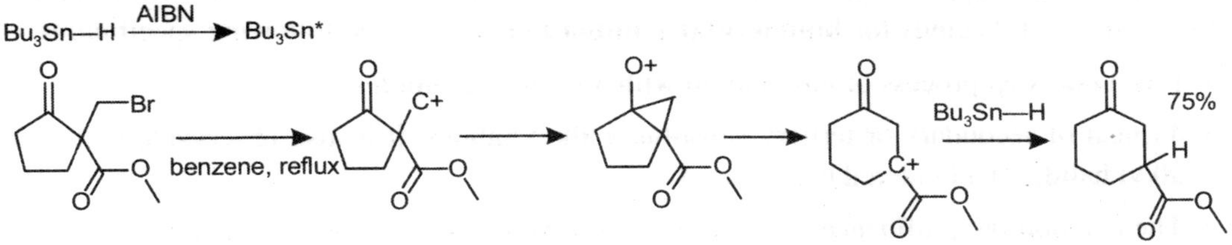

$Bu_3Sn—H \xrightarrow{AIBN} Bu_3Sn^*$

Br C+ O+ C+ $Bu_3Sn—H$ 75% H

benzene, reflux

The reaction mechanism involves a bicyclic intermediate. The reaction is initiated by thermal decomposition of AIBN. The resulting radicals abstract hydrogen from tributyltin hydride to a tributyltin radical which in turn abstracts the halogen atom to form an alkyl radical. This radical attacks the carbonyl group to an intermediate bicyclic ketyl. This intermediate then rearranges with ring expansion to a new carbon radical species which recombines with a proton radical from tributyltin hydride propagating the catalytic cycle.

## Scope

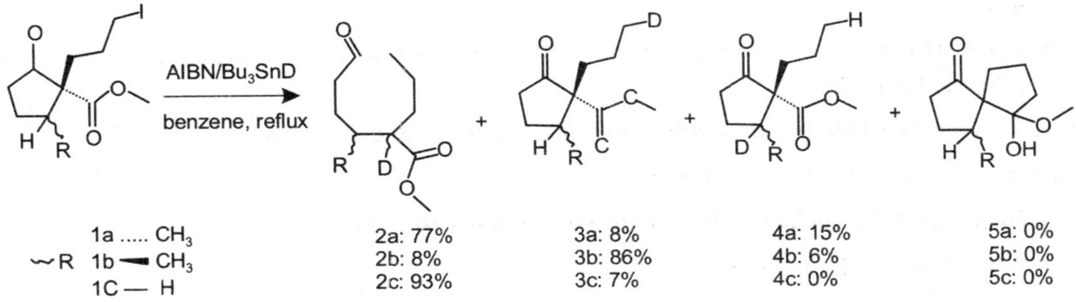

AIBN/$Bu_3SnD$

benzene, reflux

+ + +

1a ..... $CH_3$
~R 1b ◄ $CH_3$
1C — H

2a: 77%
2b: 8%
2c: 93%

3a: 8%
3b: 86%
3c: 7%

4a: 15%
4b: 6%
4c: 0%

5a: 0%
5b: 0%
5c: 0%

A side reaction accompanying this ring expansion is organic reduction of the halo alkane to a saturated alkyl group. One study [4] shows that the success depends critically on the accessibility of the carbonyl group. Deuterium experiments also show the presence of a 1,5 hydride shift. The reaction of the alkyl radical with the ester carbonyl group is also a posiibility but has an unfavorable activation energy.

## Elimination Reaction

An **Elimination reaction** is a type of organic reaction in which two substituents are removed from a molecule in either a one or two-step mechanism [1]. Either the unsaturation of the molecule increases (as in most organic elimination reactions) or the valence of an atom in the molecule decreases by two, a process known as reductive elimination. An important class of elimination reactions are those involving alkyl halides, or alkanes in general, with good leaving groups, reacting with a Lewis base to form an alkene in the reverse of an addition reaction. When the substrate is asymmetric, regioselectivity is determined by Zaitsev's rule. The one and two-step mechanisms are named and known as **E2 reaction** and **E1 reaction**, respectively.

## E2 Mechanism

In the 1920s, Sir Christopher Ingold proposed a model to explain a peculiar type of chemical reaction: the E2 mechanism. E2 stands for **bimolecular elimination** and has the following specificities.

- It is a one-step process of elimination with a single transition state.
- Typical of secondary or tertiary substituted alkyl halides. It is also observable with primary alkyl halides if a hindered base is used.
- The reaction rate, influenced by both the alkyl halide and the base, is second order.
- Because E2 mechanism results in formation of a Pi bond, the two leaving groups (often a hydrogen and a halogen) need to be coplanar. An antiperiplanar transition state has staggered conformation with lower energy and a synperiplanar transition state is in eclipsed conformation with higher energy. The reaction mechanism involving staggered conformation is more favourable for E2 reactions.
- Reaction often present with strong base.
- In order for the pi bond to be created, the hybridization of carbons need to be lowered from $sp^3$ to $sp^2$.
- The C-H bond is weakened in the rate determining step and therefore the deuterium isotope effect is larger than 1.
- This reaction type has similarities with the $S_N2$ reaction mechanism.

The reaction fundamental elements are:

- Breaking of the *carbon-hydrogen* and *carbon-halogen* bonds in one step.
- Formation of a *C=C Pi bond.*

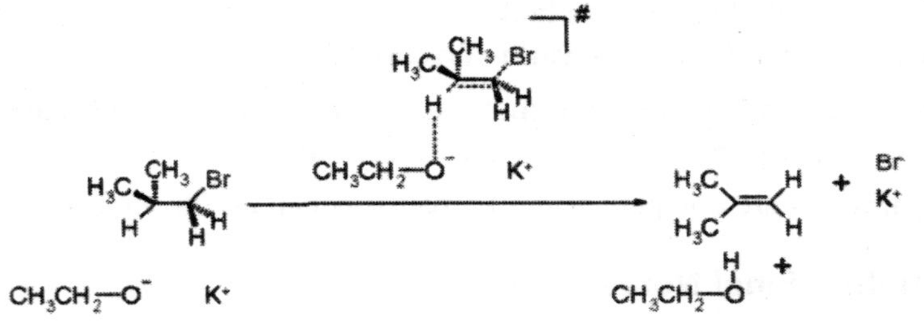

An example of this type of reaction in *scheme 1* is the reaction of isobutylbromide with potassium ethoxide in ethanol. The reaction products are isobutylene, ethanol and potassium bromide.

## E1 mechanism

E1 is a model to explain a particular type of chemical elimination reaction. E1 stands for **unimolecular elimination** and has the following specificities.

- It is a two-step process of elimination *ionization and deprotonation.*
  - Ionization, Carbon-halogen breaks to give a carbocation intermediate.
  - Deprotonation of the carbocation.
- Typical of tertiary and some secondary substituted alkyl halides.
- The reaction rate is influenced only by the concentration of the alkyl halide because carbocation formation is the slowest, rate-determining step. Therefore first order kinetics apply.
- Reaction mostly occurs in complete absence of base or presence of only weak base.
- E1 reactions are in competition with $S_N1$ reactions because they share a common carbocationic intermediate.
- Deuterium isotope effect is absent.
- accompanied by carbocationic rearrangement reactions
  - E1cB is another type of elimination reaction in which a carbanion is formed when the deprotonation happens before the ionization

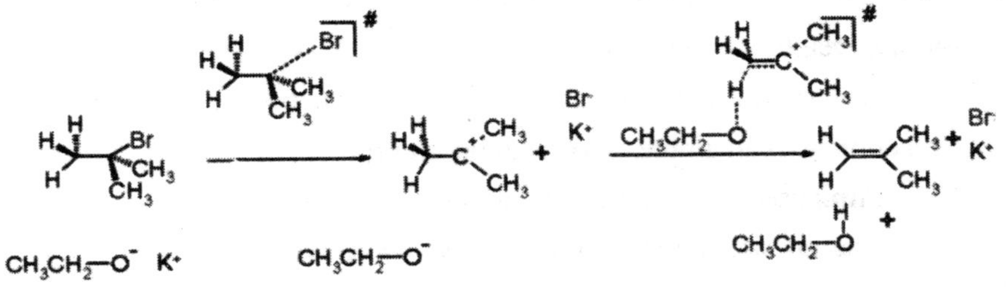

An example in *scheme 2* is the reaction of tert-butylbromide with potassium ethoxide in ethanol. E1 eliminations happen with highly substituted alkyl halides due to 2 main reasons.

- Highly substituted alkyl halides are bulky, limiting the room for the E2 one-step mechanism; therefore, the two-step E1 mechanism is favoured.
- Highly substituted carbocations are more stable than methyl or primary substituted. Such stability gives time for the two-step E1 mechanism to occur.

If $S_N1$ and E1 pathways are competing, the E1 pathway can be favoured by increasing the heat.

## E2 and E1 Elimination Final Notes

The reaction rate is influenced by halogen's reactivity; iodide and bromide being favoured. Fluoride is not a good leaving group. There is a certain level of competition between **elimination reaction** and nucleophilic substitution. More precisely, there are competitions between E2 and $S_N2$ and also between E1 and $S_N1$. Substitution generally predominates and elimination occurs only during precise circumstances. Generally, elimination is favoured over substitution when

- steric hindrance increases
- basicity increases
- temperature increases
- the steric bulk of the base increases for example Potassium tert-butoxide
- the nucleophile is poor

In one study [2] the kinetic isotope effect (KIE) was determined for the gas phase reaction of several alkyl halides with the chlorate ion. In accordance with a E2 elimination the reaction with t-butyl chloride results in a KIE of 2.3. The methyl chloride reaction (only $S_N2$ possible) on the other hand has a KIE of 0.85 consistent with a $S_N2$ reaction because in this reaction type the C-H bonds tighten in the transition state. The KIE's for the ethyl (0.99) and isopropyl (1.72) analogues suggest competition between the two reaction modes.

## Specific elimination reactions

The E1cB elimination reaction is a special type of elimination reaction involving carbanions. In an addition-elimination reaction elimination takes place after an initial addition reaction and in the Ei mechanism both substituents leave simultaneously in a syn addition.

In each of these elimination reactions the reactants have specific leaving groups:

- the dehydration reaction is one where the leaving group is water.
- the Bamford-Stevens reaction with a tosyl hydrazone leaving group assisted by alkoxide
- the Cope reaction with an amine oxide leaving group
- the Hofmann elimination with quaternary amine leaving group
- the Chugaev reaction with a methyl xanthate leaving group
- the Grieco elimination with a selenoxide leaving group
- the Shapiro reaction with a tosyl hydrazone leaving group assisted by alkyllithium
- Hydrazone iodination with a hydrazone leaving group assisted by iodine

- A Grob fragmentation with degree of unsaturation increasing in one of the leaving groups.
- the Kornblum–DeLaMare rearrangement (elimination over a (H)C-O(OR) bond) with an alcohol leaving group forming a ketone
- the Takai olefination with two bulky chromium groups.

## Friedel-Crafts Reactions

The **Friedel-Crafts reactions** are a set of reactions developed by Charles Friedel and James Crafts in 1877. There are two main types of Friedel-Crafts reactions: alkylation reactions and acylation reactions. This reaction type is part of electrophilic aromatic substitution.

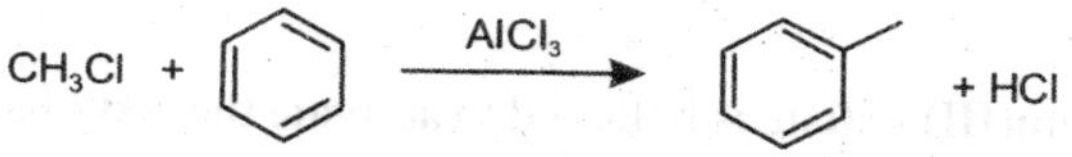

$$CH_3Cl + C_6H_6 \xrightarrow{AlCl_3} C_6H_5CH_3 + HCl$$

### Friedel-Crafts Alkylation

Friedel-Crafts alkylation involves the alkylation of an aromatic ring and an alkyl halide using a strong Lewis acid catalyst. With anhydrous ferric chloride as a catalyst, the alkyl group attaches at the former site of the chloride ion.

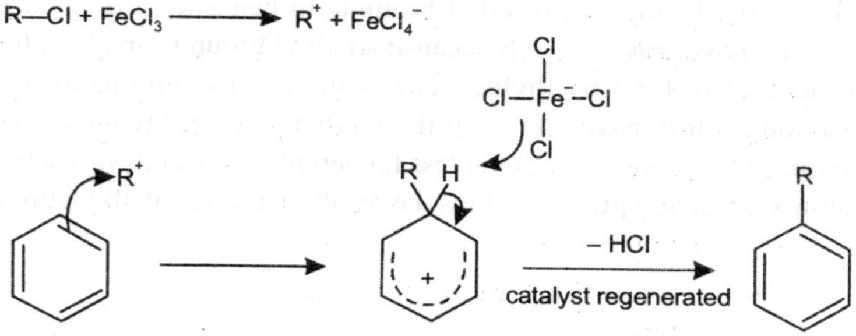

$$R{-}Cl + FeCl_3 \longrightarrow R^+ + FeCl_4^-$$

– HCl

catalyst regenerated

This reaction has one big disadvantage, namely that the product is more nucleophilic than the reactant due to the electron donating alkyl-chain. Therefore, another hydrogen is substituted with an alkyl-chain, which leads to overalkylation of the molecule. Also, if the chlorine is not on a tertiary carbon, carbocation rearrangement reaction will occur. This is due to the relative stability of the tertiary carbocation over the secondary and primary carbocations.

Steric hindrance can be exploited to limit the number of alkylations, as in the t-butylation of 1,4-dimethoxybenzene.

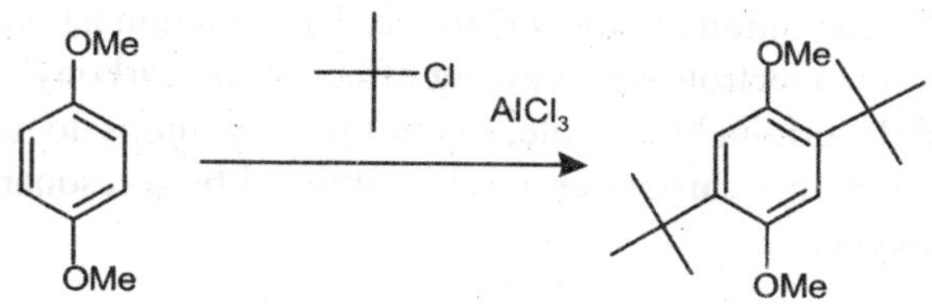

OMe OMe

$AlCl_3$

OMe OMe

Alkylations are not limited to alkyl halides: Friedel-Crafts reactions are possible with any carbocationic intermediate such as those derived from alkenes and a protic acid or lewis acid, enones and epoxides. In one study the electrophile is a bromonium ion derived from an alkene and NBS:[6]

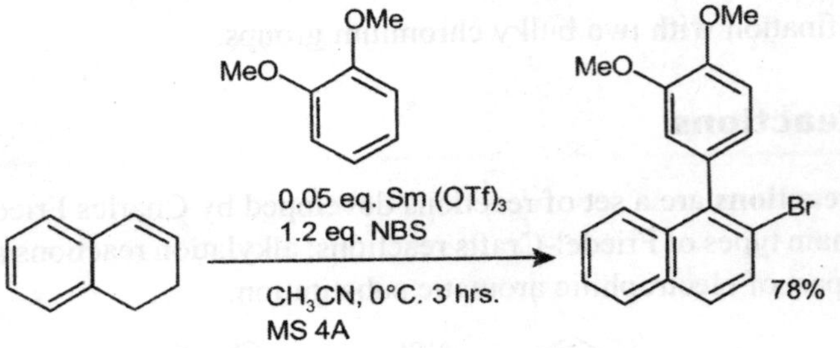

OMe
MeO
OMe
MeO
0.05 eq. $Sm(OTf)_3$
1.2 eq. NBS
Br
$CH_3CN$, 0°C. 3 hrs.
MS 4A
78%

In this reaction samarium(III) triflate is believed to activate the NBS halogen donor in halonium ion formation.

## Friedel-Crafts Dealkylation

Friedel-Crafts alkylation is a reversible reaction. In a **reversed Friedel-Crafts reaction** or **Friedel-Crafts dealkylation**, alkyl groups can be removed in the presence of protons and a Lewis acid.

For example, in a multiple addition of ethyl bromide to benzene, *ortho* and *para* substitution is expected after the first monosubstitution step because an alkyl group is an activating group. However, the actual reaction product is 1,3,5-triethylbenzene with all alkyl groups as a meta substituent.[7] Thermodynamic reaction control makes sure that thermodynamically favoured *meta* substitution with steric hindrance minimized takes prevalence over less favourable *ortho* and *para* substitution by chemical equilibration. The ultimate reaction product is thus the result of a series of alkylations and dealkylations.

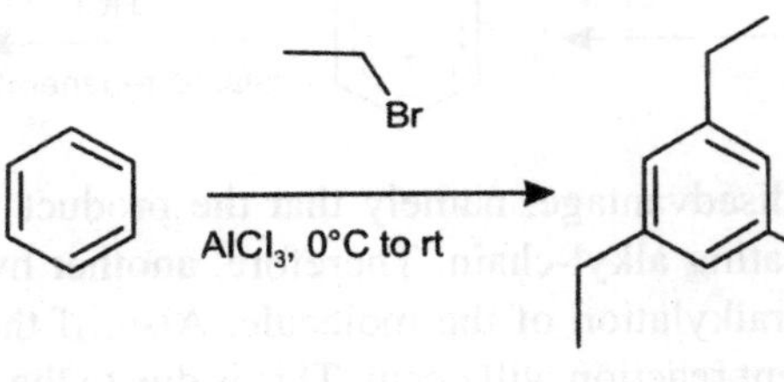

Br
$AlCl_3$, 0°C to rt

## Friedel-Crafts Acylation

Friedel-Crafts acylation is the acylation of aromatic rings with an acyl chloride using a strong Lewis acid catalyst. Friedel-Crafts acylation is also possible with acid anhydrides. Reaction conditions are similar to the Friedel-Crafts alkylation mentioned above. This reaction has several advantages over the alkylation reaction. Due to the electron-withdrawing effect of the carbonyl group, the ketone product is always less reactive than the original molecule, so multiple acylations do not occur. Also, there are no carbocation rearrangements, as the carbonium ion is stabilized by a resonance structure in which the positive charge is on the oxygen.

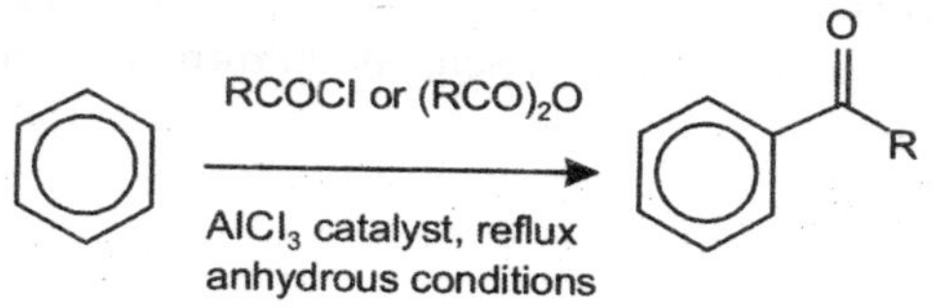

RCOCl or $(RCO)_2O$

$AlCl_3$ catalyst, reflux
anhydrous conditions

The viability of the Friedel-Crafts acylation depends on the stability of the acyl chloride reagent. Formyl chloride, for example, is too unstable to be isolated. Thus, synthesis of benzaldehyde via the Friedel-Crafts pathway requires that formyl chloride be synthesized *in situ*. This is accomplished via the Gatterman-Koch Synthesis, accomplished by reacting benzene with carbon monoxide and hydrogen chloride under high pressure, catalyzed by a mixture of aluminium chloride and cuprous chloride.

## Reaction Mechanism

In a simple mechanistic view, the first step consists of dissociation of a chlorine atom to form an acyl cation:

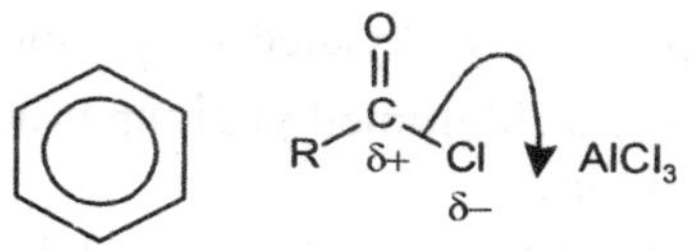

R–C(=O)$^{\delta+}$–Cl$^{\delta-}$ $AlCl_3$

This is followed by nucleophilic attack of the arene toward the acyl group:

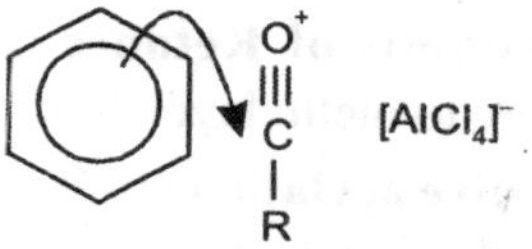

$O^+$≡C–R $[AlCl_4]^-$

Finally, a chlorine atom reacts to form HCl, and the $AlCl_3$ catalyst is regenerated:

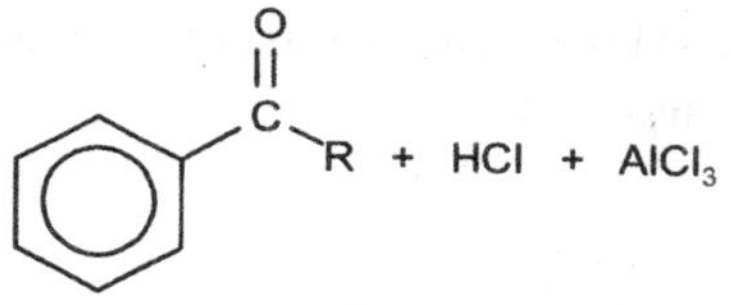

Ph–C(=O)–R + HCl + $AlCl_3$

## Friedel-Crafts Hydroxyalkylation

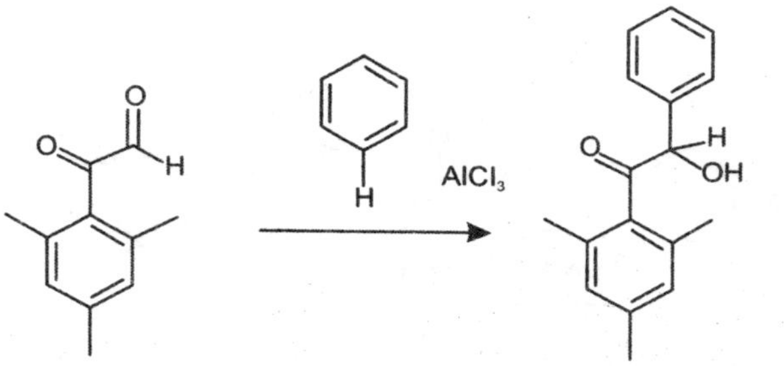

$AlCl_3$

Arenes react with certain aldehydes and ketones to the hydroxyalkylated product for example in the reaction of the mesityl derivative of glyoxal with benzene to form a benzoin with an alcohol rather than a carbonyl group:

## Scope & Variations

This reaction is related to several classic named reactions:

- The acylated reaction product can be converted into the alkylated product via a Clemmensen reduction.
- The Gattermann-Koch reaction can be used to synthesize benzaldehyde from benzene.
- The Gatterman reaction describes arene reactions with hydrocyanic acid
- The Houben-Hoesch reaction describes arene reactions with nitriles
- A reaction modification with an aromatic phenyl ester as a reactant is called the Fries rearrangement.
- In the Scholl reaction two arenes couple directly (sometimes called **Friedel-Crafts arylation**).
- In the Zincke-Suhl reaction p-cresol is alkylated to a cyclohexadienone with tetrachloromethane
- In the Blanc chloromethylation a chloromethyl group is added to an arene with formaldehyde, hydrochloric acid and zinc chloride.
- The **Bogert-Cook Synthesis** (1933) involves the dehydration and isomerization of *1-β-phenylethylcyclohexanol* to the octahydro derivative of phenanthrene
- The **Darzens-Nenitzescu Synthesis of Ketones** (1910, 1936) involves the acylation of cyclohexene with acetyl chloride to methylcyclohexenylketone.
- In the related **Nenitzescu reductive acylation** (1936) a saturated hydrocarbon is added making it a reductive acylation to methylcyclohexylketone
- In a green chemistry variation aluminium chloride is replaced by graphite in an alkylation of p-xylene with 2-bromobutane. This variation will not work with primary halides from which less carbocation involvement is inferred.

## Dyes

Friedel-Crafts reactions have been used in the synthesis of several triarylmethane and xanthene dyes [13]. Examples are the synthesis of thymolphthalein (a pH indicator) from two equivalents of thymol and phthalic anhydride:

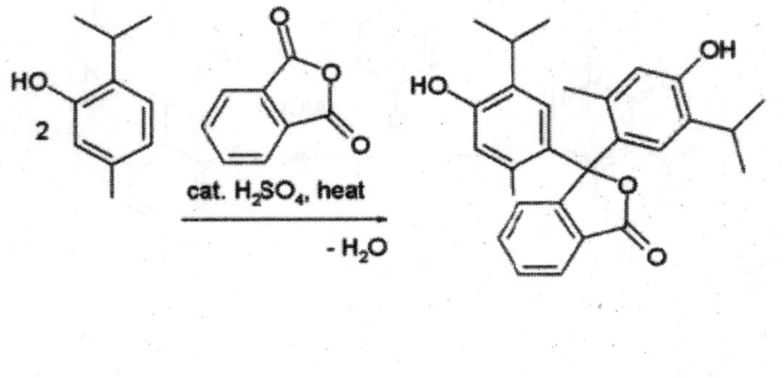

HO 2 — cat. $H_2SO_4$, heat — $- H_2O$ — HO, OH

A reaction of phthalic anhydride with resorcinol in the presence of zinc chloride gives the fluorophore Fluorescein. Replacing resorcinol by N,N-diethylaminophenol in this reaction gives rhodamine B:

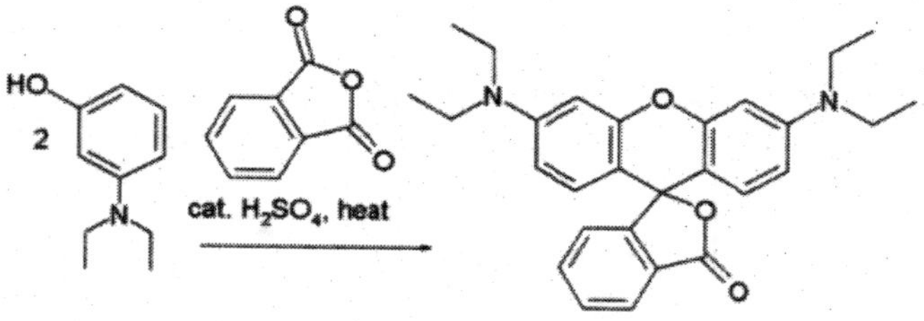

cat. $H_2SO_4$, heat

## Grignard Reaction

The **Grignard reaction**, named for the French chemist François Auguste Victor Grignard, is an organometallic chemical reaction in which alkyl- or aryl-magnesium halides (**Grignard reagents**), which act as nucleophiles, attack electrophilic carbon atoms that are present within polar bonds (e.g., a carbonyl group, see below) to yield a carbon-carbon bond (compare to Wittig reaction), thus altering hybridization about the reaction center. The Grignard reaction is an important tool in the formation of carbon-carbon bonds and for the formation of carbon-phosphorus, carbon-tin, carbon-silicon, carbon-boron and other carbon-heteroatom bonds.

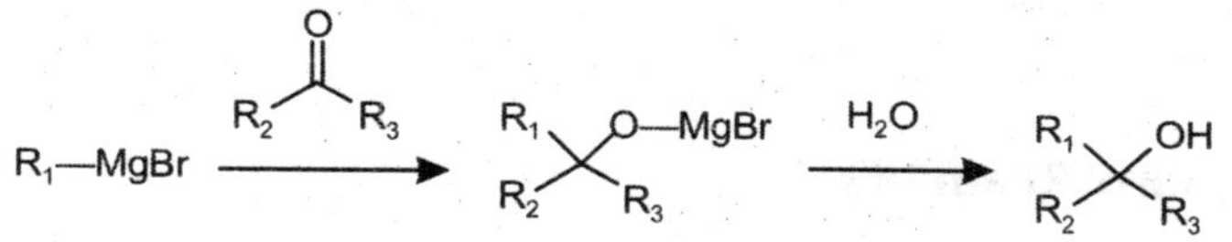

$R_1$—MgBr, $R_2$, $R_3$, O—MgBr, $H_2O$, OH

The addition to the nucleophile is irreversible due to the high $pK_a$ value of the alkyl component ($pK_a$ = ~45). Grignard reagents react with electrophilic chemical compounds. It should be noted that such reactions are **not** ionic; the Grignard reagent exists as an organometallic cluster (in ether). Victor Grignard (University Of Nancy, France) was awarded the 1912 Nobel Prize in Chemistry for the discovery of such reagents. The disadvantage of the Grignard reagents is that they readily react with protic solvents (such as water), or functional groups with acidic protons, such as alcohols and amines. In fact, atmospheric humidity in the lab can dictate one's success when trying to synthesize a Grignard reagent from magnesium turnings and an alkyl halide. To circumvent this issue, the reaction vessel is often flame-dried to evaporate all moisture, then sealed to prevent more from entering.

An example of the Grignard reaction is a key step in the industrial production of Tamoxifen:

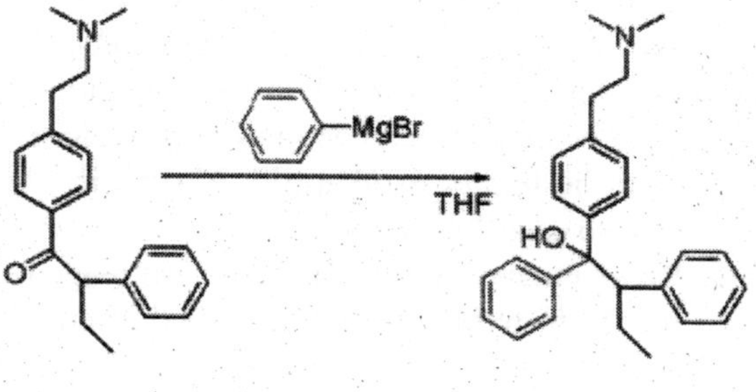

MgBr, THF, HO

## Reaction mechanism

The addition of the Grignard reagent to the carbonyl typically proceeds through a six-membered ring transition state.

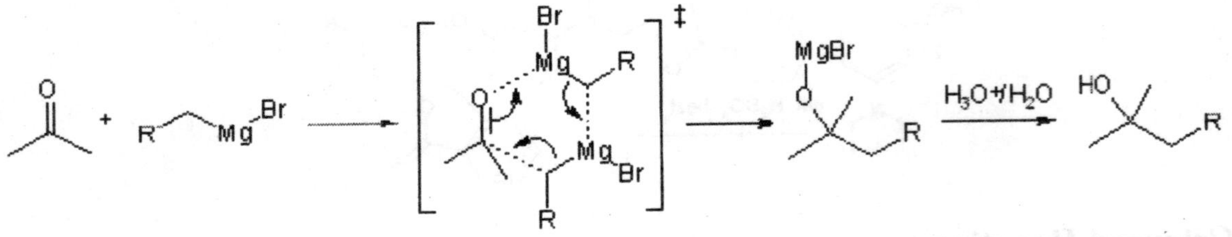

However, with hindered Grignard reagents, the reaction may proceed by single-electron transfer.

In a reaction involving Grignard reagents, it is important to ensure that no water is present, which would otherwise cause the reagent to rapidly decompose. Thus, most Grignard reactions occur in solvents such as anhydrous diethyl ether or tetrahydrofuran, because the oxygen of these solvents stabilizes the magnesium reagent. The reagent may also react with oxygen present in the atmosphere, inserting an oxygen atom between the carbon base and the magnesium halide group. Usually, this side-reaction may be limited by the volatile solvent vapors displacing air above the reaction mixture. However, it may be preferable for such reactions to be carried out in nitrogen or argon atmospheres, especially for smaller scales.

## Synthesis of Grignard Reagents

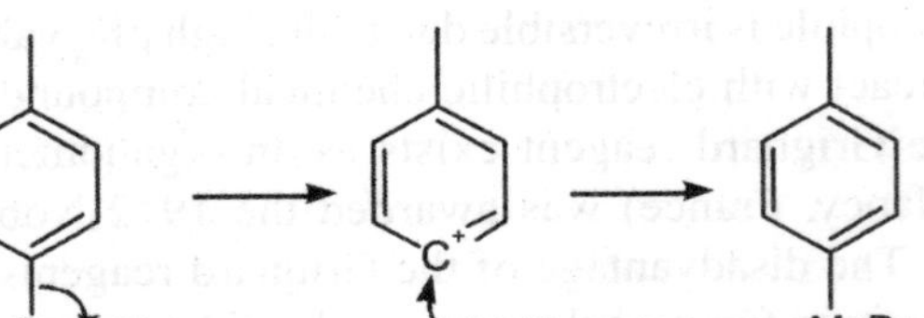

Grignard reagents are formed via the action of an alkyl or aryl halide on magnesium metal.[6] The reaction is conducted by adding the organic halide to a suspension of magnesium in an ether, which provides ligands required to stabilize the organomagnesium compound. Typical solvents are diethyl ether and tetrahydrofuran. Oxygen and protic solvents such as water or alcohols are not compatible with Grignard reagents. The reaction proceeds through single electron transfer.Grignard reactions often start slowly. As is common for reactions involving solids and solution, initiation follows an induction period during which reactive magnesium becomes exposed to the organic reagents. After this induction period, the reactions can be highly exothermic. Alkyl and aryl bromides and iodides are common substrates. Chlorides are also used, but fluorides are generally unreactive, except with specially activated magnesium, such as Rieke magnesium. Many Grignard reagents such as phenylmagnesium bromide are available commercially in tetrahydrofuran or diethyl ether solutions. Via the Schlenk equilibrium, Grignard reagents form varying amounts of diorganomagnesium compounds (R = organic group, X = halide):

$$2\ \text{RMgX} \text{ € } \text{R}_2\text{Mg} + \text{MgX}_2$$

## Practical tips

Many methods have been developed to initiate sluggish Grignard reactions. Mechanical methods include crushing of the Mg pieces in situ; rapid stirring and sonication of the suspension is also effective. Iodine, methyl iodide, and 1,2-dibromoethane are commonly employed activating agents. The use of 1,2-dibromoethane is particularly advantageous as its action can be monitored by the observation of bubbles of ethylene. Furthermore, the side-products are innocuous:

$$\text{Mg} + \text{BrC}_2\text{H}_4\text{Br} \rightarrow \text{C}_2\text{H}_4 + \text{MgBr}_2$$

The amount of Mg consumed by these activating agents is usually insignificant.

The addition of a small amount of mercuric chloride amalgamates the surface of the metal, allowing it to react. These methods weaken the passivating layer of MgO, thereby exposing highly reactive magnesium to the organic halide.

## Variations

Grignard reagents will react with a variety of carbonyl derivatives.

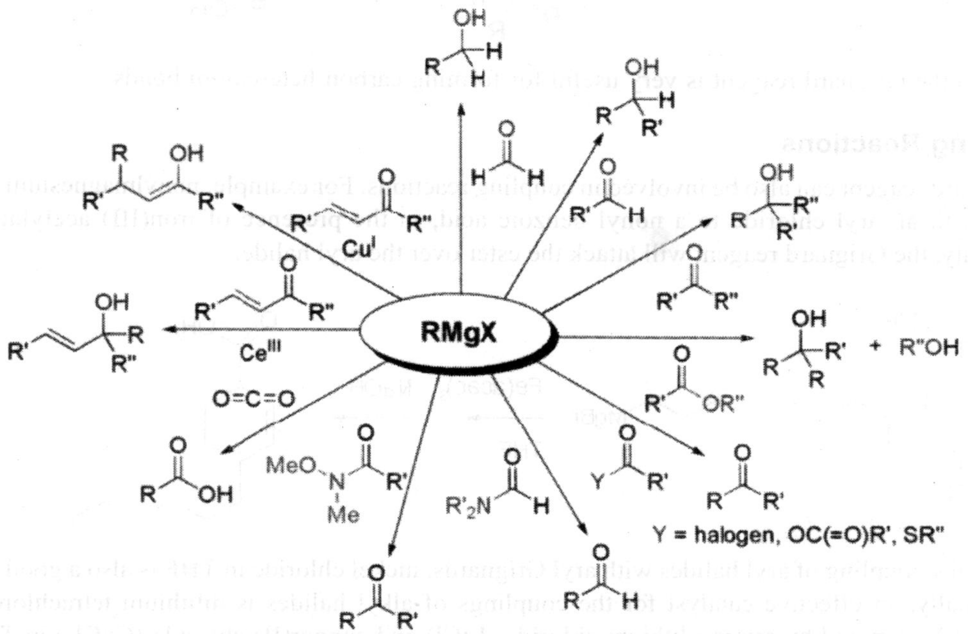

Y = halogen, OC(=O)R', SR"

In addition, Grignard reagents will react with other various electrophiles.

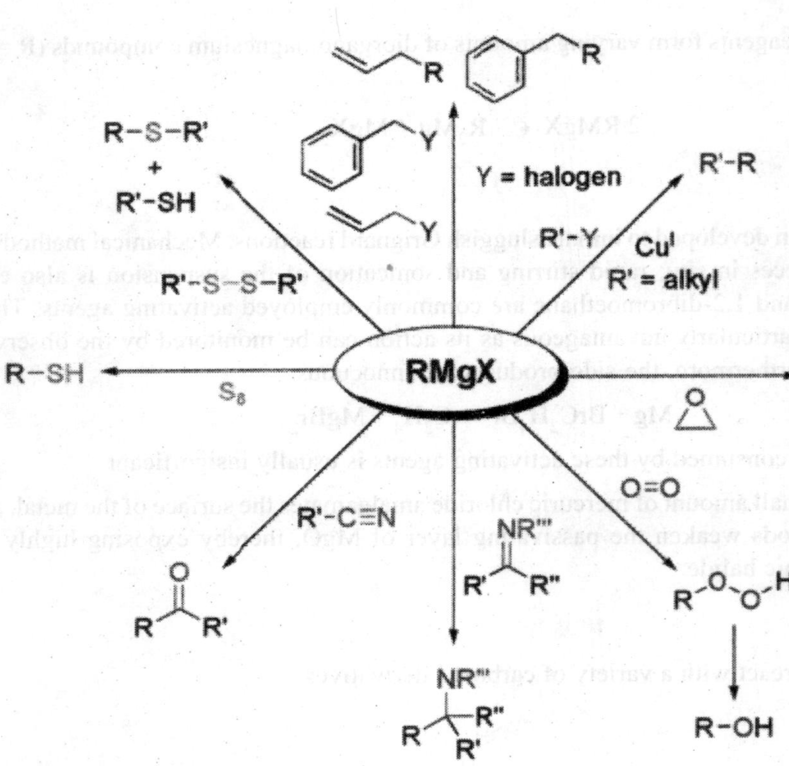

R–S–R' + R'–SH
R'–S–S–R'
Y = halogen
R'–Y Cu' R' = alkyl
R'–R
R–SH S8
RMgX
OH
R'–C≡N
NR'''
O=O
R–OH

Also the Grignard reagent is very useful for forming carbon-heteroatom bonds.

## Coupling Reactions

A Grignard reagent can also be involved in coupling reactions. For example, nonylmagnesium bromide reacts with an aryl chloride to a nonyl benzoic acid, in the presence of iron(III) acetylacetonate. Ordinarily, the Grignard reagent will attack the ester over the aryl halide.

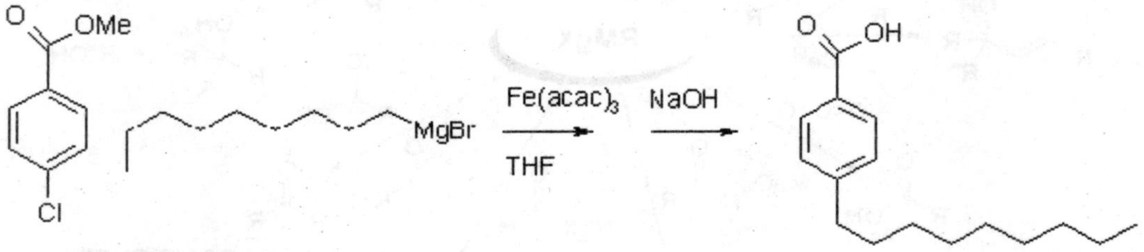

OMe Cl MgBr $Fe(acac)_3$ THF NaOH OH

For the coupling of aryl halides with aryl Grignards, nickel chloride in THF is also a good catalyst. Additionally, an effective catalyst for the couplings of alkyl halides is dilithium tetrachlorocuprate ($Li_2CuCl_4$), prepared by mixing lithium chloride (LiCl) and copper(II) chloride ($CuCl_2$) in THF. The Kumada-Corriu coupling gives access to styrenes.

## Oxidation

The oxidation of a Grignard reagent with oxygen takes place through a radical intermediate to a magnesium hydroperoxide. Hydrolysis of this complex yields hydroperoxides and reduction with an additional equivalent of Grignard reagent gives an alcohol.

$$R\text{-}MgX \xrightarrow{O_2} R^{\cdot} + O_2^{\cdot -} + MgX^{+} \longrightarrow R\text{-}O\text{-}O\text{-}MgX \xrightarrow{H_3O^+} R\text{-}O\text{-}O\text{-}H + HO\text{-}MgX + H^+$$

$$R\text{-}O\text{-}O\text{-}MgX \xrightarrow{R\text{-}MgX} R\text{-}O\text{-}MgX \xrightarrow{H_3O^+} R\text{-}O\text{-}H$$

The synthetic utility of Grignard oxidations can be increased by a reaction of Grignards with oxygen in presence of an alkene to an ethylene extended alcohol. This modification requires aryl or vinyl Grignards. Adding just the Grignard and the alkene does not result in a reaction demonstrating that the presence of oxygen is essential. Only drawback is the requirement of at least two equivalents of Grignard although this can partly be circumvented by the use of a dual Grignard system with a cheap reducing Grignard such as n-butylmagnesium bromide.

4 eq. $(CH_3)_3Si\text{-}CH_2\text{-}MgCl$, $O_2$, $(CH_3CH_2)_2O$, 25°C, 12 hrs. — 80%

## Nucleophilic Aliphatic Substitution

Grignard reagents are nucleophiles in nucleophilic aliphatic substitutions for instance with alkyl halides in a key step in industrial Naproxen production:

MgBr — Br–CH($CH_3$)–COOMgBr, THF — COOH

## Elimination

In the Boord olefin synthesis, the addition of magnesium to certain β-haloethers results in an elimination reaction to the alkene. This reaction can limit the utility of Grignard reactions.

$$\text{X-C-C-OR} \xrightarrow[\text{- ROMX}]{\text{M}} \text{C=C}$$

## Grignard Degradation

At one time was a tool in structure elucidation in which a Grignard RMgBr formed from a heteroaryl bromide HetBr reacts with water to Het-H (bromine replaced by a hydrogen atom) and MgBrOH. This hydrolysis method allows the determination of the number of halogen atoms in an organic compound. In modern usage Grignard degradation is used in the chemical analysis of certain triacylglycerols

## Hofmann Rearrangement

The **Hofmann rearrangement** is the organic reaction of a primary amide to a primary amine with one fewer carbon atom.

$$RC(=O)NH_2 \xrightarrow[NaOH]{Br_2} [R{-}N{=}C{=}O] \xrightarrow[-CO_2]{H_2O} R{-}NH_2$$

The reaction of bromine with sodium hydroxide forms sodium hypobromite in situ, which transforms the primary amide into an intermediate isocyanate. The intermediate isocyanate is hydrolyzed to a primary amine giving off carbon dioxide.The reaction is named after its discoverer: August Wilhelm von Hofmann. This reaction is also sometimes called the **Hofmann degradation**, and should not be confused with the Hofmann elimination.

### Variations

Several reagents can substitute for bromine. N-Bromosuccinimide and 1,8-Diazabicyclo[5.4.0]undec-7-ene (DBU) can effect a Hofmann rearrangement. In the following example, the intermediate isocyanate is trapped by methanol forming a carbamate.

NH2 —NBS, DBU / MeOH, reflux→ 73%

A mild alternative to bromine is also (bis(trifluoroacetoxy)iodo)benzene.

The Hofmann-Löffler reaction or Hofmann-Löffler-Freytag Reaction is an organic reaction is which a haloamine is converted to a cyclic amine such as a pyrrolidine with heat and an acid

An example is the synthesis of n-butylpyrrolidine

$Cl_2$ / NaOH aq. / ligroin; $H_2SO_4$, $H_2O$

The reaction type is a radical substitution. In the first step the halogen atom migrates from nitrogen to a gamma or delta carbon position in a rearrangement reaction.

A conceptually related reaction is the Barton reaction.

## Hunsdiecker Reaction

The **Hunsdiecker reaction** (also called the *Borodin reaction* after Alexander Borodin) is the organic reaction of silver salts of carboxylic acids with halogens to give organic halides. The reaction is named after Heinz Hunsdiecker and Cläre Hunsdiecker.

$$RCOO^- Ag^+ \xrightarrow[CCl_4]{Br_2} R{-}Br$$

Several reviews have been published. Mercuric oxide will also affect this transformation.

## Reaction Mechanism

The reaction mechanism of the Hunsdiecker reaction is believed to involve organic radical intermediates. The silver salt of the carboxylic acid **1** will quickly react with bromine to form intermediate **2**. Formation of the diradical pair **3** allows for radical decarboxylation to form the diradical pair **4**, which will quickly recombine to form the desired organic halide **5**.

$$\underset{\mathbf{1}}{RCOO^- Ag^+} \xrightarrow[-AgBr]{Br_2} \underset{\mathbf{2}}{RCOOBr} \rightleftharpoons \underset{\mathbf{3}}{RCOO^\bullet \; Br^\bullet} \xrightarrow{-CO_2} \underset{\mathbf{4}}{R^\bullet \; Br^\bullet} \longrightarrow \underset{\mathbf{5}}{R{-}Br}$$

# Variations

## Simonini Reaction

The reaction of silver salts of carboxylic acids with iodine is called the **Simonini reaction** , named after Angelo Simonini a student of Adolf Lieben at the University of Vienna.

$$2\, RCOO^- Ag^+ \xrightarrow[-CO_2]{I_2} RCOOR$$

# Hydration Reaction

In organic chemistry, a **hydration reaction** is a chemical reaction in which a hydroxyl group ($OH^-$) and a hydrogen cation (an acidic proton) are added to the two carbon atoms bonded together in the carbon-carbon double bond which makes up an alkene functional group. The reaction usually runs in a strong acidic, aqueous solution. Hydration differs from hydrolysis in that hydrolysis cleaves the non-water component in two. Hydration leaves the non-water component intact.

The general chemical equation of the reaction is the following:

$$RRC = CH_2 \text{ in } H_2O/\text{acid} \rightarrow RRC - CH_2 - OH$$

In the first step, the acidic proton bonds to the less substituted carbon of the double bond following Markovnikov's rule. In the second step an $H_2O$ molecule bonds to the other, more highly substituted carbon. The oxygen atom at this point has three bonds and carries a positive charge. Another water molecule comes along and takes up the extra proton. When carried out in the laboratory, this reaction tends to yield many undesirable side products and in its simple form described here is not considered very useful for the production of alcohol.

## Mechanism

This is an example reaction mechanism of the hydration of 1-methylcyclohexene to 1 methylcyclohexanol, using sulphuric acid as a catalyst.

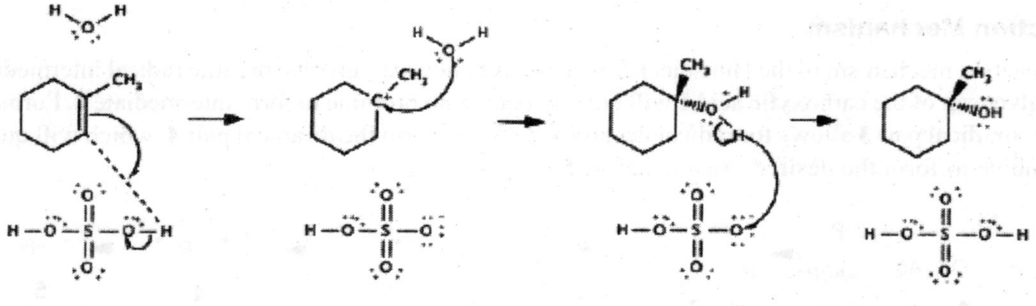

In organic chemistry, the **hydroboration-oxidation reaction** is a two-step organic chemical reaction that converts an alkene into a neutral alcohol by the net addition of water across the double bond.[1][2][3] The hydrogen and hydroxyl group are added in a syn addition leading to cis stereochemistry. Hydroboration-oxidation is an anti-Markovnikov reaction, with the hydroxyl group attaching to the less-substituted carbon.

The general form of the reaction is as follows:

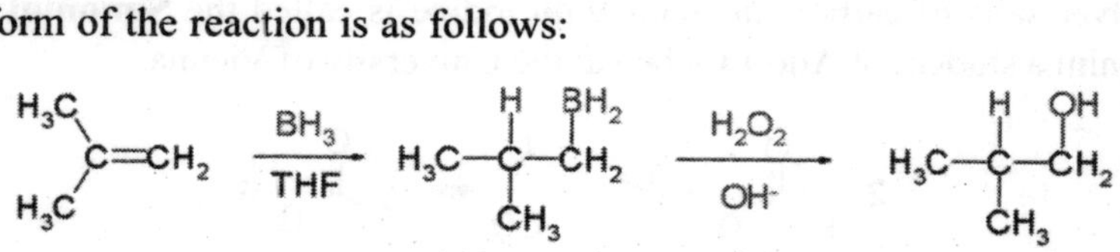

where THF is tetrahydrofuran, the archetypal solvent used for this reaction. In the first step, borane ($BH_3$) adds to the double bond, transferring one hydrogen from itself to the adjacent carbon. The second step substitutes the boron group $BH_2$ with the hydroxyl group, creating the final product.

## Hydroboration Mechanism

Borane exists as a toxic, colourless gas called diborane ($B_2H_6$). In diborane, two hydrogen atoms are each bonded to both boron atoms by single pairs of electrons ("three-center two-electron bonds"). This delocalization satisfies the octet around each boron and reduces the electrophilicity. That said, even diborane is intensely Lewis acidic, because of its vacant p orbitals. Because dimerization happens instantaneously, it is not possible to isolate pure borane. However, when diborane is treated with an ether or amine, a stable complex is formed, as the lone pair from the Lewis basic oxygen or nitrogen atom is donated to the borane. These complexes act chemically like borane. Solutions of $BH_3$ complexes in THF or diethyl ether are commercially available and more easily handled than diborane gas, and so are the more common form found in laboratories. For simplicity in illustration, borane will be used instead of the borane-ether complex in this article.

diborane + $2R-\ddot{O}-R$ ⇌ $2H_3\overset{-}{B}-\overset{+}{O}R_2$

diborane ether borane-ether complex

The addition of $BH_3$ to the alkene is a concerted reaction, with multiple bond formation and breaking occurring simultaneously. The intermediate step can be visualized more clearly by a theoretical transition state.

Knowing that the group containing the boron will be replaced by a hydroxyl group, it can be seen that the first step is the stereospecific-determining step. The hydroborane will add to the alkene so that the boron always ends up on the lesser substituted carbon. In the transition state, the more substituted carbon bears a partial positive charge (a partial carbocation). As a general rule, carbocations that are more substituted tolerate positive charge better than those that aren't. Had the hydroborane attacked with the opposite orientation, the lesser substituted carbon will bear the positive charge, which is electronically unfavourable. Until all hydrogens attached to boron have been transferred away, the boron group $BH_2$ will continue adding to more alkenes. This means that one equivalent of hydroborane will conduct the reaction with three equivalents of alkene. Furthermore, it is not necessary for the hydroborane to have more than one hydrogen. Therefore, $BH_3$ can be better represented as R-BH, where R can represents the remainder of the molecule. A widely used hydroboration reagent is 9-BBN which has just one hydrogen at boron and the same applies for catecholborane. Hydroborations also take place stereoselective in a syn mode, that is on the same face of the alkene. Thus 1-methylcyclopentene reacts with diborane predominantly to the trans-alkane.

## Hydroboration-Oxidation

In the second **hydroboration-oxidation** step, the nucleophilic hydroperoxide anion attacks the boron atom. Alkyl migration to oxygen gives the alkyl borane with retention of stereochemistry (in reality, the reaction occurs via the trialkyl borate $B(OR)_3$, rather than the monoalkly borinic ester $BH_2OR$).

A hydroboration reaction also takes place on alkynes. Again the mode of action is *syn* and secondary reaction products are aldehydes from terminal alkynes and ketones from internal alkynes. In order to prevent hydroboration across both the pi-bonds, a bulky borane like disiamyl (di-sec-iso-amyl) borane

**is used. Amines can be obtained by action of chloramine.[6] Reaction with iodine or bromine afford the corresponding alkyl halides. A carboxylic acid simply replaces the borane group by a proton.**

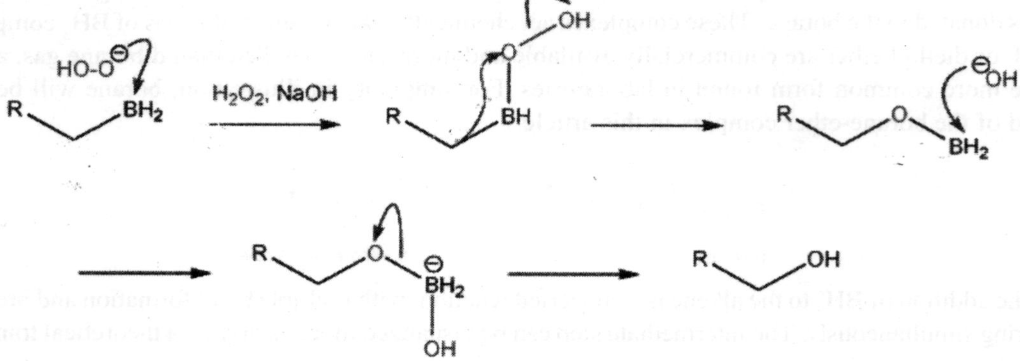

## Oxymercuration Reaction

**The oxymercuration reaction is an electrophilic addition organic reaction that transforms an alkene into a neutral alcohol. In oxymercuration, the alkene reacts with mercuric acetate (AcO-Hg-OAc) in aqueous solution to yield the addition of an acetoxymercuri (HgOAc) group and a hydroxy (OH) group across the double bond. Carbocations are not formed in this process and thus rearrangements are not observed. The reaction follows Markovnikov's rule (the hydroxy group will always be added to the more substituted carbon) and it is an anti addition (the two groups will be trans to each other).**

**Oxymercuration followed by demercuration is called an oxymercuration-reduction reaction. This reaction, which is almost always done in practice instead of oxymercuration, is treated at the conclusion of the article.**

### Mechanism

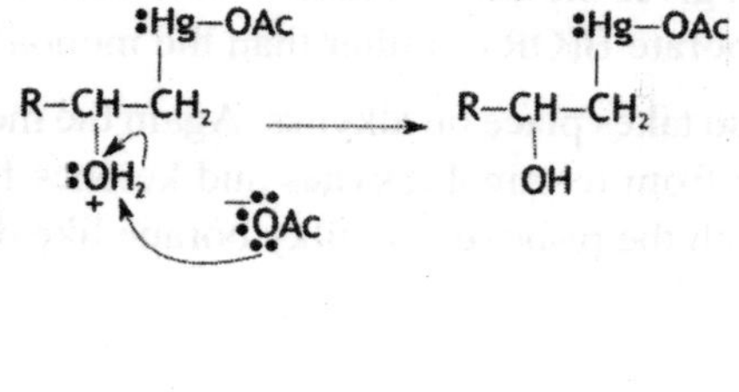

Curved-arrow mechanism, in sequential order from top to bottom.

Oxymercuration can be fully described in three steps(the whole process is sometimes called *deoxymercuration*), which is illustrated in stepwise fashion to the right. In the first step, the nucleophilic double bond attacks the mercury ion, ejecting an acetoxy group. The electron pair on the mercury ion in turn attacks a carbon on the double bond, forming a *mercurinium ion* in which the mercury atom bears a positive charge. The electrons in the highest occupied molecular orbital of the double bond are donated to mercury's empty $dz^2$ orbital and the electrons in mercury's dxz orbital are donated in the lowest unoccupied molecular orbital of the double bond.

In the second step, the nucleophilic water molecule attacks the more substituted carbon, liberating the electrons participating in its bond with mercury. The electrons collapse to the mercury ion and neutralizes it. The oxygen in the water molecule now bears a positive charge.

In the third step, the negatively charged acetoxy ion that was expelled in the first step attacks a hydrogen of the water group, forming the waste product HOAc. The two electrons participating in the bond between oxygen and the attacked hydrogen collapse into the oxygen, neutralizing its charge and creating the final alcohol product.

## Regioselectivity and Stereochemistry

Oxymercuration is very regioselective and is a textbook Markovnikov reaction; ruling out extreme cases, the water nucleophile will always preferentially attack the more substituted carbon, depositing the resultant hydroxy group there. This phenomenon is explained by examining the three resonance *structures* of the mercurinium ion formed at the end of the step one.

By inspection of these structures, it is seen that the positive charge of the mercury atom will sometimes reside on the more substituted carbon (approximately 4% of the time). This forms a temporary tertiary *carbocation*, which is a very reactive electrophile. The nucleophile will attack the mercurinium ion at this time. Therefore, the nucleophile attacks the more substituted carbon because it retains more *positive character* than the lesser substituted carbon.

Stereochemically, oxymercuration is an anti addition. As illustrated by the second step, the nucleophile cannot attack the carbon from the same face as the mercury ion because of steric hindrance. There is simply insufficient room on that face of the molecule to accommodate both a mercury ion and the attacking nucleophile. Therefore, when free rotation is impossible, the hydroxy and acetoxymercuri groups will always be trans to each other.

## Oxymercuration-reduction

In practice, the mercury adduct product created by the oxymercuration reaction is almost always treated with sodium borohydride ($NaBH_4$) in aqueous base in a reaction called *demercuration*. In demercuration, the acetoxymercuri group is replaced with a hydrogen in a stereochemically insensitive reaction. The combination of oxymercuration followed immediately by demercuration is called an **oxymercuration-reduction** reaction.

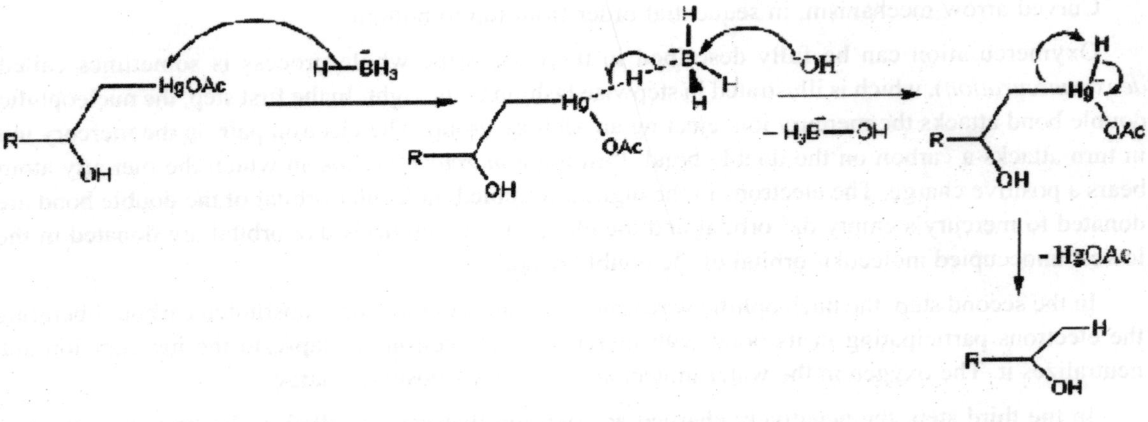

Mechanism for demercuration

Therefore, the oxymercuration-reduction reaction is the net addition of water across the double bond. Any stereochemistry set up by the oxymercuration step is scrambled by the demercuration step, so that the hydrogen and hydroxy group may be cis or trans from each other. Oxymercuration-reduction is a popular laboratory technique to create a "clean" alcohol (the acetoxymercuri group left over from the oxymercuration reaction is often undesirable).

## Prins Reaction

### History

The original reactants employed by Dutch chemist Hendrik Jacobus Prins in his 1919 publication were styrene (*scheme 2*), pinene, camphene, eugenol, isosafrole and anethole. The Prins reaction is an organic reaction consisting of an electrophilic addition of an aldehyde or ketone to an alkene or alkyne followed by capture of a nucleophile. The outcome of the reaction depends on reaction conditions (*scheme 1*). With water and a protic acid such as sulfuric acid as the reaction medium and formaldehyde the reaction product is a 1,3-diol. When water is absent dehydration takes place to an allyl alcohol. With an excess of formaldehyde and a low reaction temperature the reaction product is a dioxane. When water is replaced by acetic acid the corresponding esters are formed.

### Reaction Mechanism

The reaction mechanism for this reaction is depicted in **scheme 5**. The carbonyl reactant (2) is protonated by a protic acid and for the resulting oxonium ion **3** two resonance structures can be drawn. This electrophile engages in an electrophilic addition with the alkene to the carbocationic intermediate **4**. Exactly how much positive charge is present on the secondary carbon atom in this intermediate should be determined for each reaction set. Evidence exists for NGP of the hydroxyl oxygen or its neighbouring carbon atom. When the overall reaction has a high degree of concertedness, the charge built-up will be modest.

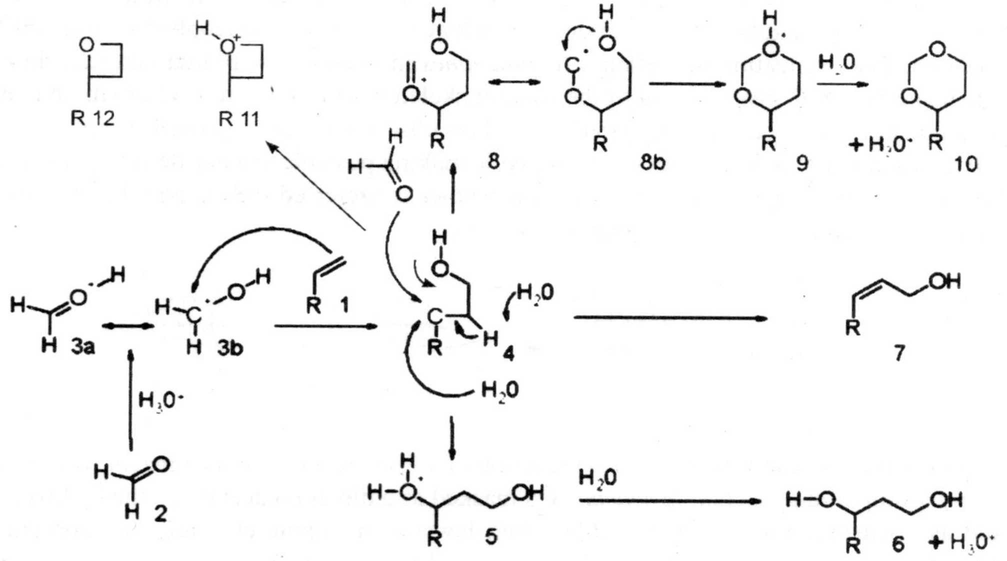

The three reaction modes open to this oxo-carbenium intermediate are:

- in blue: capture of the carbocation by water or any suitable nucleophile through **5** to the 1,3-adduct **6**.
- in black: proton abstraction in an elimination reaction to unsaturated compound **7**. When the olefin carries a methylene group, elimination and addition can be concerted with transfer of an allyl proton to the carbonyl group which in effect is an ene reaction in *scheme 6*.

carbonyl-ene reaction Prins reaction

- in green: capture of the carbocation by additional carbonyl reactant. In this mode the positive charge is dispersed over oxygen and carbon in the resonance structures **8a** and **8b**. Ring closure leads through intermediate **9** to the dioxane **10**. An example is the conversion of styrene to 4-phenyl-m-dioxane.
- in gray: only in specific reactions and when the carbocation is very stable the reaction takes a shortcut to the oxetane **12**. The photochemical Paterno-Büchi reaction between alkenes and aldehydes to oxetanes is more straightforward.

## Variations

Many variations of the Prins reaction exist because it lends itself easily to cyclization reactions and because it is possible to capture the oxo-carbenium ion with a large array of nucleophiles. The **halo-**

**Prins reaction** is one such modification with replacement of protic acids and water by lewis acids such as stannic chloride and boron tribromide. The halogen is now the nucleophile recombining with the carbocation. The cyclization of certain *allyl pulegones* in *scheme 7* with titanium tetrachloride in dichloromethane at -78°C gives access to the decalin skeleton with the hydroxyl group and chlorine group predominantly in cis configuration (91% cis). This observed cis diastereoselectivity is due to the intermediate formation of a trichlorotitanium alkoxide making possible an easy delivery of chlorine to the carbocation ion from the same face. The trans isomer is preferred (98% cis) when the switch is made to a tin tetrachloride reaction at room temperature.

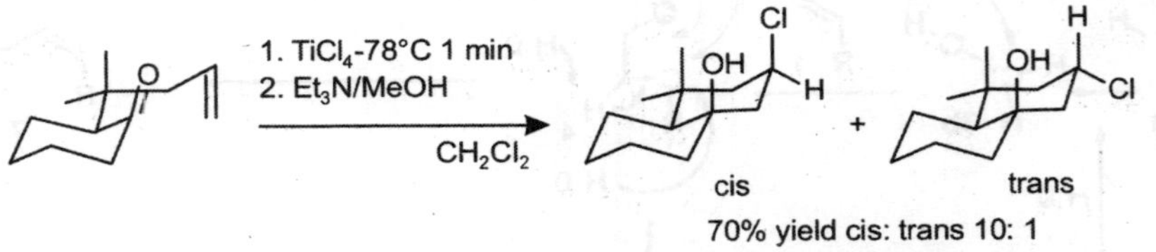

70% yield cis: trans 10: 1

The **Prins-pinacol reaction** is a cascade reaction of a Prins reaction and a pinacol rearrangement. The carbonyl group in the reactant in *scheme 8* is masked as a dimethyl acetal and the hydroxyl group is masked as a triisopropylsilyl ether (TIPS). With lewis acid stannic chloride the oxonium ion is activated and the pinacol rearrangement of the resulting Prins intermediate results in ring contraction and referral of the positive charge to the TIPS ether which eventually forms an aldehyde group in the final product as a mixture of cis and trans isomers with modest diastereoselectivity.

69% yield distrans 1.0:3.4

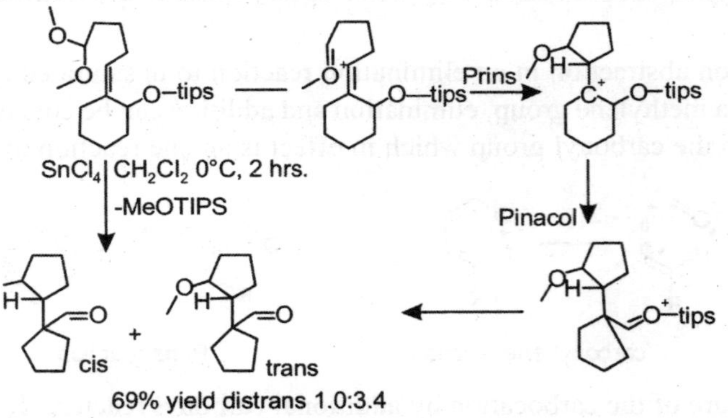

## Uses

The Prins reaction is used in total synthesis for example in that of *Exiguolide*

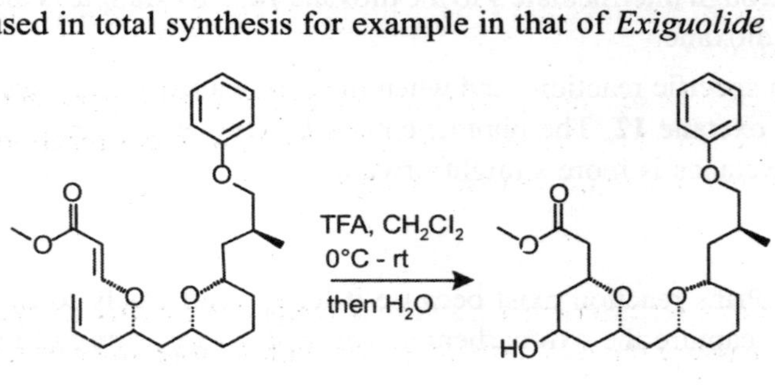

## Radical Substitution Reaction

In organic chemistry, a **radical substitution reaction** is a substitution reaction involving free radicals as a reactive intermediate.The reaction always involves at least two steps, and possibly a third. In the first step called **initiation** a free radical is created by homolysis. Homolysis can be brought about by heat or light but also by radical initiators such as organic peroxides or azo compounds. Light is used to create two free radicals from one diatomic species. The final step is called **termination** in which the radical recombines with another radical species. If the reaction is not terminated, but instead the radical group(s) go on to react further, the steps where new radicals are formed and then react is collectively known as **propagation** because a new radical is created available for secondary reactions.

### Radical Substitution Reactions

In free radical halogenation reactions radical substitution takes place with halogen reagents and alkane substrates. Another important class of radical substitutions involve aryl radicals. One example is the hydroxylation of benzene by Fenton's reagent. Many oxidation and reduction reactions in organic chemistry have free radical intermediates, for example the oxidation of aldehydes to carboxylic acids with chromic acid. Coupling reactions can also be considered radical substitutions. Certain aromatic substitutions takes place by radical-nucleophilic aromatic substitution. Auto-oxidation is a process responsible for deterioration of paints and food and lab hazards such as diethyl ether peroxide.

More radical substitutions are listed below:

- The Barton-McCombie deoxygenation is a way to substitute a hydroxyl group for a proton.
- The Wohl-Ziegler reaction involves the allylic bromination of alkenes.
- The Hunsdiecker reaction converts silver salts of carboxylic acids to alkyl halides.
- The Dowd-Beckwith reaction involves ring expansion of cyclic β-keto esters.
- The Barton reaction involves synthesis of nitrosoalcohols from nitrites.

### Wohl-Ziegler Reaction

The **Wohl-Ziegler reaction** is a chemical reaction that involves the allylic or benzylic bromination of hydrocarbons using an *N*-bromoimide and a radical initiator.[3]

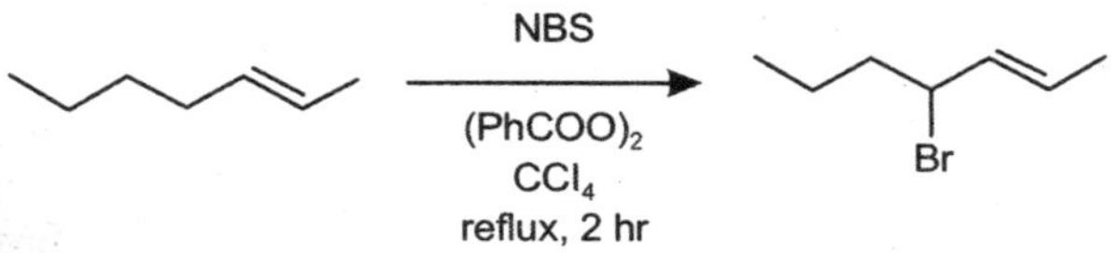

NBS

$(PhCOO)_2$

$CCl_4$

reflux, 2 hr

Br

- Best yields are achieved with *N*-bromosuccinimide in carbon tetrachloride solvent.
- Several reviews have been published.[4][5]
- In a typical setup a a stoichiometric amount of *N*-bromosuccinimide solution and a small quantity of initiator are added to a solution of the substrate in $CCl_4$, and the reaction mixture is stirred and heated to the boiling point. Initiation of the reaction is indicated by more vigorous

boiling; sometimes the heat source may need to be removed. Once all *N*-bromosuccinimide (which is denser than the solvent) has been converted to succinimide (which floats on top) the reaction has finished.

## Wurtz-Fittig Reaction

The **Wurtz-Fittig reaction** is the chemical reaction of aryl halides with alkyl halides and sodium metal to give substituted aromatic compounds.

Br R—I Na° R

- The reaction is named after Charles-Adolphe Wurtz, who discovered in 1855 a similar reaction between two alkyl halides (Wurtz reaction), and Rudolph Fittig, who discovered that also aryl halides undergo this reaction.

## Zimmermann Reaction

In the **Zimmermann reaction** the Janovski adduct is oxidized with excess base to a strongly coloured enolate with subsequent reduction of the dinitro compound to the aromatic nitro amine. This reaction is the basis of the **Zimmermann test** used for the detection of Ketosteroids.

O R H → O⁻ R
$O_2N$ $NO_2$ base → $O_2N$ H O R $NO_2$ → O C R $O_2N$ $NO_2$ + $O_2N$ $NH_2$

# Chirality*

## History and Introduction

The term optical activity derives from the interaction of chiral materials with polarized light. A solution of the (–)-form of an optical

*This subhead is written in the request and honour by Dr. Palash Gangopadhyay (author's friend) the subject chirality was speciality of Dr. Gangopadhyay.

isomer rotates the plane of polarization of a beam of plane polarized light in a counterclockwise direction, *vice versa* for the (+) optical isomer. The property was first observed by Jean-Baptiste Biot in 1815, and gained considerable importance in the sugar industry, analytical chemistry, and pharmaceuticals. Louis Pasteur deduced in 1848 that this phenomenon has a molecular basis. Artificial composite materials displaying the analog of optical activity but in the microwave region were introduced by J.C. Bose in 1898, and gained considerable attention from the mid-1980s.

The word "racemic" is derived from the Latin word for grape; the term having its origins in the work of Louis Pasteur who isolated racemic tartaric acid from wine. he term **chiral** is used to describe an object that is non-superimposable on its mirror image.

Human hands are perhaps the most universally recognized example of chirality: The left hand is a non-superimposable mirror image of the right hand; no matter how the two hands are oriented, it is impossible for all the major features of both hands to coincide. This difference in symmetry becomes obvious if someone attempts to shake the right hand of a person using his left hand, or if a left-handed glove is placed on a right hand. The term *chirality* is derived from the Greek word for hand, χετρ (/cheir/).

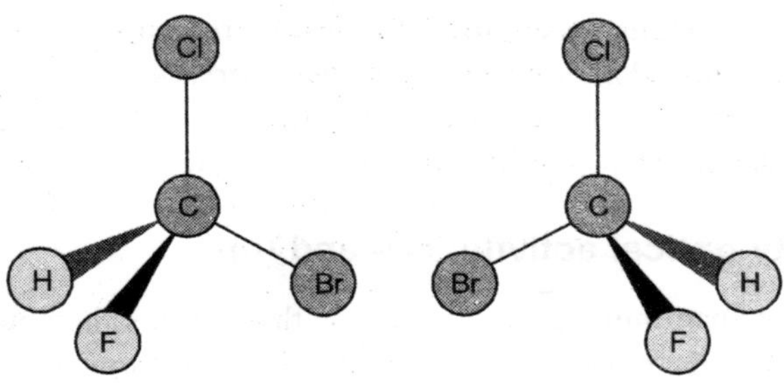


When used in the context of chemistry, chirality usually refers to molecules. Two mirror images of a molecule that cannot be superimposed onto each other are referred to as enantiomers or optical isomers. Because the difference between right and left hands is universally known and easy to observe, many pairs of enantiomers are designated as "right-" and "left-handed." A mixture of equal amounts of the two enantiomers is said to be a racemic mixture. Molecular chirality is of interest because of its application to stereochemistry in inorganic chemistry, organic chemistry, physical chemistry, biochemistry, and supramolecular chemistry.

The symmetry of a molecule (or any other object) determines whether it is chiral. A molecule is *achiral* (not chiral) if and only if it has an axis of improper rotation; that is, an n-fold rotation (rotation by 360°/n) followed by a reflection in the plane perpendicular to this axis that maps the molecule onto itself. (See chirality (mathematics).) A simplified rule applies to tetrahedrally-bonded carbon, as shown in the illustration: if all four substituents are different, the molecule is chiral. A chiral molecule is not necessarily asymmetric, that is, devoid of any symmetry elements, as it can have, for example, rotational symmetry.

## Naming Conventions

### By Configuration: *R*- and *S*-

For chemists, the *R* / *S* system is the most important nomenclature system for denoting enantiomers, which does not involve a reference molecule such as glyceraldehyde. It labels each chiral center *R* or

*S* according to a system by which its substituents are each assigned a *priority*, according to the Cahn Ingold Prelog priority rules, based on atomic number. If the center is oriented so that the lowest-priority of the four is pointed away from a viewer, the viewer will then see two possibilities: If the priority of the remaining three substituents decreases in clockwise direction, it is labelled *R* (for *Rectus*), if it decreases in counterclockwise direction, it is *S* (for *Sinister*).

This system labels each chiral center in a molecule (and also has an extension to chiral molecules not involving chiral centers). Thus, it has greater generality than the D/L system, and can label, for example, an (*R*,*R*) isomer versus an (*R*,*S*) — diastereomers.

The *R* / *S* system has no fixed relation to the (+)/(–) system. An *R* isomer can be either dextrorotatory or levorotatory, depending on its exact substituents.

The *R* / *S* system also has no fixed relation to the D/L system. For example, the side-chain one of serine contains a hydroxy group, -OH. If a thiol group, -SH, were swapped in for it, the D/L labelling would, by its definition, not be affected by the substitution. But this substitution would invert the molecule's *R* / *S* labelling, due to the fact that the CIP priority of $CH_2OH$ is lower than that for $CO_2H$ but the CIP priority of $CH_2SH$ is higher than that for $CO_2H$.

For this reason, the D/L system remains in common use in certain areas of biochemistry, such as amino acid and carbohydrate chemistry, because it is convenient to have the same chiral label for all of the commonly occurring structures of a given type of structure in higher organisms. In the D/L system, they are all L; in the *R* / *S* system, they are *mostly S* but there are some common exceptions.

## By optical activity: (+)- and (–)-

An enantiomer can be named by the direction in which it rotates the plane of polarized light. If it rotates the light clockwise (as seen by a viewer towards whom the light is travelling), that enantiomer is labelled (+). Its mirror-image is labeled (–). The (+) and (–) isomers have also been termed *d*- and *l*-, respectively (for *dextrorotatory* and *levorotatory*). This labeling is easy to confuse with D- and L-.

## By configuration: D- and L-

An optical isomer can be named by the spatial configuration of its atoms. The D/L system does this by relating the molecule to glyceraldehyde. Glyceraldehyde is chiral itself, and its two isomers are labeled D and L. Certain chemical manipulations can be performed on glyceraldehyde without affecting its configuration, and its historical use for this purpose (possibly combined with its convenience as one of the smallest commonly used chiral molecules) has resulted in its use for nomenclature. In this system, compounds are named by analogy to glyceraldehyde, which, in general, produces unambiguous designations, but is easiest to see in the small biomolecules similar to glyceraldehyde. One example is the amino acid alanine, which has two optical isomers, and they are labeled according to which isomer of glyceraldehyde they come from. On the other hand, glycine, the amino acid derived from glyceraldehyde, has no optical activity, as it is not chiral (achiral). Alanine, however, is chiral.

The D/L labelling is unrelated to (+)/(–); it does not indicate which enantiomer is dextrorotatory and which is levorotatory. Rather, it says that the compound's stereochemistry is related to that of the dextrorotatory or levorotatory enantiomer of glyceraldehyde—the dextrorotatory isomer of glyceraldehyde is, in fact, the D isomer. Nine of the nineteen L-amino acids commonly found in proteins are dextrorotatory (at a wavelength of 589 nm), and D-fructose is also referred to as *levulose* because it is levorotatory.

A rule of thumb for determining the D/L isomeric form of an amino acid is the "CORN" rule. The groups:

COOH, R, $NH_2$ and H (where R is a variant carbon chain)

are arranged around the *chiral center* carbon atom. Sighting with the hydrogen atom away from the viewer, if these groups are arranged clockwise around the carbon atom, then it is the D-form. If counter-clockwise, it is the L-form.

## Types

In general, chiral molecules have **point chirality**, centering around a single atom, usually carbon, which has four different substituents. The two enantiomers of such compounds are said to have different **absolute configurations** at this center. This center is thus stereogenic (i.e., a grouping within a molecular entity that may be considered a focus of stereoisomerism), and is exemplified by the α-carbon of amino acids. A molecule can have multiple chiral centers without being chiral overall if there is a symmetry element (a mirror plane or inversion center), which relates the two (or more) chiral centers. Such a molecule is called a meso compound. It is also possible for a molecule to be chiral without having actual point chirality. Common examples include 1,1'-bi-2-naphthol (BINOL) and 1,3-dichloro-allene, which have axial chirality, and (*E*)-cyclooctene, which has planar chirality.

It is important to keep in mind that molecules that are dissolved in solution or are in the gas phase usually have considerable flexibility, and, thus, may adopt a variety of different conformations. These various conformations are themselves almost always chiral. However, when assessing chirality, one must use a structural picture of the molecule that corresponds to just one chemical conformation - the most symmetric conformation possible.

When the optical rotation for an enantiomer is too low for practical measurement it is said to exhibit cryptochirality.

Even isotopic differences must be considered when examining chirality. Replacing one of the two $^1H$ atoms at the $CH_2$ position of benzyl alcohol with a deuterium ($^2H$) makes that carbon a stereocenter. The resulting benzyl-α-*d* alcohol exists as two distinct enantiomers, which can be assigned by the usual stereochemical naming conventions. The *S* enantiomer has $[\alpha]_D = +0.715°$.[5]

## Properties of Enantiomers

Enantiomers are identical with respect to ordinary chemical reactions, but differences arise when they are in the presence of other chiral molecules or objects. Different enantiomers of chiral compounds often taste and smell differently and have different effects as drugs - see below.

One chiral 'object' that interacts differently with the two enantiomers of a chiral compound is circularly polarised light: An enantiomer will absorb left- and right-circularly polarised light to differing degrees. This is the basis of circular dichroism (CD) spectroscopy. Usually the difference in absorptivity is relatively small (parts per thousand). CD spectroscopy is a powerful analytical technique for investigating the secondary structure of proteins and for determining the absolute configurations of chiral compounds, in particular, transition metal complexes. CD spectroscopy is replacing polarimetry as a method for characterising chiral compounds, although the latter is still popular with sugar chemists.

## In Biology

Many biologically active molecules are chiral, including the naturally occurring amino acids (the building blocks of proteins), and sugars. In biological systems, most of these compounds are of the same chirality: most amino acids are L and sugars are D. Typical naturally occurring proteins, made of L amino acids, are known as *left-handed proteins*, whereas D amino acids produce *right-handed proteins*.

The origin of this homochirality in biology is the subject of much debate. Most scientists believe that Earth life's choice of chirality was purely random, and that if carbon-based life forms exist elsewhere in the universe, their chemistry could theoretically have opposite chirality. But a few scientists are looking for fundamental reasons that favour the chirality on Earth, such as the weak nuclear force.[1]

Enzymes, which are chiral, often distinguish between the two enantiomers of a chiral substrate. Imagine an enzyme as having a glove-like cavity that binds a substrate. If this glove is right-handed, then one enantiomer will fit inside and be bound, whereas the other enantiomer will have a poor fit and is unlikely to bind.

D-form amino acids tend to taste sweet, whereas L-forms are usually tasteless. Spearmint leaves and caraway seeds, respectively, contain L-carvone and D-carvone - enantiomers of carvone. These smell different to most people because our olfactory receptors also contain chiral molecules that behave differently in the presence of different enantiomers.

## In Drugs

Many chiral drugs must be made with high enantiomeric purity due to potential side-effects of the other enantiomer. (The other enantiomer may also merely be inactive.)

- Thalidomide: Thalidomide is racemic. One enantiomer is effective against morning sickness, whereas the other is teratogenic. In this case, administering just one of the enantiomers to a pregnant patient does not help, as the two enantiomers are readily interconverted *in vivo*. Thus, if a person is given either enantiomer, both the D and L isomers will eventually be present in the patient's serum.
- Ethambutol: Whereas one enantiomer is used to treat tuberculosis, the other causes blindness.
- Naproxen: One enantiomer is used to treat arthritis pain, but the other causes liver poisoning with no analgesic effect.
- Steroid receptor sites also show stereoisomer specificity.
- Penicillin's activity is stereodependent. The antibiotic must mimic the D-alanine chains that occur in the cell walls of bacteria in order to react with and subsequently inhibit bacterial transpeptidase enzyme.

- Only L-propranolol is a powerful adrenoceptor antagonist, whereas D-propranolol is not. However, both have local anesthetic effect.
- The L-isomer of Methorphan, levomethorphan is a potent opioid analgesic, while the D-isomer, dextromethorphan is a dissiociative cough suppressant.
- S(-) isomer of carvedilol, a drug that interacts with adrenoceptors, is 100 times more potent as beta receptor blocker than R(+) isomer. However, both the isomers are approximately equipotent as alpha receptor blockers.
- The D-isomers of amphetamine and methamphetamine are strong CNS stimulants, while the L-isomers of both drugs lack appreciable CNS(central nervous system) stimulant effects, but instead stimulate the peripheral nervous system. For this reason, the Levo-isomer of methamphetamine is available as an OTC nasal inhaler in some countries, while the Dextro-isomer is banned from medical use in all but a few countries in the world, and higly regulated in those countries who do allow it to be used medically.

## In Inorganic Chemistry

Many coordination compounds are chiral; for example, the well-known $[Ru(2,2'\text{-bipyridine})_3]^{2+}$ complex in which the three bipyridine ligands adopt a chiral propeller-like arrangement [6]. In this case, the Ru atom may be regarded as a stereogenic center, with the complex having point chirality. The two enantiomers of complexes such as $[Ru(2,2'\text{-bipyridine})_3]^{2+}$ may be designated as Ë (left-handed twist of the propeller described by the ligands) and Ä (right-handed twist). Hexol is a chiral cobalt complex that was first investigated by Alfred Werner. Resolved hexol is significant as being the first compound devoid of carbon to display optical activity.

## Chirality of Amines

$R_3$ ⋯ N: ⇌ :N ⋯ $R_3$ (with $R_1$ and $R_2$ on each nitrogen)

Tertiary amines (see image) are chiral in a way similar to carbon compounds: The nitrogen atom bears four distinct substituents counting the lone pair. However, the energy barrier for the inversion of the stereocenter is, in general, about 30 kJ/mol, which means that the two stereoisomers are rapidly interconverted at room temperature. As a result, amines such as NHRR' cannot be resolved optically and NRR'R" can only be resolved when the R, R', and R" groups are constrained in cyclic structures.

# Index

### C

**F**

## N

## O

**T**

U

**V**

**Y**



**Z**